A Second Course
in Linear Algebra

A Second Course in Linear Algebra

WILLIAM C. BROWN
Michigan State University
East Lansing, Michigan

A Wiley-Interscience Publication

JOHN WILEY & SONS

New York · Chichester · Brisbane · Toronto · Singapore

89-8092

Library of Congress Cataloging-in-Publication Data

Brown, William C. (William Clough), 1943–
    A second course in linear algebra.

    "A Wiley-Interscience publication."
    Bibliography: p.
    Includes index.
    1. Algebras, Linear.   I. Title.
QA184.B765   1987   512'.5   87–23117
ISBN 0-471-62602-3

Printed in the United States of America

10 9 8 7 6 5 4 3 2 1

To Linda

Preface

For the past two years, I have been teaching a first-year graduate-level course in linear algebra and analysis. My basic aim in this course has been to prepare students for graduate-level work. This book consists mainly of the linear algebra in my lectures. The topics presented here are those that I feel are most important for students intending to do advanced work in such areas as algebra, analysis, topology, and applied mathematics.

Normally, a student interested in mathematics, engineering, or the physical sciences will take a one-term course in linear algebra, usually at the junior level. In such a course, a student will first be exposed to the theory of matrices, vector spaces, determinants, and linear transformations. Often, this is the first place where a student is required to do a mathematical proof. It has been my experience that students who have had only one such linear algebra course in their undergraduate training are ill prepared to do advanced-level work. I have written this book specifically for those students who will need more linear algebra than is normally covered in a one-term junior-level course.

This text is aimed at seniors and beginning graduate students who have had at least one course in linear algebra. The text has been designed for a one-quarter or semester course at the senior or first-year graduate level. It is assumed that the reader is familiar with such animals as functions, matrices, determinants, and elementary set theory. The presentation of the material in this text is deliberately formal, consisting mainly of theorems and proofs, very much in the spirit of a graduate-level course.

The reader will note that many familiar ideas are discussed in Chapter I. I urge the reader not to skip this chapter. The topics are familiar, but my approach, as well as the notation I use, is more sophisticated than a junior-level treatment. The material discussed in Chapters II–V is usually only touched upon (if at all) in a one-term course. I urge the reader to study these chapters carefully.

Having written five chapters for this book, I obviously feel that the reader should study all five parts of the text. However, time considerations often demand that a student or instructor do less. A shorter but adequate course could consist of Chapter I, Sections 1–6, Chapter II, Sections 1 and 2, and Chapters III and V. If the reader is willing to accept a few facts about extending scalars, then Chapters III, IV, and V can be read with no reference to Chapter II. Hence, a still shorter course could consist of Chapter I, Sections 1–6 and Chapters III and V.

It is my firm belief that any second course in linear algebra ought to contain material on tensor products and their functorial properties. For this reason, I urge the reader to follow the first version of a short course if time does not permit a complete reading of the text. It is also my firm belief that the basic linear algebra needed to understand normed linear vector spaces and real inner product spaces should not be divorced from the intrinsic topology and analysis involved. I have therefore presented the material in Chapter IV and the first half of Chapter V in the same spirit as many analysis texts on the subject. My original lecture notes on normed linear vector spaces and (real) inner product spaces were based on Loomis and Sternberg's classic text *Advanced Calculus*. Although I have made many changes in my notes for this book, I would still like to take this opportunity to acknowledge my debt to these authors and their fine text for my current presentation of this material.

One final word about notation is in order here. All important definitions are clearly displayed in the text with a number. Notation for specific ideas (e.g., $\mathbb{N}$ for the set of natural numbers) is introduced in the main body of the text as needed. Once a particular notation is introduced, it will be used (with only a few exceptions) with the same meaning throughout the rest of the text. A glossary of notation has been provided at the back of the book for the reader's convenience.

WILLIAM C. BROWN

East Lansing, Michigan
September 1987

Contents

A Second Course
in Linear Algebra

Chapter I

Linear Algebra

1. DEFINITIONS AND EXAMPLES OF VECTOR SPACES

In this book, the symbol F will denote an arbitrary field. A field is defined as follows:

Definition 1.1: A nonempty set F together with two functions $(x, y) \to x + y$ and $(x, y) \to xy$ from $F \times F$ to F is called a *field* if the following nine axioms are satisfied:

F1. $x + y = y + x$ for all x, $y \in F$.

F2. $x + (y + z) = (x + y) + z$ for all x, y, $z \in F$.

F3. There exists a unique element $0 \in F$ such that $x + 0 = x$ for all $x \in F$.

F4. For every $x \in F$, there exists a unique element $-x \in F$ such that $x + (-x) = 0$.

F5. $xy = yx$ for all x, $y \in F$.

F6. $x(yz) = (xy)z$ for all x, y, $z \in F$.

F7. There exists a unique element $1 \neq 0$ in F such that $x1 = x$ for all $x \in F$.

F8. For every $x \neq 0$ in F, there exists a unique $y \in F$ such that $xy = 1$.

F9. $x(y + z) = xy + xz$ for all x, y, $z \in F$.

Strictly speaking a field is an ordered triple $(F, (x, \ y) \to x + y, (x, y) \to xy)$ satisfying axioms F1–F9 above. The map from $F \times F \to F$ given by $(x, y) \to x + y$ is called *addition*, and the map $(x, y) \to xy$ is called *multiplication*. When referring to some field $(F, (x, y) \to x + y, (x, y) \to xy)$, references to addition and multiplication are dropped from the notation, and the letter F is used to denote both the set and the two maps satisfying axioms F1–F9. Although this procedure is somewhat ambiguous, it causes no confusion in concrete situations. In our first example below, we introduce some notation that we shall use throughout the rest of this book.

Example 1.2: We shall let $\mathbb{Q}$ denote the set of rational numbers, $\mathbb{R}$, the set of real numbers, and $\mathbb{C}$, the set of complex numbers. With the usual addition and multiplication, $\mathbb{Q}$, $\mathbb{R}$, and $\mathbb{C}$ are all fields with $\mathbb{Q} \subseteq \mathbb{R} \subseteq \mathbb{C}$. □

The fields in Example 1.1 are all infinite in the sense that the cardinal number attached to the underlying set in question is infinite. Finite fields are very important in linear algebra as well. Much of coding theory is done over finite algebraic extensions of the field $\mathbb{F}_p$ described in Example 1.3 below.

Example 1.3: Let $\mathbb{Z}$ denote the set of integers with the usual addition $x + y$ and multiplication xy inherited from $\mathbb{Q}$. Let p be a positive prime in $\mathbb{Z}$ and set $\mathbb{F}_p = \{0, 1, \ldots, p - 1\}$. $\mathbb{F}_p$ becomes a (finite) field if we define addition $\oplus$ and multiplication $\cdot$ modulo p. Thus, for elements $x, y \in \mathbb{F}_p$, there exist unique integers $k, z \in \mathbb{Z}$ such that $x + y = kp + z$ with $z \in \mathbb{F}_p$. We define $x \oplus y$ to be z. Similarly, $x \cdot y = w$ where $xy = k'p + w$ and $0 \leqslant w < p$.

The reader can easily check that $(\mathbb{F}_p, \oplus, \cdot)$ satisfies axioms F1–F9. Thus, $\mathbb{F}_p$ is a finite field of cardinality p. □

Except for some results in Section 7, the definitions and theorems in Chapter I are completely independent of the field F. Hence, we shall assume that F is an arbitrary field and study vector spaces over F.

Definition 1.4: A vector space V over F is a nonempty set together with two functions, $(\alpha, \beta) \to \alpha + \beta$ from $V \times V$ to V (called addition) and $(x, \alpha) \to x\alpha$ from $F \times V$ to V (called scalar multiplication), which satisfy the following axioms:

V1. $\alpha + \beta = \beta + \alpha$ for all $\alpha, \beta \in V$.

V2. $\alpha + (\beta + \gamma) = (\alpha + \beta) + \gamma$ for all $\alpha, \beta, \gamma \in V$.

V3. There exists an element $0 \in V$ such that $0 + \alpha = \alpha$ for all $\alpha \in V$.

V4. For every $\alpha \in V$, there exists a $\beta \in V$ such that $\alpha + \beta = 0$.

V5. $(xy)\alpha = x(y\alpha)$ for all $x, y \in F$, and $\alpha \in V$.

V6. $x(\alpha + \beta) = x\alpha + x\beta$ for all $x \in F$, and $\alpha, \beta \in V$.

V7. $(x + y)\alpha = x\alpha + y\alpha$ for all $x, y \in F$, and $\alpha \in V$.

V8. $1\alpha = \alpha$ for all $\alpha \in V$.

As with fields, we should make the comment that a vector space over F is really a triple $(V, (\alpha, \beta) \to \alpha + \beta, (x, \alpha) \to x\alpha)$ consisting of a nonempty set V together with two functions from $V \times V$ to V and $F \times V$ to V satisfying axioms V1–V8. There may be many different ways to endow a given set V with the structure of a vector space over F. Nevertheless, we shall drop any reference to addition and scalar multiplication when no confusion can arise and just use the notation V to indicate a given vector space over F.

If V is a vector space over F, then the elements of V will be called *vectors* and the elements of F *scalars*. We assume the reader is familiar with the elementary arithmetic in V, and, thus, we shall use freely such expressions as $-\alpha, \alpha - \beta$, and $\alpha_1 + \cdots + \alpha_n$ when dealing with vectors in V. Let us review some well-known examples of vector spaces.

Example 1.5: Let $\mathbb{N} = \{1, 2, 3, \ldots\}$ denote the set of natural numbers. For each $n \in \mathbb{N}$, we have the vector space $F^n = \{(x_1, \ldots, x_n) \mid x_i \in F\}$ consisting of all n-tuples of elements from F. Vector addition and scalar multiplication are defined componentwise by $(x_1, \ldots, x_n) + (y_1, \ldots, y_n) = (x_1 + y_1, \ldots, x_n + y_n)$ and $x(x_1, \ldots, x_n) = (xx_1, \ldots, xx_n)$. In particular, when $n = 1$, we see F itself is a vector space over F. $\square$

If A and B are two sets, let us denote the set of functions from A to B by B^A. Thus, $B^A = \{f \colon A \to B \mid f \text{ is a function}\}$. In Example 1.5, F^n can be viewed as the set of functions from $\{1, 2, \ldots, n\}$ to F. Thus, $\alpha = (x_1, \ldots, x_n) \in F^n$ is identified with the function $g_\alpha \in F^{\{1, \ldots, n\}}$ given by $g_\alpha(i) = x_i$ for $i = 1, \ldots, n$. These remarks suggest the following generalization of Example 1.5.

Example 1.6: Let V be a vector space over F and A an arbitrary set. Then the set V^A consisting of all functions from A to V becomes a vector space over F when we define addition and scalar multiplication pointwise. Thus, if $f, g \in V^A$, $f + g$ is the function from A to V defined by $(f + g)(a) = f(a) + g(a)$ for all $a \in A$. For $x \in F$ and $f \in V^A$, xf is defined by $(xf)(a) = x(f(a))$. $\square$

If A is a finite set of cardinality n in Example 1.6, then we shall shorten our notation for the vector space V^A and simply write V^n. In particular, if $V = F$, then $V^n = F^n$ and we recover the example in 1.5.

Example 1.7: We shall denote the set of $m \times n$ matrices (a_{ij}) with coefficients $a_{ij} \in F$ by $M_{m \times n}(F)$. The usual addition of matrices $(a_{ij}) + (b_{ij}) = (a_{ij} + b_{ij})$ and scalar multiplication $x(a_{ij}) = (xa_{ij})$ make $M_{m \times n}(F)$ a vector space over F. $\square$

Note that our choice of notation implies that F^n and $M_{1 \times n}(F)$ are the same vector space. Although we now have two different notations for the same vector space, this redundancy is useful and will cause no confusion in the sequel.

Example 1.8: We shall let F[X] denote the set of all polynomials in an indeterminate X over F. Thus, a typical element in F[X] is a finite sum of the form $a_n X^n + a_{n-1} X^{n-1} + \cdots + a_0$. Here $n \in \mathbb{N} \cup \{0\}$, and $a_0, \ldots, a_n \in F$. The usual notions of adding two polynomials and multiplying a polynomial by a constant, which the reader is familiar with from the elementary calculus, make sense over any field F. These operations give F[X] the structure of a vector space over F. □

Many interesting examples of vector spaces come from analysis. Here are some typical examples.

Example 1.9: Let I be an interval (closed, open, or half open) in ℝ. We shall let C(I) denote the set of all continuous, real valued functions on I. If $k \in \mathbb{N}$, we shall let $C^k(I)$ denote those $f \in C(I)$ that are k-times differentiable on the interior of I. Then $C(I) \supseteq C^1(I) \supseteq C^2(I) \supseteq \cdots$. These sets are all vector spaces over ℝ when endowed with the usual pointwise addition $(f + g)(x) = f(x) + g(x)$, $x \in I$, and scalar multiplication $(yf)(x) = y(f(x))$. □

Example 1.10: Let $A = [a_1, b_1] \times \cdots \times [a_n, b_n] \subseteq \mathbb{R}^n$ be a closed rectangle. We shall let $\mathscr{R}(A)$ denote the set of all real valued functions on A that are Riemann integrable. Clearly $\mathscr{R}(A)$ is a vector space over ℝ when addition and scalar multiplication are defined as in Example 1.9. □

We conclude our list of examples with a vector space, which we shall study carefully in Chapter III.

Example 1.11: Consider the following system of linear differential equations:

$$f_1' = a_{11}f_1 + \cdots + a_{1n}f_n$$
$$\vdots$$
$$f_n' = a_{n1}f_1 + \cdots + a_{nn}f_n$$

Here $f_1, \ldots, f_n \in C^1(I)$, where I is some open interval in ℝ. f_i' denotes the derivative of f_i, and the a_{ij} are scalars in ℝ. Set $A = (a_{ij}) \in M_{n \times n}(\mathbb{R})$. A is called the *matrix* of the system. If B is any matrix, we shall let B^t denote the transpose of B. Set $f = (f_1, \ldots, f_n)^t$. We may think of f as a function from $\{1, \ldots, n\}$ to $C^1(I)$, that is, $f \in C^1(I)^n$. With this notation, our system of differential equations becomes $f' = Af$. The set of solutions to our system is $V = \{f \in C^1(I)^n \,|\, f' = Af\}$. Clearly, V is a vector space over ℝ if we define addition and scalar multiplication componentwise as in Example 1.9. □

Now suppose V is a vector space over F. One rich source of vector spaces associated with V is the set of subspaces of V. Recall the following definition:

Definition 1.12: A nonempty subset W of V is a subspace of V if W is a vector space under the same vector addition and scalar multiplication as for V.

Thus, a subset W of V is a subspace if W is closed under the operations of V. For example, C([a, b]), $C^k([a, b])$, ℝ[X], and $\mathscr{R}([a, b])$ are all subspaces of $\mathbb{R}^{[a,b]}$.

If we have a collection $\mathscr{S} = \{W_i | i \in \Delta\}$ of subspaces of V, then there are some obvious ways of forming new subspaces from $\mathscr{S}$. We gather these constructions together in the following example:

Example 1.13: Let $\mathscr{S} = \{W_i | i \in \Delta\}$ be an indexed collection of subspaces of V. In what follows, the indexing set Δ of $\mathscr{S}$ can be finite or infinite. Certainly the intersection, $\bigcap_{i \in \Delta} W_i$, of the subspaces in $\mathscr{S}$ is a subspace of V. The set of all finite sums of vectors from $\bigcup_{i \in \Delta} W_i$ is also a subspace of V. We shall denote this subspace by $\sum_{i \in \Delta} W_i$. Thus, $\sum_{i \in \Delta} W_i = \{\sum_{i \in \Delta} \alpha_i | \alpha_i \in W_i \text{ for all } i \in \Delta\}$. Here and throughout the rest of this book, if Δ is infinite, then the notation $\sum_{i \in \Delta} \alpha_i$ means that all α_i are zero except possibly for finitely many $i \in \Delta$. If Δ is finite, then without any loss of generality, we can assume $\Delta = \{1, \ldots, n\}$ for some $n \in \mathbb{N}$. (If $\Delta = \phi$, then $\sum_{i \in \Delta} W_i = (0)$.) We shall then write $\sum_{i \in \Delta} W_i = W_1 + \cdots + W_n$.

If $\mathscr{S}$ has the property that for every i, $j \in \Delta$ there exists a $k \in \Delta$ such that $W_i \cup W_j \subseteq W_k$, then clearly $\bigcup_{i \in \Delta} W_i$ is a subspace of V. $\square$

In general, the union of two subspaces of V is not a subspace of V. In fact, if W_1 and W_2 are subspaces of V, then $W_1 \cup W_2$ is a subspace if and only if $W_1 \subseteq W_2$ or $W_2 \subseteq W_1$. This fact is easy to prove and is left as an exercise. In our first theorem, we discuss one more important fact about unions.

Theorem 1.14: Let V be a vector space over an infinite field F. Then V cannot be the union of a finite number of proper subspaces.

Proof: Suppose $W_1, \ldots, W_n$ are proper subspaces of V such that $V = W_1 \cup \cdots \cup W_n$. We shall show that this equation is impossible. We remind the reader that a subspace W of V is proper if $W \neq V$. Thus, $V - W \neq \phi$ for a proper subspace W of V.

We may assume without loss of generality that $W_1 \not\subseteq W_2 \cup \cdots \cup W_n$. Let $\alpha \in W_1 - \bigcup_{i=2}^{n} W_i$. Let $\beta \in V - W_1$. Since F is infinite, and neither α nor β is zero, $\Delta = \{\alpha + x\beta | x \in F\}$ is an infinite subset of V. Since there are only finitely many subspaces W_i, there exists a $j \in \{1, \ldots, n\}$ such that $\Delta \cap W_j$ is infinite.

Suppose $j \in \{2, \ldots, n\}$. Then there exist two nonzero scalars x, $x' \in F$ such that $x \neq x'$, and $\alpha + x\beta$, $\alpha + x'\beta \in W_j$. Since W_j is a subspace, $(x' - x)\alpha = x'(\alpha + x\beta) - x(\alpha + x'\beta) \in W_j$. Since $x' - x \neq 0$, we conclude $\alpha \in W_j$. But this is contrary to our choice of $\alpha \notin W_2 \cup \cdots \cup W_n$. Thus, $j = 1$.

Now if $j = 1$, then again there exist two nonzero scalars x, $x' \in F$ such that $x \neq x'$, and $\alpha + x\beta$, $\alpha + x'\beta \in W_1$. Then $(x - x')\beta = (\alpha + x\beta) - (\alpha + x'\beta) \in W_1$. Since $x - x' \neq 0$, $\beta \in W_1$. This is impossible since β was chosen in $V - W_1$. We conclude that V cannot be equal to the union of $W_1, \ldots, W_n$. This completes the proof of Theorem 1.14. $\square$

If F is finite, then Theorem 1.14 is false in general. For example, let $V = (\mathbb{F}_2)^2$. Then $V = W_1 \cup W_2 \cup W_3$, where $W_1 = \{(0, 0), (1, 1)\}$, $W_2 = \{(0, 0), (0, 1)\}$, and $W_3 = \{(0, 0), (1, 0)\}$.

Any subset S of a vector space V determines a subspace $L(S) = \cap \{W \mid W$ a subspace of V, $W \supseteq S\}$. We shall call $L(S)$ the *linear span* of S. Clearly, $L(S)$ is the smallest subspace of V containing S. Thus, in Example 1.13, for instance, $L(\bigcup_{i\in\Delta} W_i) = \sum_{i\in\Delta} W_i$.

Let $\mathscr{P}(V)$ denote the set of all subsets of V. If $\mathscr{S}(V)$ denotes the set of all subspaces of V, then $\mathscr{S}(V) \subseteq \mathscr{P}(V)$, and we have a natural function $L: \mathscr{P}(V) \to \mathscr{S}(V)$, which sends a subset $S \in \mathscr{P}(V)$ to its linear span $L(S) \in \mathscr{S}(V)$. Clearly, L is a surjective map whose restriction to $\mathscr{S}(V)$ is the identity. We conclude this section with a list of the more important properties of the function $L(\cdot)$.

Theorem 1.15: The function $L: \mathscr{P}(V) \to \mathscr{S}(V)$ satisfies the following poperties:

(a) For $S \in \mathscr{P}(V)$, $L(S)$ is the subspace of V consisting of all finite linear combinations of vectors from S. Thus,

$$L(S) = \left\{ \sum_{i=1}^{n} x_i\alpha_i \mid x_i \in F, \ \alpha_i \in S, \ n \geqslant 0 \right\}$$

(b) If $S_1 \subseteq S_2$, then $L(S_1) \subseteq L(S_2)$.

(c) If $\alpha \in L(S)$, then there exists a finite subset $S' \subseteq S$ such that $\alpha \in L(S')$.

(d) $S \subseteq L(S)$ for all $S \in \mathscr{P}(V)$.

(e) For every $S \in \mathscr{P}(V)$, $L(L(S)) = L(S)$.

(f) If $\beta \in L(S \cup \{\alpha\})$ and $\beta \notin L(S)$, then $\alpha \in L(S \cup \{\beta\})$. Here α, $\beta \in V$ and $S \in \mathscr{P}(V)$.

Proof: Properties (a)–(e) follow directly from the definition of the linear span. We prove (f). If $\beta \in L(S \cup \{\alpha\}) - L(S)$, then β is a finite linear combination of vectors from $S \cup \{\alpha\}$. Furthermore, α must occur with a nonzero coefficient in any such linear combination. Otherwise, $\beta \in L(S)$. Thus, there exist vectors $\alpha_1, \ldots, \alpha_n \in S$ and nonzero scalars $x_1, \ldots, x_n$, $x_{n+1} \in F$ such that $\beta = x_1\alpha_1 + \cdots + x_n\alpha_n + x_{n+1}\alpha$. Since $x_{n+1} \neq 0$, we can write α as a linear combination of β and $\alpha_1, \ldots, \alpha_n$. Namely, $\alpha = x_{n+1}^{-1}\beta - x_{n+1}^{-1}x_1\alpha_1 - \cdots - x_{n+1}^{-1}x_n\alpha_n$. Thus, $\alpha \in L(S \cup \{\beta\})$. $\square$

EXERCISES FOR SECTION 1

(1) Complete the details in Example 1.3 and argue $(\mathbb{F}_p, \oplus, \cdot)$ is a field.

(2) Let $\mathbb{R}(X) = \{f(x)/g(x) \mid f, g \in \mathbb{R}[X]$ and $g \neq 0\}$ denote the set of rational functions on $\mathbb{R}$. Show that $\mathbb{R}(X)$ is a field under the usual definition of addition $f/g + h/k = (kf + gh)/gk$ and multiplication $(f/g)(h/k) = fh/gk$. $\mathbb{R}(X)$ is called the field of rational functions over $\mathbb{R}$. Does $F(X)$ make sense for any field F?

(3) Set $F = \{a + b\sqrt{-5}\,|\,a,\ b \in \mathbb{Q}\}$. Show that F is a subfield of $\mathbb{C}$, that is, F is a field under complex addition and multiplication. Show that $\{a + b\sqrt{-5}\,|\,a, b \text{ integers}\}$ is not a subfield of $\mathbb{C}$.

(4) Let I be an open interval in $\mathbb{R}$. Let $a \in I$. Let $V_a = \{f \in \mathbb{R}^I\,|\,f \text{ has a derivative at } a\}$. Show that V_a is a subspace of $\mathbb{R}^I$.

(5) The vector space $\mathbb{R}^\mathbb{N}$ is just the set of all sequences $\{a_i\} = (a_1, a_2, a_3, \dots)$ with $a_i \in \mathbb{R}$. What are vector addition and scalar multiplication here?

(6) Show that the following sets are subspaces of $\mathbb{R}^\mathbb{N}$:
   (a) $W_1 = \{\{a_i\} \in \mathbb{R}^\mathbb{N}\,|\,\lim_{i \to \infty} a_i = 0\}$.
   (b) $W_2 = \{\{a_i\} \in \mathbb{R}^\mathbb{N}\,|\,\{a_i\} \text{ is a bounded sequence}\}$.
   (c) $W_3 = \{\{a_i\} \in \mathbb{R}^\mathbb{N}\,|\,\sum_{i=1}^{\infty} a_i^2 < \infty\}$.

(7) Let $(a_1, \dots, a_n) \in F^n - (0)$. Show that $\{(x_1, \dots, x_n) \in F^n\,|\,\sum_{i=1}^{n} a_i x_i = 0\}$ is a proper subspace of F^n.

(8) Identify all subspaces of $\mathbb{R}^2$. Find two subspaces W_1 and W_2 of $\mathbb{R}^2$ such that $W_1 \cup W_2$ is not a subspace.

(9) Let V be a vector space over F. Suppose W_1 and W_2 are subspaces of V. Show that $W_1 \cup W_2$ is a subspace of V if and only if $W_1 \subseteq W_2$ or $W_2 \subseteq W_1$.

(10) Consider the following subsets of $\mathbb{R}[X]$:
   (a) $W_1 = \{f \in \mathbb{R}[X]\,|\,f(0) = 0\}$.
   (b) $W_2 = \{f \in \mathbb{R}[X]\,|\,2f(0) = f(1)\}$.
   (c) $W_3 = \{f \in \mathbb{R}[X]\,|\,\text{the degree of } f \leqslant n\}$.
   (d) $W_4 = \{f \in \mathbb{R}[X]\,|\,f(t) = f(1 - t) \text{ for all } t \in \mathbb{R}\}$.
   In which of these cases is W_i a subspace of $\mathbb{R}[X]$?

(11) Let K, L, and M be subspaces of a vector space V. Suppose $K \supseteq L$. Prove Dedekind's modular law: $K \cap (L + M) = L + (K \cap M)$.

(12) Let $V = \mathbb{R}^3$. Show that $\delta_1 = (1, 0, 0)$ is not in the linear span of α, β, and γ where $\alpha = (1, 1, 1)$, $\beta = (0, 1, -1)$, and $\gamma = (1, 0, 2)$.

(13) If S_1 and S_2 are subsets of a vector space V, show that $L(S_1 \cup S_2) = L(S_1) + L(S_2)$.

(14) Let S be any subset of $\mathbb{R}[X] \subseteq \mathbb{R}^\mathbb{R}$. Show that $e^x \notin L(S)$.

(15) Let $\alpha_i = (a_{i1},\ a_{i2}) \in F^2$ for $i = 1, 2$. Show that $F^2 = L(\{\alpha_1, \alpha_2\})$ if and only if the determinant of the 2×2 matrix $M = (a_{ij})$ is nonzero. Generalize this result to F^n.

(16) Generalize Example 1.8 to $n + 1$ variables $X_0, \dots, X_n$. The resulting vector space over F is called the *ring of polynomials* in $n + 1$ variables (over F). It is denoted $F[X_0, \dots, X_n]$. Show that this vector space is spanned by all monomials $X_0^{m_0}, \dots, X_n^{m_n}$ as $(m_0, \dots, m_n) \in (\mathbb{N} \cup \{0\})^{n+1}$.

(17) A polynomial $f \in F[X_0, \ldots, X_n]$ is said to be *homogeneous of degree d* if f is a finite linear combination of monomials $X_0^{m_0}, \ldots, X_n^{m_n}$ of degree d (i.e., $m_0 + \cdots + m_n = d$). Show that the set of homogeneous polynomials of degree d is a subspace of $F[X_0, \ldots, X_n]$. Show that any polynomial f can be written uniquely as a finite sum of homogeneous polynomials.

(18) Let $V = \{A \in M_{n \times n}(F) \mid A = A^t\}$. Show that V is a subspace of $M_{n \times n}(F)$. V is the subspace of symmetric matrices of $M_{n \times n}(F)$.

(19) Let $W = \{A \in M_{n \times n}(F) \mid A^t = -A\}$. Show that W is a subspace of $M_{n \times n}(F)$. W is the subspace of all skew-symmetric matrices in $M_{n \times n}(F)$.

(20) Let W be a subspace of V, and let $\alpha, \beta \in V$. Set $A = \alpha + W$ and $B = \beta + W$. Show that $A = B$ or $A \cap B = \varnothing$.

2. BASES AND DIMENSION

Before proceeding with the main results of this section, let us recall a few facts from set theory. If A is any set, we shall denote the cardinality of A by $|A|$. Thus, A is a finite set if and only if $|A| < \infty$. If A is not finite, we shall write $|A| = \infty$. The only fact from cardinal arithmetic that we shall need in this section is the following:

2.1: Let A and B be sets, and suppose $|A| = \infty$. If for each $x \in A$, we have some finite set $\Delta_x \subseteq B$, then $|A| \geqslant |\bigcup_{x \in A} \Delta_x|$.

A proof of 2.1 can be found in any standard text in set theory (e.g., [1]), and, consequently, we omit it.

A relation R on a set A is any subset of the crossed product $A \times A$. Suppose R is a relation on a set A. If $x, y \in A$ and $(x, y) \in R$, then we shall say x relates to y and write $x \sim y$. Thus, $x \sim y \Leftrightarrow (x, y) \in R$. We shall use the notation $(A, \sim)$ to indicate the composite notion of a set A and a relation $R \subseteq A \times A$. This notation is a bit ambiguous since the symbol $\sim$ has no reference to R in it. However, the use of $\sim$ will always be clear from the context. In fact, the only relation R we shall systematically exploit in this section is the inclusion relation $\subseteq$ among subsets of $\mathscr{P}(V)$ [V some vector space over a field F].

A set A is said to be partially ordered if A has a relation $R \subseteq A \times A$ such that (1) $x \sim x$ for all $x \in A$, (2) if $x \sim y$, and $y \sim x$, then $x = y$, and (3) if $x \sim y$, and $y \sim z$, then $x \sim z$. A typical example of a partially ordered set is $\mathscr{P}(V)$ together with the relation $A \sim B$ if and only if $A \subseteq B$. If $(A, \sim)$ is a partially ordered set, and $A_1 \subseteq A$, then we say A_1 is totally ordered if for any two elements $x, y \in A_1$, we have at least one of the relations $x \sim y$ or $y \sim x$. If $(A, \sim)$ is a partially ordered set, and $A_1 \subseteq A$, then an element $x \in A$ is called an upper bound for A_1 if $y \sim x$ for all $y \in A_1$. Finally, an element $x \in (A, \sim)$ is a maximal element of A if $x \sim y$ implies $x = y$.

We say a partially ordered set $(A, \sim)$ is inductive if every totally ordered subset of A has an upper bound in A. The crucial point about inductive sets is the following result, which is called Zorn's lemma:

2.2: If a partially ordered set $(A, \sim)$ is inductive, then a maximal element of A exists.

We shall not give a proof of Zorn's lemma here. The interested reader may consult [3, p. 33] for more details.

Now suppose V is an arbitrary vector space over a field F. Let S be a subset of V.

Definition 2.3: S is linearly dependent over F if there exists a finite subset $\{\alpha_1, \ldots, \alpha_n\} \subseteq S$ and nonzero scalars $x_1, \ldots, x_n \in F$ such that $x_1\alpha_1 + \cdots + x_n\alpha_n = 0$. S is linearly independent (over F) if S is not linearly dependent.

Thus, if S is linearly independent, then whenever $\sum_{i=1}^{n} x_i\alpha_i = 0$ with $\{\alpha_1, \ldots, \alpha_n\} \subseteq S$ and $\{x_1, \ldots, x_n\} \subseteq F$, then $x_1 = \cdots x_n = 0$. Note that our definition implies the empty set ϕ is linearly independent over F. When considering questions of dependence, we shall drop the words "over F" whenever F is clear from the context. It should be obvious, however, that if more than one field is involved, a given set S could be dependent over one field and independent over another. The following example makes this clear.

Example 2.4: Suppose $V = \mathbb{R}$, the field of real numbers. Let $F_1 = \mathbb{Q}$, and $F_2 = \mathbb{R}$. Then V is a vector space over both F_1 and F_2. Let $S = \{\alpha_1 = 1, \alpha_2 = \sqrt{2}\}$. S is a set of two vectors in V. Using the fact that every integer factors uniquely into a product of primes, one sees easily that S is independent over F_1. But, clearly S is dependent over F_2 since $(\sqrt{2})\alpha_1 + (-1)\alpha_2 = 0$. $\square$

Definition 2.5: A subset S of V is called a *basis* of V if S is linearly independent over F and $L(S) = V$.

If S is a basis of a vector space V, then every nonzero vector $\alpha \in V$ can be written uniquely in the form $\alpha = x_1\alpha_1 + \cdots + x_n\alpha_n$, where $\{\alpha_1, \ldots, \alpha_n\} \subseteq S$ and $x_1, \ldots, x_n$ are nonzero scalars in F. Every vector space has a basis. In fact, any given linearly independent subset S of V can be expanded to a basis.

Theorem 2.6: Let V be a vector space over F, and suppose S is a linearly independent subset of V. Then there exists a basis B of V such that $B \supseteq S$.

Proof: Let $\mathscr{S}$ denote the set of all independent subsets of V that contain S. Thus, $\mathscr{S} = \{A \in \mathscr{P}(V) \mid A \supseteq S$ and A is linearly independent over $F\}$. We note that $\mathscr{S} \neq \phi$ since $S \in \mathscr{S}$. We partially order $\mathscr{S}$ by inclusion. Thus, for $A_1, A_2 \in \mathscr{S}$,

$A_1 \sim A_2$ if and only if $A_1 \subseteq A_2$. The fact that $(\mathscr{S}, \subseteq)$ is a partially ordered set is clear.

Suppose $\mathscr{T} = \{A_i \,|\, i \in \Delta\}$ is an indexed collection of elements from $\mathscr{S}$ that form a totally ordered subset of $\mathscr{S}$. We show $\mathscr{T}$ has an upper bound. Set $A = \bigcup_{i \in \Delta} A_i$. Clearly, $A \in \mathscr{P}(V)$, $S \subseteq A$, and $A_i \subseteq A$ for all $i \in \Delta$. If A fails to be linearly independent, then there exists a finite subset $\{\alpha_1, \ldots, \alpha_n\} \subseteq A$ and nonzero scalars $x_1, \ldots, x_n \in F$ such that $x_1 \alpha_1 + \cdots + x_n \alpha_n = 0$. Since $\mathscr{T}$ is totally ordered, there exists an index $i_0 \in \Delta$ such that $\{\alpha_1, \ldots, \alpha_n\} \subseteq A_{i_0}$. But then A_{i_0} is dependent, which is impossible since $A_{i_0} \in \mathscr{S}$. We conclude that A is linearly independent, and, consequently, $A \in \mathscr{S}$. Thus, $\mathscr{T}$ has an upper bound A in $\mathscr{S}$.

Since $\mathscr{T}$ was arbitrary, we can now conclude that $(\mathscr{S}, \subseteq)$ is an inductive set. Applying 2.2, we see that $\mathscr{S}$ has a maximal element B. Since $B \in \mathscr{S}$, $B \supseteq S$ and B is linearly independent. We claim that B is in fact a basis of V. To prove this assertion, we need only argue $L(B) = V$. Suppose $L(B) \neq V$. Then there exists a vector $\alpha \in V - L(B)$. Since $\alpha \notin L(B)$, the set $B \cup \{\alpha\}$ is clearly linearly independent. But then $B \cup \{\alpha\} \in \mathscr{S}$, and $B \cup \{\alpha\}$ is strictly larger than B. This is contrary to the maximality of B in $\mathscr{S}$. Thus, $L(B) = V$, and B is a basis of V containing S. $\square$

Let us look at a few concrete examples of bases before continuing.

Example 2.7: The empty set ϕ is a basis for the zero subspace (0) of any vector space V. If we regard a field F as a vector space over itself, then any nonzero element α of F forms a basis of F. $\square$

Example 2.8: Suppose $V = F^n$, $n \in \mathbb{N}$. For each $i = 1, \ldots, n$, let $\delta_i = (0, \ldots, 1, \ldots, 0)$. Thus, δ_i is the n-tuple whose entries are all zero except for a 1 in the ith position. Set $\underline{\delta} = \{\delta_1, \ldots, \delta_n\}$. Since $(x_1, \ldots, x_n) = x_1 \delta_1 + \cdots + x_n \delta_n$, we see $\underline{\delta}$ is a basis of F^n. We shall call $\underline{\delta}$ the canonical (standard) basis of F^n. $\square$

Example 2.9: Let $V = M_{m \times n}(F)$. For any $i = 1, \ldots, m$, and $j = 1, \ldots, n$, let e_{ij} denote the $m \times n$ matrix whose entries are all zero except for a 1 in the $(i\,j)$th position. Since $(a_{ij}) = \sum_{i,j} a_{ij} e_{ij}$, we see $B = \{e_{ij} \,|\, 1 \leqslant i \leqslant m, 1 \leqslant j \leqslant n\}$ is a basis for V. The elements e_{ij} in B are called the matrix units of $M_{m \times n}(F)$. $\square$

Example 2.10: Let $V = F[X]$. Let B denote the set of all monic monomials in X. Thus, $B = \{1 = X^0, X, X^2, \ldots\}$. Clearly, B is a basis of $F[X]$. $\square$

A specific basis for the vector space $C^k(I)$ in Example 1.9 is hard to write down. However, since $\mathbb{R}[X] \subseteq C^k(I)$, Theorem 2.6 guarantees that one basis of $C^k(I)$ contains the monomials $1, X, X^2, \ldots$.

Theorem 2.6 says that any linearly independent subset of V can be expanded to a basis of V. There is a companion result, which we shall need in Section 3. Namely, if some subset S of V spans V, then S contains a basis of V.

Theorem 2.11: Let V be a vector space over F, and suppose $V = L(S)$. Then S contains a basis of V.

Proof: If $S = \phi$ or $\{0\}$, then $V = (0)$. In this case, ϕ is a basis of V contained in S. So, we can sume S contains a nonzero vector α. Let $\mathcal{S} = \{A \subseteq S \mid A$ linearly independent over F$\}$. Clearly, $\{\alpha\} \in \mathcal{S}$. Partially order $\mathcal{S}$ by inclusion. If $\mathcal{T} = \{A_i \mid i \in \Delta\}$ is a totally ordered subset of $\mathcal{S}$, then $\bigcup_{i \in \Delta} A_i$ is an upper bound for $\mathcal{T}$ in $\mathcal{S}$. Thus, $(\mathcal{S}, \subseteq)$ is inductive. Applying 2.2, we see that $\mathcal{S}$ has a maximal element B.

We claim B is a basis for V. Since $B \in \mathcal{S}$, $B \subseteq S$ and B is linearly independent over F. If $L(B) = V$, then B is a basis of V, and the proof is complete. Suppose $L(B) \neq V$. Then $S \not\subseteq L(B)$, for otherwise $V = L(S) \subseteq L(L(B)) = L(B)$. Hence there exists a vector $\beta \in S - L(B)$. Clearly, $B \cup \{\beta\}$ is linearly independent over F. Thus, $B \cup \{\beta\} \in \mathcal{S}$. But $\beta \notin L(B)$ implies $\beta \notin B$. Hence, $B \cup \{\beta\}$ is strictly larger than B in $\mathcal{S}$. Since B is maximal, this is a contradiction. Therefore, $L(B) = V$ and our proof is complete. $\square$

A given vector space V has many different bases. For example, $\underline{\alpha} = \{(0, \ldots, x, \ldots, 0) + \delta_i \mid i = 1, \ldots, n\}$ is clearly a basis for F^n for any $x \neq -1$ in F. What all bases of V have in common is their cardinality. We prove this fact in our next theorem.

Theorem 2.12: Let V be a vector space over F, and suppose B_1 and B_2 are two bases of V. Then $|B_1| = |B_2|$.

Proof: We divide this proof into two cases.

CASE 1: Suppose V has a basis B that is finite.
In this case, we shall argue $|B_1| = |B| = |B_2|$. Suppose $B = \{\alpha_1, \ldots, \alpha_n\}$. It clearly suffices to show $|B_1| = n$. We suppose $|B_1| \neq n$ and derive a contradiction. There are two possibilities to consider here. Either $|B_1| = m < n$ or $|B_1| > n$. Let us first suppose $B_1 = \{\beta_1, \ldots, \beta_m\}$ with $m < n$. Since $\beta_1 \in L(B)$, $\beta_1 = x_1\alpha_1 + \cdots + x_n\alpha_n$. At least one x_i here is nonzero since $\beta_1 \neq 0$. Relabeling the α_i if need be, we can assume $x_1 \neq 0$. Since B is linearly independent over F, we conclude that $\beta_1 \in L(\{\alpha_2, \ldots, \alpha_n\} \cup \{\alpha_1\}) - L(\{\alpha_2, \ldots, \alpha_n\})$. It now follows from Theorem 1.15(f) that $\alpha_1 \in L(\{\beta_1, \alpha_2, \ldots, \alpha_n\})$. Since $\{\alpha_2, \ldots, \alpha_n\}$ is linearly independent over F, and $\beta_1 \notin L(\{\alpha_2, \ldots, \alpha_n\})$, we see that $\{\beta_1, \alpha_2, \ldots, \alpha_n\}$ is linearly independent over F. Since $\alpha_1 \in L(\{\beta_1, \alpha_2, \ldots, \alpha_n\})$, $V = L(\{\beta_1, \alpha_2, \ldots, \alpha_n\})$. Thus, $\{\beta_1, \alpha_2, \ldots, \alpha_n\}$ is a basis of V.

Now we can repeat this argument m times. We get after possibly some permutation of the α_i that $\{\beta_1, \ldots, \beta_m, \alpha_{m+1}, \ldots, \alpha_n\}$ is a basis of V. But $\{\beta_1, \ldots, \beta_m\}$ is already a basis of V. Thus, $\alpha_{m+1} \in L(\{\beta_1, \ldots, \beta_m\})$. This implies $\{\beta_1, \ldots, \beta_m, \alpha_{m+1}, \ldots, \alpha_n\}$ is linearly dependent which is a contradiction. Thus, $|B_1|$ cannot be less than n.

Now suppose $|B_1| > n$ ($|B_1|$ could be infinite here). By an argument similar to that given above, we can exchange n vectors of B_1 with $\alpha_1, \ldots, \alpha_n$. Thus, we construct a basis of V of the form $B \cup S$, where S is some nonempty subset of B_1. But B is already a basis of V. Since $S \neq \phi$, $B \cup S$ must then be linearly dependent. This is impossible. Thus, if V has a basis consisting of n vectors, then any basis of V has cardinality n, and the proof of the theorem is complete in case 1.

CASE 2: Suppose no basis of V is finite.

In this case, both B_1 and B_2 are infinite sets. Let $\alpha \in B_1$. Since B_2 is a basis of V, there exists a unique, finite subset $\Delta_\alpha \subseteq B_2$ such that $\alpha \in L(\Delta_\alpha)$ and $\alpha \notin L(\Delta')$ for any proper subset Δ' of Δ_α. Thus, we have a well-defined function $\varphi: B_1 \to \mathscr{P}(B_2)$ given by $\varphi(\alpha) = \Delta_\alpha$. Since B_1 is infinite, we may apply 2.1 and conclude that $|B_1| \geq |\bigcup_{\alpha \in B_1} \Delta_\alpha|$. Since $\alpha \in L(\Delta_\alpha)$ for all $\alpha \in B_1$, $V = L(\bigcup_{\alpha \in B_1} \Delta_\alpha)$. Thus $\bigcup_{\alpha \in B_1} \Delta_\alpha$ is a subset of B_2 that spans all of V. Since B_2 is a basis of V, we conclude $\bigcup_{\alpha \in B_1} \Delta_\alpha = B_2$. In particular, $|B_1| \geq |B_2|$. Reversing the roles of B_1 and B_2 gives $|B_2| \geq |B_1|$. This completes the proof of Theorem 2.12. $\square$

We shall call the common cardinality of any basis of V the *dimension of V.* We shall write dim V for the dimension of V. If we want to stress what field we are over, then we shall use the notation $\dim_F V$ for the dimension of the F-vector space V. Thus, dim V = |B|, where B is any basis of V when the base field F is understood.

Let us check the dimensions of some of our previous examples. In Example 2.4, $\dim_{F_2} V = 1$, and $\dim_{F_1}(V) = |\mathbb{R}|$, the cardinality of $\mathbb{R}$. In Example 2.7, $\dim_F(0) = 0$. In Example 2.8, $\dim F^n = n$. In Example 2.9, $\dim M_{m \times n}(F) = mn$. In Example 2.10, $\dim V = |\mathbb{N}|$, the cardinality of $\mathbb{N}$.

If the dimension of a vector space V is infinite, as in Examples 2.4 and 2.10, we shall usually make no attempt to distinguish which cardinal number gives dim V. Instead, we shall merely write dim V = ∞. If V has a finite basis $\{\alpha_1, \ldots, \alpha_n\}$, we shall call V a finite-dimensional vector space and write dim V < ∞, or, more precisely, dim V = n < ∞. Thus, for example, $\dim_{\mathbb{R}} C^k(I) = \infty$, whereas $\dim_{\mathbb{R}} \mathbb{R}^n = n < \infty$. In our next theorem, we gather together some of the more elementary facts about dim V.

Theorem 2.13: Let V be a vector space over F.

(a) If W is a subspace of V, then dim W $\leqslant$ dim V.

(b) If V is finite dimensional and W is a subspace of V such that dim W = dim V, then W = V.

(c) If W is a subspace of V, then there exists a subspace W' of V such that $W + W' = V$ and $W \cap W' = (0)$.

(d) If V is finite dimensional and W_1 and W_2 are subspaces of V, then $\dim(W_1 + W_2) + \dim(W_1 \cap W_2) = \dim W_1 + \dim W_2$.

Proof: It follows from Theorem 2.6 that any basis of a subspace W of V can be enlarged to a basis of V. This immediately proves (a) and (b). Suppose W is a subspace of V. Let B be a basis of W. By Theorem 2.6, there exists a basis C of V such that $B \subseteq C$. Let $W' = L(C - B)$. Since $C = B \cup (C - B)$, $V = L(C) = L(B) + L(C - B) = W + W'$. Since C is linearly independent and $B \cap (C - B) = \phi$, $L(B) \cap L(C - B) = (0)$. Thus, $W \cap W' = (0)$, and the proof of (c) is complete.

To prove (d), let $B_0 = \{\alpha_1, \ldots, \alpha_n\}$ be a basis of $W_1 \cap W_2$. If $W_1 \cap W_2 = (0)$, then we take B_0 to be the empty set ϕ. We can enlarge B_0 to a basis $B_1 = \{\alpha_1, \ldots, \alpha_n, \beta_1, \ldots, \beta_m\}$ of W_1. We can also enlarge B_0 to a basis $B_2 = \{\alpha_1, \ldots, \alpha_n, \gamma_1, \ldots, \gamma_p\}$ of W_2. Thus, $\dim W_1 \cap W_2 = n$, $\dim W_1 = n + m$, and $\dim W_2 = n + p$. We claim that $B = \{\alpha_1, \ldots, \alpha_n, \beta_1, \ldots, \beta_m, \gamma_1, \ldots, \gamma_p\}$ is a basis of $W_1 + W_2$. Clearly $L(B) = W_1 + W_2$. We need only argue B is linearly independent. Suppose $\sum_{i=1}^{n} x_i \alpha_i + \sum_{i=1}^{m} y_i \beta_i + \sum_{i=1}^{p} z_i \gamma_i = 0$ for some x_i, y_i, $z_i \in F$. Then $\sum_{i=1}^{p} z_i \gamma_i \in W_1 \cap W_2 = L(\{\alpha_1, \ldots, \alpha_n\})$. Thus, $\sum_{i=1}^{p} z_i \gamma_i = \sum_{i=1}^{n} w_i \alpha_i$ for some $w_i \in F$. Since B_2 is a basis of W_2, we conclude that $z_1 = \cdots = z_p = 0$. Since B_1 is a basis of W_1, $x_1 = \cdots = x_n = y_1 = \cdots y_m = 0$. In particular, B is linearly independent. Thus, $\dim(W_1 + W_2) = |B| = n + m + p$, and the proof of (d) follows.   □

A few comments about Theorem 2.13 are in order here. Part (d) is true whether V is finite dimesional or not. The proof is the same as that given above when $\dim(W_1 + W_2) < \infty$. If $\dim(W_1 + W_2) = \infty$, then either W_1 or W_2 is an infinite-dimensional subspace with the same dimension as $W_1 + W_2$. Thus, the result is still true but rather uninteresting.

If V is not finite dimensional, then (b) is false in general. A simple example illustrates this point.

Example 2.14: Let $V = F[X]$, and let W be the subspace of V consisting of all even polynomials. Thus, $W = \{\sum a_i X^{2i} \mid a_i \in F\}$. A basis of W is clearly all even powers of X. Thus, $\dim V = \dim W$, but $W \neq V$.   □

The subspace W' of V constructed in part (c) of Theorem 2.13 is called a complement of W. Note that W' is not in general unique. For example, if $V = \mathbb{R}^2$ and $W = L((1, 0))$, then any subspace of the form $L((a, b))$ with $b \neq 0$ is a complement of W.

Finally, part (d) of Theorem 2.13 has a simple extension to finitely many subspaces $W_1, \ldots, W_k$ of V. We record this extension as a corollary.

Corollary 2.15: Let V be a finite-dimensional vector space of dimension n. Suppose $W_1, \ldots, W_k$ are subspaces of V. For each $i = 1, \ldots, k$, set $f_i = n - \dim W_i$. Then

(a) $\dim(W_1 \cap \cdots \cap W_k) = n - \sum_{i=1}^{k} f_i + \sum_{j=1}^{k-1}$
$\{n - \dim((W_1 \cap \cdots \cap W_j) + W_{j+1})\}$.

(b) $\dim(W_1 \cap \cdots \cap W_k) \geqslant n - \sum_{i=1}^{k} f_i$.

(c) $\dim(W_1 \cap \cdots \cap W_k) = n - \sum_{i=1}^{k} f_i$ if and only if for all $i = 1, \ldots, k$, $W_i + (\bigcap_{j \neq i} W_j) = V$.

Proof: Part (a) follows from Theorem 2.13 (d) by induction. Parts (b) and (c) are easy consequences of (a). We leave the technical details for an exercise at the end of this section. □

Before closing this section, let us develop some useful notation concerning bases. Suppose V is a finite-dimensional vector space over F. If $\alpha = \{\alpha_1, \ldots, \alpha_n\}$ is a basis of V, then we have a natural function $[\cdot]_\alpha V \to M_{n \times 1}(F)$ defined as follows.

Definition 2.16: If $\alpha = \{\alpha_1, \ldots, \alpha_n\}$ is a basis of V, then $[\beta]_\alpha = (x_1, \ldots, x_n)^t \in M_{n \times 1}(F)$ if and only if $\sum_{i=1}^{n} x_i \alpha_i = \beta$.

Since α is a basis of V, the representation of a given vector β as a linear combination of $\alpha_1, \ldots, \alpha_n$ is unique. Thus, Definition 2.16 is unambiguous. The function $[\cdot]_\alpha: V \to M_{n \times 1}(F)$ is clearly bijective and preserves vector addition and scalar multiplication. Consequently, $[x\beta + y\delta]_\alpha = x[\beta]_\alpha + y[\delta]_\alpha$ for all x, y $\in$ F and $\beta, \delta \in V$. The column vector $[\beta]_\alpha$ is often called the α skeleton of β.

Suppose $\alpha = \{\alpha_1, \ldots, \alpha_n\}$ and $\delta = \{\delta_1, \ldots, \delta_n\}$ are two bases of V. Then there is a simple relationship between the α and δ skeletons of a given vector β. Let $M(\delta, \alpha)$ denote the n × n matrix whose columns are defined by the following equation:

2.17: $M(\delta, \alpha) = ([\delta_1]_\alpha | \cdots | [\delta_n]_\alpha)$

In equation 2.17, the ith column of $M(\delta, \alpha)$ is the n × 1 matrix $[\delta_i]_\alpha$. Multiplication by $M(\delta, \alpha)$ induces a map from $M_{n \times 1}(F)$ to $M_{n \times 1}(F)$ that connects the α and δ skeletons. Namely:

Theorem 2.18: $M(\delta, \alpha)[\beta]_\delta = [\beta]_\alpha$ for all $\beta \in V$.

Proof: Let us denote the ith column of any matrix M by $\text{Col}_i(M)$. Then for each $i = 1, \ldots, n$, we have $M(\delta, \alpha)[\delta_i]_\delta = M(\delta, \alpha)(0, \ldots, 0, 1, 0, \ldots, 0)^t$ $= \text{Col}_i(M(\delta, \alpha)) = [\delta_i]_\alpha$. Thus, the theorem is correct for $\beta \in \delta$.

Now we have already noted that $[\cdot]_\delta$ and $[\cdot]_\alpha$ preserve vector addition and scalar multiplication. So does multiplication by $M(\delta, \alpha)$ as a map on $M_{n \times 1}(F)$. Since any $\beta \in V$ is a linear combination of the vectors in δ, we conclude that $M(\delta, \alpha)[\beta]_\delta = [\beta]_\alpha$ for every $\beta \in V$. □

The matrix $M(\delta, \alpha)$ defined in 2.17 is called the change of basis matrix (between δ and α). It is often convenient to think of Theorem 2.18 in terms of the following commutative diagram:

2.19:

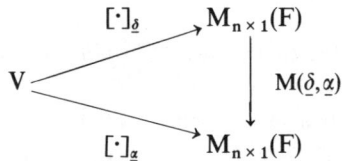

By a *diagram*, we shall mean a collection of vector spaces and maps (represented by arrows) between these spaces. A diagram is said to be commutative if any two sequences of maps (i.e., composites of functions in the diagram) that originate at the same space and end at the same space are equal. Thus, 2.19 is commutative if and only if the two paths from V to $M_{n \times 1}(F)$, clockwise and counterclockwise, are the same maps. This is precisely what Theorem 2.18 says.

Most of the maps or functions that we shall encounter in the diagrams in this book will be linear transformations. We take up the formal study of linear transformations in Section 3.

EXERCISES FOR SECTION 2

(1) Let $V_n = \{f(X) \in F[X] \mid \text{degree } f \leqslant n\}$. Show that each V_n is a finite-dimensional subspace of $F[X]$ of dimension $n + 1$. Since $F[X] = \bigcup_{n=1}^{\infty} V_n$, observe that Theorem 1.14 is false when the word "finite" is taken out of the theorem.

(2) Let V_n be as in Exercise 1 with $F = \mathbb{F}_2$. Find a basis of V_5 containing $1 + x$ and $x^2 + x + 1$.

(3) Show that any set of nonzero polynomials in $F[X]$, no two of which have the same degree, is linearly independent over F.

(4) Let $V = \{(a_1, a_2, \ldots) \in \mathbb{R}^N \mid a_i = 0 \text{ for all } i \text{ sufficiently large}\}$. Show that V is an infinite-dimensional subspace of $\mathbb{R}^N$. Find a basis for V.

(5) Prove Theorem 2.13(d) when $\dim(W_1 + W_2) = \infty$.

(6) Prove Corollary 2.15.

(7) Find the dimension of the subspace $V = L(\{\alpha, \beta, \gamma, \delta\}) \subseteq \mathbb{R}^4$, where $\alpha = (1, 2, 1, 0)$, $\beta = (-1, 1, -4, 3)$, $\gamma = (2, 3, 3, -1)$, and $\delta = (0, 1, -1, 1)$.

(8) Compute the following dimensions:
   (a) $\dim_{\mathbb{R}}(\mathbb{C})$.
   (b) $\dim_{\mathbb{Q}}(\mathbb{R})$.
   (c) $\dim_{\mathbb{Q}}(F)$, where F is the field given in Exercise 3 of Section 1.

(9) Suppose V is an n-dimensional vector space over the finite field $\mathbb{F}_p$. Argue that V is a finite set and find $|V|$.

(10) Suppose V is a vector space over a field F for which $|V| > 2$. Show that V has more than one basis.

(11) Let F be a subfield of the field F′. This means that the operations of addition and multiplication on F′ when restricted to F make F a field.
   (a) Show that F′ is a vector space over F.
   (b) Suppose $\dim_F(F') = n$. Let V be an m-dimensional vector space over F′. Show that V is an mn-dimensional vector space over F.

(12) Show that $\dim(V^n) = n \dim(V)$.

(13) Return to the space V_n in Exercise 1. Let $p_i(X) = \sum_{j=0}^{n} a_{ji} X^j$ for $i = 1, \ldots, r$. Set $A = (a_{ji}) \in M_{(n+1) \times r}(F)$. Show that the dimension of $L(\{p_1, \ldots, p_r\})$ is precisely the rank of A.

(14) Show that the dimension of the subspace of homogeneous polynomials of degree d in $F[X_0, \ldots, X_n]$ is the binomial coefficient $\binom{n+d}{d}$.

(15) Find the dimensions of the vector spaces in Exercises 18 and 19 of Section 1.

(16) Let $A \in M_{m \times n}(F)$. Set $CS(A) = \{AX \mid X \in M_{n \times 1}(F)\}$. $CS(A)$ is called the *column space* of A. Set $NS(A) = \{X \in M_{n \times 1}(F) \mid AX = 0\}$. $NS(A)$ is called the *null space* of A. Show that $CS(A)$ is a subspace of $M_{m \times 1}(F)$, and $NS(A)$ is a subspace of $M_{n \times 1}(F)$. Show that $\dim(CS(A)) + \dim(NS(A)) = n$.

(17) With the same notation as in Exercise 16, show the linear system $AX = B$ has a solution if and only if $\dim(CS(A)) = \dim(CS(A \mid B))$. Here $B \in M_{m \times 1}(F)$, and $(A \mid B)$ is the $m \times (n+1)$ augmented matrix obtained from A by adjoining the column B.

(18) Suppose V and W are two vector spaces over a field F such that $|V| = |W|$. Is $\dim V = \dim W$?

(19) Consider the set W of 2×2 matrices of the form

$$\begin{pmatrix} x & -x \\ y & z \end{pmatrix}$$

and the set Y of 2×2 matrices of the form

$$\begin{pmatrix} x & y \\ -x & z \end{pmatrix}$$

Show that W and Y are subspaces of $M_{2 \times 2}(F)$ and compute the numbers $\dim(W)$, $\dim(Y)$, $\dim(W + Y)$, and $\dim(W \cap Y)$.

3. LINEAR TRANSFORMATIONS

Let V and W be vector spaces over a field F.

Definition 3.1: A function T: $V \to W$ is called a *linear transformation* (linear map, homomorphism) if $T(x\alpha + y\beta) = xT(\alpha) + yT(\beta)$ for all x, $y \in F$ and $\alpha, \beta \in V$.

Before we state any general theorems about linear transformations, let us consider a few examples.

Example 3.2: The map that sends every vector in V to $0 \in W$ is clearly a linear map. We shall call this map the *zero map* and denote it by 0. If T: $V \to W$ and S: $W \to Z$ are linear transformations, then clearly the composite map ST: $V \to Z$ is a linear transformation. $\square$

Example 3.3: If V is finite dimensional with basis $\underline{\alpha} = \{\alpha_1, \ldots, \alpha_n\}$, then $[\cdot]_{\underline{\alpha}}$: $V \to M_{n \times 1}(F)$ is a linear transformation that is bijective. $\square$

Example 3.4: Taking the transpose, $A \to A^t$, is clearly a linear map from $M_{m \times n}(F) \to M_{n \times m}(F)$. $\square$

Example 3.5: Suppose $V = M_{m \times n}(F)$ and $A \in M_{m \times m}(F)$. Then multiplication by A (necessarily on the left) induces a linear transformation $T_A: V \to V$ given by $T_A(B) = AB$ for all $B \in V$. $\square$

Example 3.3 and 3.5 show that the commutative diagram in 2.19 consists of linear transformations.

Example 3.6: Suppose $V = C^k(I)$ with $k \geqslant 2$. Then ordinary differentiation $f \to f'$ is a linear transformation from $C^k(I)$ to $C^{k-1}(I)$. $\square$

Example 3.7: Suppose $V = F[X]$. We can formally define a derivative $f \to f'$ on V as follows: If $f(X) = \sum_{i=0}^{n} a_i X^i$, then $f'(X) = \sum_{i=1}^{n} i a_i X^{i-1}$. The reader can easily check that this map, which is called the *canonical derivative* on F[X], is a linear transformation. $\square$

Example 3.8: Suppose $V = \mathscr{R}(A)$ as in Example 1.10. Then $T(f) = \int_A f$ is a linear transformation from V to $\mathbb{R}$. $\square$

We shall encounter many more examples of linear transformations as we proceed. At this point, let us introduce a name for the collection of all linear transformations from V to W.

Definition 3.9: Let V and W be vector spaces over F. The set of all linear transformations from V to W will be denoted by $\text{Hom}_F(V, W)$.

When the base field F is clear from the context, we shall often write Hom(V, W) instead of $\text{Hom}_F(V, W)$. Thus, Hom(V, W) is the subset of the vector space W^V (Example 1.6) consisting of all linear transformations from V to W. If T, S ∈ Hom(V, W) and x, y ∈ F, then the function $xT + yS \in W^V$ is in fact a linear transformation. For if a, b ∈ F and α, β ∈ V, then $(xT + yS)(a\alpha + b\beta) = xT(a\alpha + b\beta) + yS(a\alpha + b\beta) = xaT(\alpha) + xbT(\beta) + yaS(\alpha) + ybS(\beta) = a(xT(\alpha) + yS(\alpha)) + b(xT(\beta) + yS(\beta)) = a(xT + yS)(\alpha) + b(xT + yS)(\beta)$. Therefore, $xT + yS \in \text{Hom}(V, W)$. We have proved the following theorem:

Theorem 3.10: Hom(V, W) is a subspace of W^V. ☐

Since any T ∈ Hom(V, W) has the property that T(0) = 0, we see that Hom(V, W) is always a proper subspace of W^V whenever W ≠ (0).

At this point, it is convenient to introduce the following terminology.

Definition 3.11: Let T ∈ Hom(V, W). Then,

(a) ker $T = \{\alpha \in V \mid T(\alpha) = 0\}$.

(b) Im $T = \{T(\alpha) \in W \mid \alpha \in V\}$.

(c) T is injective (monomorphism, 1 − 1) if ker T = (0).

(d) T is surjective (epimorphism, onto) if Im T = W.

(e) T is bijective (isomorphism) if T is both injective and surjective.

(f) We say V and W are isomorphic and write $V \cong W$ if there exists an isomorphism T ∈ Hom(V, W).

The set ker T is called the kernel of T and is clearly a subspace of V. Im T is called the image of T and is a subspace of W. Before proceeding further, let us give a couple of important examples of isomorphisms between vector spaces.

Example 3.12: $M_{n \times 1}(F) \cong M_{1 \times n}(F)$ via the transpose $A \to A^t$. We have already mentioned that $F^n = M_{1 \times n}(F)$. Thus, all three of the vector spaces $M_{n \times 1}(F)$, $M_{1 \times n}(F)$, and F^n are isomorphic to each other. ☐

Example 3.13: Suppose V is a finite-dimensional vector space over F. Then every basis $\alpha = \{\alpha_1, \ldots, \alpha_n\}$ of V determines a linear transformation $T(\alpha)$: $V \to F^n$ given by $T(\alpha)(\beta) = (x_1, \ldots, x_n)$ if and only if $\sum_{i=1}^n x_i \alpha_i = \beta$. $T(\alpha)$ is just the composite of the coordinate map $[\cdot]_\alpha: V \to M_{n \times 1}(F)$ and the transpose $M_{n \times 1}(F) \to M_{1 \times n}(F) = F^n$. Since both of these maps are isomorphisms, we see $T(\alpha)$ is an isomorphism. ☐

It is often notationally convenient to switch back and forth from column vectors to row vectors. For this reason, we give a formal name to the isomorphism $T(\alpha)$ introduced in Example 3.13.

Definition 3.14: Let V be a finite-dimensional vector space over F. If α is a basis of V, then $(\cdot)_\alpha\colon V \to F^n$ is the linear transformation defined by $(\beta)_\alpha = ([\beta]_\alpha)^t$ for all $\beta \in V$.

Thus, $(\beta)_\alpha = T(\alpha)(\beta)$ for all $\beta \in V$. We can now state the following theorem, whose proof is given in Example 3.13:

Theorem 3.15: Let V be a finite-dimensional vector space over F and suppose dim V = n. Then every basis α of V determines an isomorphism $(\cdot)_\alpha\colon V \to F^n$. $\square$

We now have two isomorphisms $[\cdot]_\alpha\colon V \to M_{n \times 1}(F)$ and $(\cdot)_\alpha\colon V \to F^n$ for every choice of basis α of a (finite-dimensional) vector space V. We shall be careful to distinguish between these two maps although they only differ by an isomorphism from $M_{n \times 1}(F)$ to $M_{1 \times n}(F)$. Notationally, F^n is easier to write than $M_{n \times 1}(F)$, and so most of our subsequent theorems will be written using the map $(\cdot)_\alpha$. With this in mind, let us reinterpret the commutative diagram given in 2.19.

If A is any $n \times n$ matrix with coefficients in F, then A induces a linear transformation $S_A\colon F^n \to F^n$ given by the following equation:

3.16:

$$S_A((x_1,\ldots,x_n)) = (A(x_1,\ldots,x_n)^t)^t = (x_1,\ldots,x_n)A^t$$

Using the notation in Example 3.5, we see S_A is the linear transformation that makes the following diagram commutative:

3.17:

The vertical arrows in 3.17 are isomorphisms. Clearly, T_A is an isomorphism if and only if A is invertible. Thus, S_A is an isomorphism if and only if A is invertible.

We shall replace the notation S_A (or T_A) with A^t (or A) and simply write

$$F^n \xrightarrow{\ A^t\ } F^n \quad \text{or} \quad M_{n \times 1}(F) \xrightarrow{\ A\ } M_{n \times 1}(F)$$

Now suppose α and δ are two bases of a finite-dimensional vector space V of dimension n. If we combine diagrams 2.19 and 3.17, we have the following commutative diagram:

3.18:

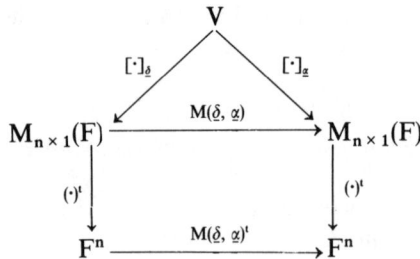

Since $(\cdot)_{\underline{\delta}} = ([\cdot]_{\underline{\delta}})^t$ and $(\cdot)_{\underline{\alpha}} = ([\cdot]_{\underline{\alpha}})^t$, we get the following corollary to Theorem 3.15:

Corollary 3.19: Suppose V is a finite-dimensional vector space of dimension n over F. If $\underline{\alpha}$ and $\underline{\delta}$ are two bases of V, then

3.20:

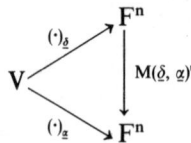

is a commutative diagram of isomorphisms.

Proof: We have already noted from 3.18 that 3.20 is commutative. Both $(\cdot)_{\underline{\alpha}}$ and $(\cdot)_{\underline{\delta}}$ are isomorphisms from Theorem 3.15. We need only argue $M(\underline{\delta}, \underline{\alpha})$ is an invertible matrix. Then the map $M(\underline{\delta}, \underline{\alpha})^t = S_{M(\underline{\delta}, \underline{\alpha})} : F^n \to F^n$ is an isomorphism.

Now change of basis matrices $M(\underline{\delta}, \underline{\alpha})$ are always invertible. This follows from Theorem 2.18. For any $\beta \in V$, we have $M(\underline{\alpha}, \underline{\delta})M(\underline{\delta}, \underline{\alpha})[\beta]_{\underline{\delta}} = M(\underline{\alpha}, \underline{\delta})[\beta]_{\underline{\alpha}} = [\beta]_{\underline{\delta}}$. This equation easily implies $M(\underline{\alpha}, \underline{\delta})M(\underline{\delta}, \underline{\alpha}) = I_n$, the n × n identity matrix. $\square$

In our next theorem, we shall need an isomorphic description of the vector space V^n introduced in Example 1.6.

Example 3.21: In this example, we construct a vector space isomorphic to V^n. Let V be a vector space over F, and let $n \in \mathbb{N}$. Consider the Cartesian product $V \times \cdots \times V$ (n times) $= \{(\alpha_1, \ldots, \alpha_n) | \alpha_i \in V\}$. Clearly, $V \times \cdots \times V$ is a vector space over F when we define vector addition and scalar multiplication by $(\alpha_1, \ldots, \alpha_n) + (\beta_1, \ldots, \beta_n) = (\alpha_1 + \beta_1, \ldots, \alpha_n + \beta_n)$, and $x(\alpha_1, \ldots, \alpha_n) = (x\alpha_1, \ldots, x\alpha_n)$.

Suppose A is any finite set with $|A| = n$. We can without any loss of generality assume $A = \{1, \ldots, n\}$. Then $V^A = V^n$. There is a natural isomorphism $T: V \times \cdots \times V \to V^n$ given by $T((\alpha_1, \ldots, \alpha_n)) = f \in V^n$, where $f(i) = \alpha_i$ for all $i = 1, \ldots, n$. The fact that T is an isomorphism is an easy exercise, which we leave to the reader. $\square$

Henceforth, we shall identify the vector spaces $V \times \cdots \times V$ (n times), V^n and V^A with $|A| = n$ and write just V^n to represent any one of these spaces. Using this notation, we have the following theorem:

Theorem 3.22: Let V and W be vector spaces over F, and suppose V is finite dimensional. Let $\dim V = n$.

(a) If $\underset{\sim}{\alpha} = \{\alpha_1, \ldots, \alpha_n\}$ is a basis of V, then for every $(\beta_1, \ldots, \beta_n) \in W^n$, there exists a unique $T \in \mathrm{Hom}(V, W)$ such that $T(\alpha_i) = \beta_i$ for $i = 1, \ldots, n$.

(b) Every basis $\underset{\sim}{\alpha}$ of V determines an isomorphism $\Psi(\underset{\sim}{\alpha})$: $\mathrm{Hom}(V, W) \cong W^n$.

Proof: (a) Let $\underset{\sim}{\alpha} = \{\alpha_1, \ldots, \alpha_n\}$ be a basis for V. Then $(\cdot)_{\underset{\sim}{\alpha}} \in \mathrm{Hom}(V, F^n)$ is an isomorphism. Let $\beta = (\beta_1, \ldots, \beta_n) \in W^n$. The n-tuple β determines a linear transformation $L_\beta \in \mathrm{Hom}(F^n, W)$ given by $L_\beta((x_1, \ldots, x_n)) = \sum_{i=1}^n x_i \beta_i$. The fact that L_β is a linear transformation is obvious. Set $T = L_\beta(\cdot)_{\underset{\sim}{\alpha}}$. Then $T \in \mathrm{Hom}(V, W)$ and $T(\alpha_i) = \beta_i$ for all $i = 1, \ldots, n$. The fact that T is the only linear transformation from V to W for which $T(\alpha_i) = \beta_i$ for $i = 1, \ldots, n$ is an easy exercise left to the reader.

(b) Fix a basis $\underset{\sim}{\alpha} = \{\alpha_1, \ldots, \alpha_n\}$ of V. Define $\Psi(\underset{\sim}{\alpha})$: $\mathrm{Hom}(V, W) \to W^n$ by $\Psi(\underset{\sim}{\alpha})(T) = (T(\alpha_1), \ldots, T(\alpha_n))$. The fact that $\Psi(\underset{\sim}{\alpha})$ is a linear transformation is obvious. We can define an inverse map χ: $W^n \to \mathrm{Hom}(V, W)$ by $\chi((\beta_1, \ldots, \beta_n)) = L_\beta(\cdot)_{\underset{\sim}{\alpha}}$. Here $\beta = (\beta_1, \ldots, \beta_n)$. Hence, $\Psi(\underset{\sim}{\alpha})$ is an isomorphism. $\square$

Theorem 3.22(a) implies that a given linear transformation $T \in \mathrm{Hom}(V, W)$ is completely determined by its values on a basis $\underset{\sim}{\alpha}$ of V. This remark is true whether V is finite dimensional or infinite dimensional. To define a linear transformation T from V to W, we need only define T on some basis $B = \{\alpha_i | i \in \Delta\}$ of V and then extend the definition of T linearly to all of $L(B) = V$. Thus, if $T(\alpha_i) = \beta_i$ for all $i \in \Delta$, then $T(\sum_{i \in \Delta} x_i \alpha_i)$ is defined to be $\sum_{i \in \Delta} x_i \beta_i$. These remarks provide a proof of the following generalization of Theorem 3.22(a):

3.23: Let V and W be vector spaces over F and suppose $B = \{\alpha_i | i \in \Delta\}$ is a basis of V. If $\{\beta_i | i \in \Delta\}$ is any subset of W, then there exists a unique $T \in \mathrm{Hom}(V, W)$ such that $T(\alpha_i) = \beta_i$ for all $i \in \Delta$. $\square$

Now suppose V and W are both finite-dimensional vector spaces over F. Let $\dim V = n$ and $\dim W = m$. If $\underset{\sim}{\alpha} = \{\alpha_1, \ldots, \alpha_n\}$ is a basis of V and

$\beta = \{\beta_1, \ldots, \beta_m\}$ is a basis of W, then the pair (α, β) determines a linear transformation $\Gamma(\alpha, \beta)$: Hom(V, W) $\to$ $M_{m \times n}(F)$ defined by the following equation:

3.24:

$$\Gamma(\alpha, \beta)(T) = ([T(\alpha_1)]_{\beta} | \cdots | [T(\alpha_n)]_{\beta})$$

In equation 3.24, $T \in$ Hom(V, W), and $\Gamma(\alpha, \beta)(T)$ is the m $\times$ n matrix whose ith column is the m $\times$ 1 matrix $[T(\alpha_i)]_{\beta}$. If $T_1, T_2 \in$ Hom(V, W) and x, y $\in$ F, then

$$
\begin{aligned}
\Gamma(\alpha, \beta)(xT_1 + yT_2) &= ([[(xT_1 + yT_2)(\alpha_1)]_{\beta} | \cdots | [(xT_1 + yT_2)(\alpha_n)]_{\beta}) \\
&= (x[T_1(\alpha_1)]_{\beta} + y[T_2(\alpha_1)]_{\beta} | \cdots | x[T_1(\alpha_n)]_{\beta} + y[T_2(\alpha_n)]_{\beta}) \\
&= x([T_1(\alpha_1)]_{\beta} | \cdots | [T_1(\alpha_n)]_{\beta}) + y([T_2(\alpha_1)]_{\beta} | \cdots | [T_2(\alpha_n)]_{\beta}) \\
&= x\Gamma(\alpha, \beta)(T_1) + y\Gamma(\alpha, \beta)(T_2)
\end{aligned}
$$

Thus $\Gamma(\alpha, \beta)$ is indeed a linear transformation from Hom(V, W) to $M_{m \times n}(F)$.

Suppose $T \in$ ker $\Gamma(\alpha, \beta)$. Then $\Gamma(\alpha, \beta)(T) = 0$. In particular, $[T(\alpha_i)]_{\beta} = 0$ for all $i = 1, \ldots, n$. But then $0 = [T(\alpha_i)]_{\beta}^t = (T(\alpha_i))_{\beta}$, and Theorem 3.15 implies $T(\alpha_i) = 0$. Thus, $T = 0$, and we conclude that $\Gamma(\alpha, \beta)$ is an injective linear transformation.

$\Gamma(\alpha, \beta)$ is surjective as well. To see this, let $A = (x_{ij}) \in M_{m \times n}(F)$. Let $\gamma_i = \sum_{j=1}^m x_{ji}\beta_j$ for $i = 1, \ldots, n$. Then $\{\gamma_1, \ldots, \gamma_n\} \subseteq W$, and $[\gamma_i]_{\beta} = (x_{1i}, \ldots, x_{mi})^t = \text{Col}_i(A)$ for all $i = 1, \ldots, n$. It follows from Theorem 3.22 that there exists a (necessarily unique) $T \in$ Hom(V, W) such that $T(\alpha_i) = \gamma_i$ for $i = 1, \ldots, n$. Thus, $\Gamma(\alpha, \beta)(T) = A$ and $\Gamma(\alpha, \beta)$ is surjective. We have, now proved the first statement in the following theorem:

Theorem 3.25: Let V and W be finite-dimensional vector spaces over F of dimensions n and m, respectively. Let α be a basis of V and β a basis of W. Then the map $\Gamma(\alpha, \beta)$: Hom(V, W) $\to$ $M_{m \times n}(F)$ defined by equation 3.24 is an isomorphism. For every $T \in$ Hom(V, W), the following diagram is commutative:

3.26:

$$
\begin{array}{ccc}
V & \xrightarrow{\quad T \quad} & W \\
{\scriptstyle (\cdot)_{\alpha}} \downarrow & & \downarrow {\scriptstyle (\cdot)_{\beta}} \\
F^n & \xrightarrow[\Gamma(\alpha, \beta)(T)^t]{} & F^m
\end{array}
$$

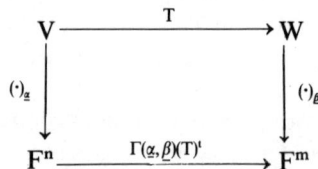

Proof: We need only argue that the diagram in 3.26 is commutative. Using the same notation as in 3.17, we have the following diagram:

3.27:

$$
\begin{array}{ccc}
V & \xrightarrow{\;\;T\;\;} & W \\
\left[\cdot\right]_{\underline{\alpha}} \downarrow & & \downarrow \left[\cdot\right]_{\underline{\beta}} \\
M_{n \times 1}(F) & \xrightarrow{\;\Gamma(\underline{\alpha},\,\underline{\beta})(T)\;} & M_{m \times 1}(F) \\
(\cdot)^{t} \downarrow & & \downarrow (\cdot)^{t} \\
F^{n} & \xrightarrow{\;\Gamma(\underline{\alpha},\,\underline{\beta})(T)^{t}\;} & F^{m}
\end{array}
$$

Since all the maps in 3.27 are linear and the bottom square commutes, we need only check $[\cdot]_{\underline{\beta}} T = \Gamma(\underline{\alpha}, \underline{\beta})(T)[\cdot]_{\underline{\alpha}}$ on a basis of V. Then the top square of 3.27 is commutative, and the commutativity of 3.26 follows. For any $\alpha_i \in \underline{\alpha}$, we have $([\cdot]_{\underline{\beta}} T)(\alpha_i) = [T(\alpha_i)]_{\underline{\beta}} = \mathrm{Col}_i(\Gamma(\underline{\alpha}, \underline{\beta})(T)) = \Gamma(\underline{\alpha}, \underline{\beta})(T)(0,\ldots, 1,\ldots, 0)^t = \Gamma(\underline{\alpha}, \underline{\beta})(T)[\alpha_i]_{\underline{\alpha}}$. $\square$

$\Gamma(\underline{\alpha}, \underline{\beta})(T)$ is called the *matrix representation* of the linear transformation T with respect to the bases $\underline{\alpha}$ and β. Since the vertical arrows in 3.26 and $\Gamma(\underline{\alpha}, \beta)$ are isomorphisms, V, W, Hom(V, W), and T are often identified with F^n, F^m, $M_{m \times n}(F)$, and $A = \Gamma(\underline{\alpha}, \underline{\beta})(T)$. Thus, the distinction between a linear transformation and a matrix is often blurred in the literature.

The matrix representation $\Gamma(\underline{\alpha}, \underline{\beta})(T)$ of T of course depends on the particular bases $\underline{\alpha}$ and β chosen. It is an easy matter to keep track of how $\Gamma(\underline{\alpha}, \underline{\beta})(T)$ changes with $\underline{\alpha}$ and $\underline{\beta}$.

Theorem 3.28: Let V and W be finite-dimensional vector spaces over F of dimensions n and m, respectively. Suppose $\underline{\alpha}$ and $\underline{\alpha}'$ are two bases of V and β and β' two bases of W. Then for every $T \in \mathrm{Hom}(V, W)$, we have

3.29:

$$
\Gamma(\underline{\alpha}', \beta')(T) = M(\beta, \beta')\Gamma(\underline{\alpha}, \underline{\beta})(T)M(\underline{\alpha}, \underline{\alpha}')^{-1}
$$

Proof: Before proving equation 3.29, we note that $M(\beta, \beta')$ (and $M(\underline{\alpha}, \underline{\alpha}')$) is the $m \times m$ (and $n \times n$) change of basis matrix given in equation 2.17. We have already noted that change of bases matrices are invertible and consequently all the terms in equation 3.29 make sense.

To see that 3.29 is in fact a valid equation, we merely combine the commutative diagrams 2.19 and 3.27. Consider the following diagram:

3.30:

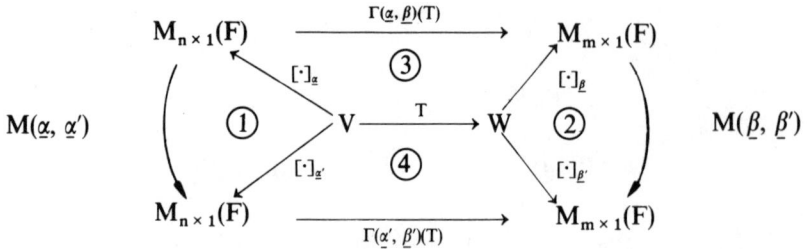

The diagram 3.30 is made up of four parts, which we have labeled ①, ②, ③, and ④. By Theorem 2.18, diagrams ① and ② are commutative. By Theorem 3.25, diagrams ③ and ④ are commutative. It follows that the entire diagram 3.30 is commutative. In particular, $M(\beta, \beta')\Gamma(\alpha, \beta)(T) = \Gamma(\alpha', \beta')(T)M(\alpha, \alpha')$. Solving this equation for $\Gamma(\alpha', \beta')(T)$ gives 3.29. □

Recall that two $m \times n$ matrices $A, B \in M_{m \times n}(F)$ are said to be equivalent if there exist invertible matrices $P \in M_{m \times m}(F)$ and $Q \in M_{n \times n}(F)$ such that $A = PBQ$. Equation 3.29 says that a given matrix representation $\Gamma(\alpha, \beta)(T)$ of T relative to a pair of bases (α, β) changes to an equivalent matrix when we replace (α, β) by new bases (α', β'). This leads to the following question: What is the simplest representation of a given linear transformation T? If we set $A = \Gamma(\alpha, \beta)(T)$, then we are asking, What is the simplest matrix B equivalent to A?

Recalling a few facts from elementary matrix theory gives us an easy answer to that question. Any invertible matrix P is a product, $P = E_r \cdots E_1$, of elementary matrices $E_1, \ldots, E_r$. $PA = E_r(\cdots (E_1 A) \cdots)$ is the $m \times n$ matrix obtained from A by preforming the elementary row operations on A represented by $E_1, \ldots, E_r$. Similarly $(PA)Q$ is the $m \times n$ matrix obtained from PA by preforming a finite number of elementary column operations on PA. Let us denote the rank of any $m \times n$ matrix A (i.e., the number of linearly independent rows or columns of A) by rk(A). If rk(A) = s, then we can clearly find invertible matrices P and Q such that

$$PAQ = \left(\begin{array}{c|c} I_s & 0 \\ \hline 0 & 0 \end{array} \right)$$

Here our notation

$$\left(\begin{array}{c|c} I_s & 0 \\ \hline 0 & 0 \end{array} \right)$$

means PAQ will have the $s \times s$ identity matrix, I_s, in its upper left-hand corner and zeros everywhere else.

If we apply these remarks to our situation in Theorem 3.28, we get the following corollary:

Corollary 3.31: Let V and W be finite-dimensional vector spaces over F of dimensions n and m, respectively. Let $\underline{\alpha}$ and β be bases of V and W. Let $T \in \text{Hom}(V, W)$, and set $A = \Gamma(\underline{\alpha}, \beta)(T)$. If $rk(A) = s$, then there exist bases $\underline{\alpha}'$ and β' of V and W, respectively, such that

3.32:

$$\Gamma(\underline{\alpha}', \underline{\beta}')(T) = \left(\begin{array}{c|c} I_s & 0 \\ \hline 0 & 0 \end{array} \right) \qquad \square$$

There is another representation problem that naturally arises when considering Theorem 3.28. Suppose $V = W$. If $\underline{\alpha}$ is a basis of V, then any $T \in \text{Hom}(V, V)$ is represented in terms of $\underline{\alpha}$ by an $n \times n$ matrix $A = \Gamma(\underline{\alpha}, \underline{\alpha})(T)$. If we change $\underline{\alpha}$ to a new basis $\underline{\alpha}'$ of V, then the representation of T changes to $B = \Gamma(\underline{\alpha}', \underline{\alpha}')(T)$. Equation 3.29 implies that $B = PAP^{-1}$, where $P = M(\underline{\alpha}, \underline{\alpha}')$. Recall that two $n \times n$ matrices A and B are similar if there exists an invertible $n \times n$ matrix P such that $B = PAP^{-1}$. Thus, different representations of the same $T \in \text{Hom}(V, V)$ with respect to different bases of V are similar matrices.

Now we can ask, What is the simplest representation of T? If we choose any basis $\underline{\alpha}$ of V and set $A = \Gamma(\underline{\alpha}, \underline{\alpha})(T)$, then our question becomes, What is the simplest matrix B similar to A? That question is not so easy to answer as the previous equivalence problem. We shall present some solutions to this question in Chapter III of this book.

Theorem 3.25 implies that $\dim \text{Hom}(V, W) = (\dim V)(\dim W)$ when V and W are finite dimensional. In our next theorem, we gather together some miscellaneous facts about linear transformations and the $\dim(\cdot)$ function.

Theorem 3.33: Let V and W be vector spaces over F and suppose $T \in \text{Hom}(V, W)$. Then

(a) If T is surjective, $\dim V \geqslant \dim W$.
(b) If $\dim V = \dim W < \infty$, then T is an isomorphism if and only if either T is injective or T is surjective.
(c) $\dim(\text{Im } T) + \dim(\ker T) = \dim V$.

Proof: (a) follows immediately from Theorem 2.11. In (b), if T is an isomorphism, then T is both injective and surjective. Suppose T is injective, and $n = \dim V = \dim W$. Let $\underline{\alpha} = \{\alpha_1, \ldots, \alpha_n\}$ be a basis of V. Since T is injective, $T\underline{\alpha} = \{T(\alpha_1), \ldots, T(\alpha_n)\}$ is a linearly independent set in W. Then $\dim W = n$ implies $T\underline{\alpha}$ is a basis of W. In particular, $W = L(T\underline{\alpha}) = T(L(\underline{\alpha})) = T(V)$. Thus, T is surjective, and hence, an isomorphism.

Suppose T is surjective. If $\underline{\alpha} = \{\alpha_1, \ldots, \alpha_n\}$ is a basis of V, then

$W = T(V) = T(L(\underline{\alpha})) = L(T\underline{\alpha})$. By Theorem 2.11, $T\underline{\alpha}$ contains a basis of W. Since dim $W = n$, $T\underline{\alpha}$ is a basis of W. Now let $\alpha \in \ker T$. Write $\alpha = \sum x_i \alpha_i$. Then $0 = \sum x_i T(\alpha_i)$. Since $T\underline{\alpha}$ is a basis of W, $x_1 = \cdots = x_n = 0$. Thus, $\alpha = 0$ and T is injective. This competes the proof of (b).

We prove (c) in the case that dim $V = n < \infty$. The infinite-dimensional case is left as an exercise at the end of this section. Let $\underline{\alpha} = \{\alpha_1, \ldots, \alpha_r\}$ be a basis of ker T. We take $r = 0$, and $\underline{\alpha} = \phi$ if T is injective. By Theorem 2.6, we can expand $\underline{\alpha}$ to a basis $\underline{\Delta} = \{\alpha_1, \ldots, \alpha_r, \beta_1, \ldots, \beta_s\}$ of V. Here $r + s = n$. We complete the proof of (c) by arguing that $T\underline{\beta} = \{T(\beta_1), \ldots, T(\beta_s)\}$ is a basis of Im T.

Suppose $\delta \in \text{Im } T$. Then $\delta = T(\gamma)$ for some $\gamma \in V$. Since $V = L(\underline{\Delta})$, $\gamma = x_1 \alpha_1 + \cdots + x_r \alpha_r + y_1 \beta_1 + \cdots + y_s \beta_s$ for some $x_i, y_i \in F$. Applying T to this equation, gives $\delta \in L(T\underline{\beta})$. Thus, $T\underline{\beta}$ spans Im T.

Suppose $\sum_{i=1}^{s} y_i T(\beta_i) = 0$ for some $y_i \in F$. Then $\sum_{i=1}^{s} y_i \beta_i \in \ker T$. Thus, $\sum_{i=1}^{s} y_i \beta_i = \sum_{i=1}^{r} x_i \alpha_i$ for some $x_i \in F$. Since $\underline{\Delta}$ is a basis of V, we conclude that $x_1 = \cdots = x_r = y_1 = \cdots = y_s = 0$. In particular, $\{T(\beta_1), \ldots, T(\beta_s)\}$ is linearly independent. Thus, $T\underline{\beta}$ is a basis of Im T, and the proof of (c) is complete. $\square$

We finish this section with a generalization of Theorem 3.33(c). We shall need the following definition.

Definition 3.34: By a chain complex $C = \{(V_i, d_i) \mid i \in \mathbb{Z}\}$ of vector spaces over F, we shall mean an infinite sequence $\{V_i\}$ of vector spaces V_i, one for each integer $i \in \mathbb{Z}$, together with a sequence $\{d_i\}$ of linear transformations, $d_i \in \text{Hom}(V_i, V_{i-1})$ for each $i \in \mathbb{Z}$, such that $d_{i+1} d_i = 0$ for all $i \in \mathbb{Z}$.

We usually draw a chain complex as an infinite sequence of spaces and maps as follows:

3.35:

$$C: \cdots \to V_{i+1} \xrightarrow{\ d_{i+1}\ } V_i \xrightarrow{\ d_i\ } V_{i-1} \to \cdots$$

If a chain complex C has only finitely many nonzero terms, then we can change notation and write C as

3.36:

$$C: 0 \to V_n \xrightarrow{\ d_n\ } V_{n-1} \xrightarrow{\ d_{n-1}\ } \cdots \to V_1 \xrightarrow{\ d_1\ } V_0 \to 0$$

It is understood here that all other vector spaces and maps not explicitly appearing in 3.36 are zero.

Definition 3.37: A chain complex

$$C: \cdots \to V_{i+1} \xrightarrow{\ d_{i+1}\ } V^i \xrightarrow{\ d_i\ } V_{i-1} \to \cdots$$

is said to be exact if $\text{Im } d_{i+1} = \ker d_i$ for every $i \in \mathbb{Z}$.

Let us consider an important example.

Example 3.38: Let V and W be vector space over F, and let $T \in \text{Hom}(V, W)$. Then

$$C: 0 \to \ker T \xrightarrow{\ i\ } V \xrightarrow{\ T\ } \text{Im } T \longrightarrow 0$$

is an exact chain complex. Here i denotes the inclusion of ker T into V.   □

We can generalize Example 3.38 slightly as follows:

Definition 3.39: By a short exact sequence, we shall mean an exact chain complex C of the following form:

3.40:

$$C: 0 \to V_2 \xrightarrow{\ d_2\ } V_1 \xrightarrow{\ d_1\ } V_0 \to 0$$

Thus, the example in 3.38 is a short exact sequence with $V_2 = \ker T$, $d_2 = i$, $V_1 = V$, etc. Clearly, a chain complex C of the form depicted in 3.40 is a short exact sequence if and only if d_2 is injective, d_1 is surjective, and $\text{Im } d_2 = \ker d_1$. Theorem 3.33(c) implies that if C is a short exact sequence, then $\dim V_2 - \dim V_1 + \dim V_0 = 0$. We can now prove the following generalization of this result:

Theorem 3.41: Suppose

$$C: 0 \to V_n \xrightarrow{\ d_n\ } V_{n-1} \xrightarrow{\ d_{n-1}\ } \cdots \to V_1 \xrightarrow{\ d_1\ } V_0 \to 0$$

is an exact chain complex. Then $\sum_{i=0}^{n} (-1)^i \dim V_i = 0$.

Proof: The chain complex C can be decomposed into the following short exact sequences

$$C_1: 0 \to \ker d_1 \to V_1 \xrightarrow{\ d_1\ } V_0 \to 0$$

$$C_2: 0 \to \ker d_2 \to V_2 \xrightarrow{\ d_2\ } \ker d_1 \to 0$$

$$\vdots$$

$$C_n: 0 \to \ker d_n \to V_n \xrightarrow{\ d_n\ } \ker d_{n-1} \to 0$$

If we now apply Theorem 3.33(c) to each C_i and add the results, we get $\sum_{i=0}^{n}(-1)^i \dim V_i = 0.$ □

EXERCISES FOR SECTION 3

(1) Let V and W be vector spaces over F.

    (a) Show that the Cartesian product $V \times W = \{(\alpha, \beta) \mid \alpha \in V, \beta \in W\}$ is a vector space under componentwise addition and scalar multiplication.

    (b) Compute $\dim(V \times W)$ when V and W are finite dimensional.

    (c) Suppose $T: V \to W$ is a function. Show that $T \in \mathrm{Hom}(V, W)$ if and only if the graph $G_T = \{(\alpha, T(\alpha)) \in V \times W \mid \alpha \in V\}$ of T is a subspace of $V \times W$.

(2) Let $T \in \mathrm{Hom}(V, W)$ and $S \in \mathrm{Hom}(W, V)$. Prove the following statements:

    (a) If ST is surjective, then S is surjective.

    (b) If ST is injective, then T is injective.

    (c) If $ST = I_V$ (the identity map on V) and $TS = I_W$, then T is an isomorphism.

    (d) If V and W have the same finite dimension n, then $ST = I_V$ implies T is an isomorphism. Similarly, $TS = I_W$ implies T is an isomorphism.

(3) Show that Exercise 2(d) is false in general. (*Hint:* Let $V = W$ be the vector space in Exercise 4 of Section 2.)

(4) Show that $F^n \cong F^m \Rightarrow n = m$.

(5) Let $T \in \mathrm{Hom}(V, V)$. If T is not injective, show there exists a nonzero $S \in \mathrm{Hom}(V, V)$ with $TS = 0$. If T is not surjective, show there exists a nonzero $S \in \mathrm{Hom}(V, V)$ such that $ST = 0$.

(6) In the proof of Corollary 3.19, we claimed that $M(\underline{\alpha}, \underline{\delta})M(\underline{\delta}, \underline{\alpha})[\beta]_{\underline{\delta}} = [\beta]_{\underline{\delta}}$ for all $\beta \in V$ implies $M(\underline{\alpha}, \underline{\delta})M(\underline{\delta}, \underline{\alpha}) = I_n$. Give a proof of this fact.

(7) When considering diagram 3.17 we claimed T_A is an isomorphism if and only if A is an invertible matrix. Give a proof of this fact.

(8) Show that Theorem 3.33(c) is correct for any vector spaces V and W. Some knowledge of cardinal arithmetic is needed for this exercise.

(9) Let $T \in \mathrm{Hom}(V, V)$. Show that $T^2 = 0$ if and only if there exist two subspaces M and N of V such that

    (a) $M + N = V$.

    (b) $M \cap N = (0)$.

    (c) $T(N) = 0$.

    (d) $T(M) \subseteq N$.

(10) Let $T \in \text{Hom}(V, V)$ be an involution, that is, $T^2 = I_V$. Show that there exists two subspaces M and N of V such that

(a) $M + N = V$.

(b) $M \cap N = (0)$.

(c) $T(\alpha) = \alpha$ for every $\alpha \in M$.

(d) $T(\alpha) = -\alpha$ for every $\alpha \in N$.

In Exercise 10, we assume $2 \neq 0$ in F. If $F = \mathbb{F}_2$, are there subspaces M and N satisfying (a)–(d)?

(11) Let $T \in \text{Hom}_F(V, V)$. If $f(X) = a_n X^n + \cdots + a_1 X + a_0 \in F[X]$, then $f(T) = a_n T^n + \cdots + a_1 T + a_0 I_V \in \text{Hom}(V, V)$. Show that $\dim_F V = m < \infty$ implies there exists a nonzero polynomial $f(X) \in F[X]$ such that $f(T) = 0$.

(12) If S, $T \in \text{Hom}_F(V, F)$ such that $S(\alpha) = 0$ implies $T(\alpha) = 0$, prove that $T = xS$ for some $x \in F$.

(13) Let W be a subspace of V with $m = \dim W \leqslant \dim V = n < \infty$. Let $Z = \{T \in \text{Hom}(V, V) \mid T(\alpha) = 0 \text{ for all } \alpha \in W\}$. Show that Z is a subspace of $\text{Hom}(V, V)$ and compute its dimension.

(14) Suppose V is a finite-dimensional vector space over F, and let S, $T \in \text{Hom}_F(V, V)$. If $ST = I_V$, show there exists a polynomial $f(X) \in F[X]$ such that $S = f(T)$.

(15) Use two appropriate diagrams as in 3.27 to prove the following theorem: Let V, W, Z be finite-dimensional vector spaces of dimensions n, m, and p, respectively. Let $\underline{\alpha}$, $\underline{\beta}$, and $\underline{\gamma}$ be bases of V, W, and Z. If $T \in \text{Hom}(V, W)$ and $S \in \text{Hom}(W, Z)$, then $\Gamma(\underline{\alpha}, \underline{\gamma})(ST) = \Gamma(\underline{\beta}, \underline{\gamma})(S)\Gamma(\underline{\alpha}, \underline{\beta})(T)$.

(16) Suppose

$$C: \cdots \to V_{i+1} \xrightarrow{d_{i+1}} V_i \xrightarrow{d_i} \cdots \to V_1 \xrightarrow{d_1} V_0 \to 0$$

and

$$C': \cdots \to V'_{i+1} \xrightarrow{d'_{i+1}} V'_i \xrightarrow{d'_i} \cdots \to V'_1 \xrightarrow{d'_1} V'_0 \to 0$$

are chain complexes with C' exact. Let $T_0 \in \text{Hom}_F(V_0, V'_0)$. Show that there exists $T_i \in \text{Hom}_F(V_i, V'_i)$ such that $T_{i-1} d_i = d'_i T_i$ for all $i = 1, \dots$. The collection of linear transformations $\{T_i\}$ is called a chain map from C to C'.

(17) Suppose $C = \{(V_i, d_i) \mid i \in \mathbb{Z}\}$ and $C' = \{(V'_i, d'_i) \mid i \in \mathbb{Z}\}$ are two chain complexes. Let $T = \{T_i\}_{i \in \mathbb{Z}}$ be a chain map from C to C'. Thus, $T_i: V_i \to V'_i$, and $T_{i-1} d_i = d'_i T_i$ for all $i \in \mathbb{Z}$. For each $i \in \mathbb{Z}$, set $V''_i = V_{i-1} \times V'_i = \{(\alpha, \beta) \mid \alpha \in V_{i-1}, \beta \in V'_i\}$. Define a map $d''_i: V''_i \to V''_{i-1}$ by

$d_i''(\alpha, \beta) = (-d_{i-1}(\alpha), T_{i-1}(\alpha) + d_i'(\beta).)$ Show that $C'' = \{(V_i'', d_i'') \,|\, i \in \mathbb{Z}\}$ is a chain complex. The complex C'' is called the *mapping cone* of T.

(18) Use Theorem 3.33(c) to give another proof of Exercise 16 in Section 2.

(19) Find a $T \in \text{Hom}_{\mathbb{R}}(\mathbb{C}, \mathbb{C})$ that is not $\mathbb{C}$-linear.

(20) Let V be a finite-dimensional vector space over F. Suppose $T \in \text{Hom}_F(V, V)$ such that $\dim(\text{Im}(T^2)) = \dim(\text{Im}(T))$. Show that $\text{Im}(T) \cap \ker(T) = \{0\}$.

(21) The special case of equation 3.29 where $V = W$, $\underline{\alpha} = \underline{\beta}$, and $\underline{\alpha}' = \underline{\beta}'$ is very important. Write out all the matrices and verify equation 3.29 in the following example: $T: \mathbb{R}^3 \to \mathbb{R}^3$ is the linear transformation given by $T(\delta_1) = 2\delta_1 + 2\delta_2 + \delta_3$, $T(\delta_2) = \delta_1 + 3\delta_2 - \delta_3$, and $T(\delta_3) = -\delta_1 + 2\delta_2$. Let $\underline{\alpha} = \{\alpha_1, \alpha_2, \alpha_3\}$, where $\alpha_1 = (1, 2, 1)$, $\alpha_2 = (1, 0, -1)$ and $\alpha_3 = (0, 1, -1)$. Compute $\Gamma(\underline{\delta}, \underline{\delta})(T)$, $\Gamma(\underline{\alpha}, \underline{\alpha})(T)$, and the change of bases matrices in 3.29.

(22) Let V be a finite-dimensional vector space over F. Suppose $TS = ST$ for every $S \in \text{Hom}_F(V, V)$. Show that $T = xI_V$ for some $x \in F$.

(23) Let $A, B \in M_{n \times n}(F)$ with at least one of these matrices nonsingular. Show that AB and BA are similar. Does this remain true if both A and B are singular?

4. PRODUCTS AND DIRECT SUMS

Let $\{V_i \,|\, i \in \Delta\}$ be a collection of vector spaces over a common field F. Our indexing set Δ may be finite or infinite. We define the product $\prod_{i \in \Delta} V_i$ of the V_i as follows:

Definition 4.1: $\prod_{i \in \Delta} V_i = \{f: \Delta \to \bigcup_{i \in \Delta} V_i \,|\, f \text{ is a function with } f(i) \in V_i \text{ for all } i \in \Delta\}$.

We can give the set $\prod_{i \in \Delta} V_i$ the structure of a vector space (over F) by defining addition and scalar multiplication pointwise. Thus, if f, $g \in \prod_{i \in \Delta} V_i$, then $f + g$ is defined by $(f + g)(i) = f(i) + g(i)$. If $f \in \prod_{i \in \Delta} V_i$ and $x \in F$, then xf is defined by $(xf)(i) = x(f(i))$. The fact that $\prod_{i \in \Delta} V_i$ is a vector space with these operations is straightforward. Henceforth, the symbol $\prod_{i \in \Delta} V_i$ will denote the vector space whose underlying set is given in 4.1 and whose vector operations are pointwise addition and scalar multiplication.

Suppose $V = \prod_{i \in \Delta} V_i$ is a product. It is sometimes convenient to identify a given vector $f \in V$ with its set of values $\{f(i) \,|\, i \in \Delta\}$. $f(i)$ is called the ith coordinate of f, and we think of f as the "Δ-tuple" $(f(i))_{i \in \Delta}$. Addition and scalar multiplication in V are given in terms of Δ-tuples as follows: $(f(i))_{i \in \Delta} + (g(i))_{i \in \Delta} = (f(i) + g(i))_{i \in \Delta}$, and $x(f(i))_{i \in \Delta} = (xf(i))_{i \in \Delta}$. This particular viewpoint is especially fruitful when $|\Delta| = n < \infty$. In this case, we can assume $\Delta = \{1, 2, \ldots, n\}$. Each $f \in V$ is then identified with the n-tuple $(f(1), \ldots, f(n))$. When $|\Delta| = n$, we shall use the notation $V_1 \times \cdots \times V_n$ instead of $\prod_{i \in \Delta} V_i$. Thus, the examples given in 1.5, 3.21, and Exercise 1 of Section 3 are all special cases of finite products. Example 1.5 is a product in which every V_i is the same vector space V.

If $V = \prod_{i \in \Delta} V_i$, then for every pair of indices $(p, q) \in \Delta \times \Delta$, there exist linear transformations $\pi_p \in \text{Hom}(V, V_p)$, and $\theta_q \in \text{Hom}(V_q, V)$ defined as follows:

Definition 4.2:

(a) $\pi_p: V \to V_p$ is given by $\pi_p(f) = f(p)$ for all $f \in V$.

(b) $\theta_q: V_q \to V$ is given by

$$\theta_q(\alpha)(i) = \begin{cases} \alpha & \text{if } i = q \\ 0 & \text{if } i \neq q \end{cases}$$

In Definition 4.2(b), $\alpha \in V_q$. $\theta_q(\alpha)$ is that function in V whose only nonzero value is α taken on at $i = q$. The fact that π_p and θ_q are linear transformations is obvious. Our next theorem lists some of the interesting properties these two sets of maps have.

Theorem 4.3: Let $V = \prod_{i \in \Delta} V_i$. Then

(a) $\pi_p \theta_p = I_{V_p}$, the identity map on V_p, for all $p \in \Delta$.

(b) $\pi_p \theta_q = 0$ for all $p \neq q$ in Δ.

(c) If Δ is finite, $\sum_{p \in \Delta} \theta_p \pi_p = I_V$, the identity map on V.

(d) π_p is surjective and θ_p is injective for all $p \in \Delta$.

(e) Let W be a second vector space over F. A function $T: W \to V$ is a linear transformation if and only if $\pi_p T \in \text{Hom}(W, V_p)$ for all $p \in \Delta$.

(f) The vector space V together with the set $\{\pi_p | p \in \Delta\}$ of linear transformations satisfies the following universal mapping property: Suppose W is any vector space over F and $\{T_p \in \text{Hom}(W, V_p) | p \in \Delta\}$ a set of linear transformations. Then there exists a unique $T \in \text{Hom}(W, V)$ such that for every $p \in \Delta$ the following diagram is commutative:

4.4:

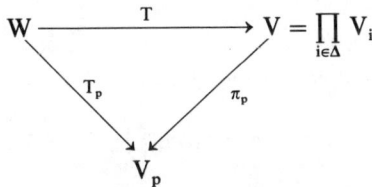

Proof: (a), (b), and (c) follow immediately from the definitions. π_p is surjective and θ_p is injective since $\pi_p \theta_p = I_{V_p}$. Thus, (d) is clear. As for (e), we need only argue that T is linear provided $\pi_p T$ is linear for all $p \in \Delta$. Let $\alpha, \beta \in W$ and $x, y \in F$.

Then for every $p \in \Delta$, we have $\pi_p(T(x\alpha + y\beta)) = \pi_p T(x\alpha + y\beta) = x\pi_p T(\alpha) + y\pi_p T(\beta) = \pi_p(xT(\alpha) + yT(\beta))$. Now it is clear from our definitions that two functions f, $g \in V$ are equal if and only if $\pi_p(f) = \pi_p(g)$ for all $p \in \Delta$. Consequently, $T(x\alpha + y\beta) = xT(\alpha) + yT(\beta)$, and T is linear.

Finally, we come to the proof of (f). We shall have no use for this fact in this text. We mention this result only because in general category theory products are defined as the unique object satisfying the universal mapping property given in (f). The map $T: W \to V$ making 4.4 commute is given by $T(\alpha) = (T_i(\alpha))_{i \in \Delta}$. We leave the details for an exercise at the end of this section. $\square$

The map $\pi_p: V \to V_p$ in Definition 4.2(a) is called the pth projection or pth coordinate map of V. The map $\theta_q: V_q \to V$ is often called the qth injection of V_q into V. These maps can be used to analyze linear transformations to and from products. We begin first with the case where $|\Delta| < \infty$.

Theorem 4.5: Suppose $V = V_1 \times \cdots \times V_n$ is a finite product of vector spaces, and let W be another vector space. If $T_i \in \mathrm{Hom}(W, V_i)$ for $i = 1, \ldots, n$, then there exists a unique $T \in \mathrm{Hom}(W, V_1 \times \cdots \times V_n)$ such that $\pi_i T = T_i$ for all $i = 1, \ldots, n$.

Proof: Set $T = \sum_{i=1}^{n} \theta_i T_i$ and apply Theorem 4.3. $\square$

As an immediate corollary to Theorem 4.5, we get the following result:

Corollary 4.6: If $|\Delta| = n < \infty$, then $\mathrm{Hom}(W, \prod_{i \in \Delta} V_i) \cong \prod_{i \in \Delta} \mathrm{Hom}(W, V_i)$.

Proof: Define a map $\Psi: \mathrm{Hom}(W, V_1 \times \cdots \times V_n) \to \mathrm{Hom}(W, V_1) \times \cdots \times \mathrm{Hom}(W, V_n)$ by $\Psi(T) = (\pi_1 T, \ldots, \pi_n T)$. One easily checks that Ψ is an injective, linear transformation. Theorem 4.5 implies Ψ is surjective. $\square$

We have a similar result for products in the first slot of Hom.

Theorem 4.7: Suppose $V = V_1 \times \cdots \times V_n$ is a finite product of vector spaces, and let W be another vector space. If $T_i \in \mathrm{Hom}(V_i, W)$ for $i = 1, \ldots, n$, then there exists a unique $T \in \mathrm{Hom}(V_1 \times \cdots \times V_n, W)$ such that $T\theta_i = T_i$ for all $i = 1, \ldots, n$.

Proof: Set $T = \sum_{i=1}^{n} T_i \pi_i$ and apply Theorem 4.3. $\square$

Corollary 4.8: If $|\Delta| = n < \infty$, then $\mathrm{Hom}(\prod_{i \in \Delta} V_i, W) \cong \prod_{i \in \Delta} \mathrm{Hom}(V_i, W)$.

Proof: Define a map $\Psi: \mathrm{Hom}(\prod_{i \in \Delta} V_i, W) \to \prod_{i \in \Delta} \mathrm{Hom}(V_i, W)$ by $\Psi(T) = (T\theta_1, \ldots, T\theta_n)$. Again the reader can easily check that Ψ is an injective, linear transformation. Theorem 4.7 implies that Ψ is surjective. $\square$

Suppose $V = V_1 \times \cdots \times V_n$ is a finite product of vector spaces over F. Let B_i be a basis of V_i, $i = 1, \ldots, n$. We can think of the vectors in V as n-tuples $(\alpha_1, \ldots, \alpha_n)$ with $\alpha_i \in V_i$. For any i and $\alpha \in V_i$, $\theta_i(\alpha) = (0, \ldots, \alpha, 0, \ldots, 0)$. Thus, $\theta_i(\alpha)$ is the n-tuple of V that is zero everywhere except for an α in the ith slot. Since θ_i is injective, $\theta_i \colon V_i \cong \theta_i(V_i)$. In particular, $\theta_i(B_i)$ is a basis of the subspace $\theta_i(V_i)$. Since $\theta_i(B_i) \cap L(\bigcup_{j \neq i} \theta_j(B_j)) = (0)$, $B = \bigcup_{i \in \Delta} \theta_i(B_i)$ is a linearly independent set. Clearly, $V = \sum_{i=1}^n \theta_i(V_i)$. Consequently, B is a basis of V. We have now proved the following theorem:

Theorem 4.9: Let $V = V_1 \times \cdots \times V_n$ be a finite product of vector spaces. If B_i is a basis of V_i, $i = 1, \ldots, n$, then $B = \bigcup_{i=1}^n \theta_i(B_i)$ is a basis of V. In particular, if each V_i is finite dimensional, then so is V. In this case, we have $\dim V = \sum_{i=1}^n \dim V_i$. $\square$

At this point, let us say a few words about our last three theorems when $|\Delta| = \infty$. Corollary 4.6 is true for any indexing set Δ. The map $\Psi(T) = (\pi_i T)_{i \in \Delta}$ is an injective, linear transformation as before. We cannot use Theorem 4.5 to conclude Ψ is surjective, since $\sum_{i \in \Delta} \theta_i T_i$ makes no sense when $|\Delta| = \infty$. However, we can argue directly that Ψ is surjective. Let $(T_i)_{i \in \Delta} \in \prod_{i \in \Delta} \text{Hom}(W, V_i)$. Define $T \in \text{Hom}(W, \prod_{i \in \Delta} V_i)$ by $T(\alpha) = (T_i(\alpha))_{i \in \Delta}$. Clearly $\Psi(T) = (T_i)_{i \in \Delta}$. Thus, we have the following generalization of 4.6:

4.10: For any indexing set Δ, $\text{Hom}(W, \prod_{i \in \Delta} V_i) \cong \prod_{i \in \Delta} \text{Hom}(W, V_i)$.

In general, Corollary 48 is false when $|\Delta| = \infty$. For example, if $W = F$ and $V_i = F$ for all $i \in \Delta$, then the reader can easily see that $|\text{Hom}_F(\prod_{i \in \Delta} F, F)| > |\prod_{i \in \Delta} F|$ when Δ is infinite. Since $\text{Hom}_F(F, F) \cong F$, we see that $\text{Hom}(\prod_{i \in \Delta} F, F)$ cannot be isomorphic to $\prod_{i \in \Delta} \text{Hom}(F, F)$.

If $V = \prod_{i \in \Delta} V_i$ with $|\Delta| = \infty$ and B_i is a basis of V_i, then $\bigcup_{i \in \Delta} \theta_i(B_i)$ is a linearly independent subset of V. But in general, $V \neq \sum_{i \in \Delta} \theta_i(V_i)$. For a concrete example, consider $V = \mathbb{R}^{\mathbb{N}}$ in Exercise 5 of Section 1. Thus, $\bigcup_{i \in \Delta} \theta_i(B_i)$ is not in general a basis for V. In particular, Theorem 4.9 is false when $|\Delta| = \infty$.

Let us again suppose $V = \prod_{i \in \Delta} V_i$ with Δ an arbitrary set. There is an important subspace of V that we wish to study.

Definition 4.11: Let $\bigoplus_{i \in \Delta} V_i = \{f \in \prod_{i \in \Delta} V_i \mid f(i) = 0 \text{ except possibly for finitely many } i \in \Delta\}$.

Clearly $\bigoplus_{i \in \Delta} V_i$ is a subspace of V under pointwise addition and scalar multiplication. In terms of Δ-tuples, the vector $f = (\alpha_i)_{i \in \Delta}$ lies in $\bigoplus_{i \in \Delta} V_i$ if and only if there exists some finite subset Δ_0 (possibly empty) of Δ such that $\alpha_i = 0$ for all $i \in \Delta - \Delta_0$. If $|\Delta| < \infty$, then $\bigoplus_{i \in \Delta} V_i = \prod_{i \in \Delta} V_i$. If $|\Delta| = \infty$, then $\bigoplus_{i \in \Delta} V_i$ is usually a proper subspace of V. Consider the following example:

Example 4.12: Let $F = \mathbb{R}$, $\Delta = \mathbb{N}$, and $V_i = \mathbb{R}$ for all $i \in \Delta$. Then the $\mathbb{N}$-tuple $(1)_{i \in \mathbb{N}}$, that is, the function $f: \mathbb{N} \to \bigcup_{i \in \mathbb{N}} \mathbb{R}$ given by $f(i) = 1$ for all $i \in \mathbb{N}$, is a vector in $V = \prod_{i \in \mathbb{N}} \mathbb{R}$ but not a vector in $\bigoplus_{i \in \mathbb{N}} \mathbb{R}$. $\square$

The vector space $\bigoplus_{i \in \Delta} V_i$ is called the direct sum of the V_i. It is also called the subdirect product of the V_i and written $\coprod_{i \in \Delta} V_i$. In this text, we shall consistently use the notation $\bigoplus_{i \in \Delta} V_i$ to indicate the direct sum of the V_i. If $|\Delta| = n < \infty$, then we can assume $\Delta = \{1, 2, \ldots, n\}$. In this case we shall write $V_1 \oplus \cdots \oplus V_n$ or $\bigoplus_{i=1}^{n} V_i$ instead of $\bigoplus_{i \in \Delta} V_i$. Thus, $V_1 \oplus \cdots \oplus V_n$, $\bigoplus_{i=1}^{n} V_i$, $V_1 \times \cdots \times V_n$, $\prod_{i \in \Delta} V_i$, and $\bigoplus_{i \in \Delta} V_i$ are all the same space when $|\Delta| = n < \infty$.

Since $\bigoplus_{i \in \Delta} V_i = \sum_{i \in \Delta} \theta_i(V_i)$, our comments after 4.10 imply the following theorem:

Theorem 4.13: Suppose $V = \bigoplus_{i \in \Delta} V_i$ is the direct sum of vector spaces V_i. Let B_i be a basis of V_i. Then $B = \bigcup_{i \in \Delta} \theta_i(B_i)$ is a basis of V. $\square$

The subspace $\bigoplus_{i \in \Delta} V_i$ constructed in Definition 4.11 is sometimes called the external direct sum of the V_i because the vector spaces $\{V_i | i \in \Delta\}$ a priori have no relationship to each other. We finish this section with a construction that is often called an internal direct sum.

Suppose V is a vector space over F. Let $\{V_i | i \in \Delta\}$ be a collection of subspaces of V. Here our indexing set Δ may be finite or infinite. We can construct the (external) direct sum $\bigoplus_{i \in \Delta} V_i$ of the V_i as in Definition 4.11 and consider the natural linear transformation $S: \bigoplus_{i \in \Delta} V_i \to V$ given by $S((\alpha_i)_{i \in \Delta}) = \sum_{i \in \Delta} \alpha_i$. Since $(\alpha_i)_{i \in \Delta} \in \bigoplus_{i \in \Delta} V_i$, only finitely many of the α_i are nonzero. Therefore, $\sum_{i \in \Delta} \alpha_i$ is a well defined finite sum in V. Thus, S is well defined and clearly linear.

Definition 4.14: Let $\{V_i | i \in \Delta\}$ be a collection of subspaces of V. We say these subspaces are independent if the linear transformation $S: \bigoplus_{i \in \Delta} V_i \to V$ defined above is injective.

Note Im $S = \sum_{i \in \Delta} V_i$. Thus, the subspaces V_i, $i \in \Delta$, are indepedent if and only if $\bigoplus_{i \in \Delta} V_i \cong \sum_{i \in \Delta} V_i$ via S. A simple example of independent subspaces is provided by Theorem 2.13(c).

Example 4.15: Let V be a vector space over F and W a subspace of V. Let W' be any complement of W. Then W, W' are independent. The direct sum $W \oplus W'$ is just the product $W \times W'$, and $S: W \times W' \to W + W'$ is given by $S((\alpha, \beta)) = \alpha + \beta$. If $(\alpha, \beta) \in \ker S$, then $\alpha + \beta = 0$. But $W \cap W' = 0$. Therefore, $\alpha = -\beta \in W \cap W'$ implies $\alpha = \beta = 0$. Thus, S is injective, and W, W' are independent. $\square$

In our next theorem, we collect a few simple facts about independent subspaces.

Theorem 4.16: Let $\{V_i | i \in \Delta\}$ be a collection of subspaces of V. Then the following statements are equivalent:

   (a) The V_i, $i \in \Delta$ are independent.
   (b) Every vector $\alpha \in \sum_{i \in \Delta} V_i$ can be written uniquely in the form $\alpha = \sum_{i \in \Delta} \alpha_i$ with $\alpha_i \in V_i$ for all $i \in \Delta$.
   (b') If $\sum_{i \in \Delta} \alpha_i = 0$ with $\alpha_i \in V_i$, then $\alpha_i = 0$ for all $i \in \Delta$.
   (c) For every $j \in \Delta$, $V_j \cap (\sum_{i \neq j} V_i) = (0)$.

Proof: In statements (b) and (b'), $\sum_{i \in \Delta} \alpha_i$ means $\alpha_i = 0$ for all but possibly finitely many $i \in \Delta$. It is obvious that (b) and (b') are equivalent. So, we argue (a) $\Leftrightarrow$ (b') $\Leftrightarrow$ (c).

Suppose the V_i are independent. If $\sum_{i \in \Delta} \alpha_i = 0$ with $\alpha_i \in V_i$ for all $i \in \Delta$, then $S((\alpha_i)_{i \in \Delta}) = \sum_{i \in \Delta} \alpha_i = 0$. Since S is injective, we conclude that $\alpha_i = 0$ for all $i \in \Delta$. Thus, (a) implies (b'). Similarly, (b') implies (a).

Suppose we assume (b'). Fix $j \in \Delta$. Let $\alpha \in V_j \cap (\sum_{i \neq j} V_i)$. Then $\alpha = \alpha_j$ for some $\alpha_j \in V_j$, and $\alpha = \sum_{i \in \Delta - \{j\}} \alpha_i$ for some $\alpha_i \in V_i$. As usual, all the α_i here are zero except possibly for finitely many indices $i \neq j$. Thus, $0 = \sum_{i \in \Delta - \{j\}} \alpha_i + (-1)\alpha_j$. (b') then implies $\alpha_j = \alpha_i = 0$ for all $i \in \Delta - \{j\}$. In particular, $\alpha = 0$, and (c) is established.

Suppose we assume (c). Let $\sum_{i \in \Delta} \alpha_i = 0$ with $\alpha_i \in V_i$. If every $\alpha_i = 0$, there is nothing to prove. Suppose some α_i, say α_j, is not zero. Then $\alpha_j = -\sum_{i \neq j} \alpha_i \in V_j \cap (\sum_{i \neq j} V_i)$ implies $V_j \cap (\sum_{i \neq j} V_i) \neq 0$. This is contrary to our assumption. Thus, (c) implies (b'), and our proof is complete. $\square$

If $\{V_i | i \in \Delta\}$ is a collection of independent subspaces of V such that $\sum_{i \in \Delta} V_i = V$, then we say V is the internal direct sum of the V_i. In this case, $V \cong \bigoplus_{i \in \Delta} V_i$ via S, and we often just identify V with $\bigoplus_{i \in \Delta} V_i$. If $|\Delta| = n < \infty$, we shall simply write $V = V_1 \oplus \cdots \oplus V_n$ when V is an internal direct sum of subspaces $V_1, \ldots, V_n$.

The reader will note that there is no difference in notation between an external direct sum and an internal direct sum. This deliberate ambiguity will cause no real confusion in the future.

Finally, suppose $V = V_1 \oplus \cdots \oplus V_n$ is an internal direct sum of independent subspaces $V_1, \ldots, V_n$. Then by Theorem 4.16(b), every vector $\alpha \in V$ can be written uniquely in the form $\alpha = \alpha_1 + \cdots + \alpha_n$ with $\alpha_i \in V_i$. Thus, the map $P_i: V \to V$, which sends α to α_i, is a well-defined function. Theorem 4.16(b) implies that each P_i is a linear transformation such that $\text{Im } P_i = V_i$. We give formal names to these maps $P_1, \ldots, P_n$.

Definition 4.17: Let $V = V_1 \oplus \cdots \oplus V_n$ be the internal direct sum of independent subspaces $V_1, \ldots, V_n$. For each $i = 1, \ldots, n$, the linear transformation P_i defined above is called the *ith projection map* of V relative to the decomposition $V_1 \oplus \cdots \oplus V_n$.

Our next theorem is an immediate consequence of Theorem 4.16(b).

Theorem 4.18: Let $V = V_1 \oplus \cdots \oplus V_n$ be an internal direct sum of the independent subspaces $V_1, \ldots, V_n$. Suppose $P_1, \ldots, P_n \in \text{Hom}(V, V)$ are the associated projection maps. Then

  (a) $P_i P_j = 0$ if $i \neq j$.
  (b) $P_i P_i = P_i$.
  (c) $\sum_{i=1}^{n} P_i = I_V$, the identity map on V. □

Theorem 4.18 says that every internal direct sum decomposition $V = V_1 \oplus \cdots \oplus V_n$ determines a set $\{P_1, \ldots, P_n\}$ of pairwise orthogonal [4.18(a)] idempotents [4.18(b)] whose sum is I_V [4.18(c)] in the algebra of endomorphisms $\mathscr{E}(V) = \text{Hom}_F(V, V)$. Let us take this opportunity to define some of the words in our last sentence.

Definition 4.19: By an *associative algebra* A over F, we shall mean a vector space $(A, (\alpha, \beta) \to \alpha + \beta, (x, \alpha) \to x\alpha)$ over F together with a second function $(\alpha, \beta) \to \alpha\beta$ from $A \times A$ to A satisfying the following axioms:

  A1. $\alpha(\beta\gamma) = (\alpha\beta)\gamma$ for all $\alpha, \beta, \gamma \in A$.
  A2. $\alpha(\beta + \gamma) = \alpha\beta + \alpha\gamma$ for all $\alpha, \beta, \gamma \in A$.
  A3. $(\beta + \gamma)\alpha = \beta\alpha + \gamma\alpha$ for all $\alpha, \beta, \gamma \in A$.
  A4. $x(\alpha\beta) = (x\alpha)\beta = \alpha(x\beta)$ for all $\alpha, \beta \in A$, $x \in F$.
  A5. There exists an element $1 \in A$ such that $1\alpha = \alpha 1 = \alpha$ for all $\alpha \in A$.

We have seen several examples of (associative) algebras in this book already. Any field F is an associative alebra over F. $M_{n \times n}(F)$ and $F[X]$ with the usual multiplication of matrices or polynomials is an algebra over F. If V is any vector space over F, then $\text{Hom}_F(V, V)$ becomes an (associative) algebra over F when we define the product of two linear transformations T_1 and T_2 to be their composite $T_1 T_2$. Clearly axioms A1–A5 are satisfied. Here 1 is the identity map from V to V. Linear transformations from V to V are called *endomorphisms* of V. The algebra $\mathscr{E}(V) = \text{Hom}_F(V, V)$ is called the *algebra of endomorphisms* of V.

Suppose A is any algebra over F. An element $\alpha \in A$ is idempotent if $\alpha\alpha = \alpha$. In F or $F[X]$, for example, the only idempotents are 0 and 1. In $M_{n \times n}(F)$, $e_{11}, \ldots, e_{nn}$ are all idempotents different from 0 or 1. Idempotents $\{\alpha_1, \ldots, \alpha_n\}$ in an algebra A are said to be pairwise orthogonal if $\alpha_i \alpha_j = 0$ whenever $i \neq j$. Thus, $\{e_{11}, \ldots, e_{nn}\}$ is a set of pairwise orthogonal idempotents in $M_{n \times n}(F)$.

Theorem 4.18 says that every internal direct sum decomposition $V = V_1 \oplus \cdots \oplus V_n$ determines a set of pairwise orthogonal idempotents whose sum is 1 in $\mathscr{E}(V)$. Our last theorem of this section is the converse of this result.

Theorem 4.20: Let V be a vector space over F, and suppose $\{P_1, \ldots, P_n\}$ is a set of pairwise orthogonal idempotents in $\mathscr{E}(V)$ such that $P_1 + \cdots + P_n = 1$. Let $V_i = \text{Im } P_i$. Then $V = V_1 \oplus \cdots \oplus V_n$.

Proof: We must show $V = V_1 + \cdots + V_n$ and $V_j \cap (\sum_{i \neq j} V_i) = 0$. Let $\alpha \in V$ and set $\alpha_i = P_i(\alpha)$. Then $\alpha = 1(\alpha) = (P_1 + \cdots + P_n)(\alpha) = P_1(\alpha) + \cdots + P_n(\alpha) = \alpha_1 + \cdots \alpha_n$. Since $\alpha_i \in \operatorname{Im} P_i = V_i$, we conclude $V \subseteq V_1 + \cdots + V_n$. Thus, $V = V_1 + \cdots + V_n$.

Fix J, and suppose $\delta \in V_j \cap (\sum_{i \neq j} V_i)$. Then $\delta = P_j(\beta) = \sum_{i \neq j} P_i(\beta_i)$ for some $\beta \in V$ and $\beta_i \in V$ ($i \neq j$). Then $\delta = P_j(\beta) = P_j P_j(\beta) = P_j(\sum_{i \neq j} P_i(\beta_i)) = \sum_{i \neq j} P_j P_i(\beta_i) = 0$. Thus, $V_j \cap (\sum_{i \neq j} V_i) = (0)$, and the proof is complete. $\square$

EXERCISES FOR SECTION 4

(1) Let $B = \{\delta_i \mid i \in \Delta\}$ be a basis of V. Show that V is the internal direct sum of $\{F\delta_i \mid i \in \Delta\}$.

(2) Show $\operatorname{Hom}_F(\bigoplus_{i \in \Delta} V_i, W) \cong \prod_{i \in \Delta} \operatorname{Hom}_F(V_i, W)$.

(3) Give a careful proof of Theorem 4.3(f).

(4) Let $V = V_1 \times \cdots \times V_n$, and for each $i = 1, \ldots, n$, set $T_i = \theta_i \pi_i$. Show that $\{T_1, \ldots, T_n\}$ is a set of pairwise orthogonal idempotents in $\mathscr{E}(V)$ whose sum is 1.

(5) Let $V = V_1 \times \cdots \times V_n$. Show that V has a collection of subspaces $\{W_1, \ldots, W_n\}$ such that $V_i \cong W_i$ for $i = 1, \ldots, n$ and $V = \bigoplus_{i=1}^n W_i$.

(6) Give a combined version of Corollaries 4.6 and 4.8 by showing directly that $\psi \colon \operatorname{Hom}_F(V_1 \times \cdots \times V_n, \quad W_1 \times \cdots \times W_m) \to \prod_{i=1}^n \prod_{j=1}^m \operatorname{Hom}(V_i, W_j)$ given by $\psi(T) = (\pi_j T \theta_i)_{i=1,\ldots,n, j=1,\ldots,m}$ is an isomorphism.

(7) Suppose $V = V_1 \oplus \cdots \oplus V_n$. Let $T \in \operatorname{Hom}(V, V)$ such that $T(V_i) \subseteq V_i$ for all $i = 1, \ldots, n$. Find a basis $\underline{\alpha}$ of V such that

$$\Gamma(\underline{\alpha}, \underline{\alpha})(T) = \begin{pmatrix} M_1 & \cdot & & 0 \\ 0 & \cdot & \cdot & M_n \end{pmatrix}$$

where M_i describes the action of T on V_i.

(8) If X, Y, Z are subspaces of V such that $X \oplus Y = X \oplus Z = V$, is $Y = Z$? Is $Y \cong Z$?

(9) Find three subspaces V_1, V_2, V_3 of $V = F[X]$ such that $V = V_1 \oplus V_2 \oplus V_3$.

(10) If $V = V_1 + V_2$, show that there exists a subspace W of V such that $W \subseteq V_2$ and $V = V_1 \oplus W$.

(11) Let A be an algebra over F. A linear transformation $T \in \operatorname{Hom}_F(A, A)$ is called an algebra homomorphism if $T(\alpha\beta) = T(\alpha)T(\beta)$ for all $\alpha, \beta \in A$. Exhibit a nontrivial algebra homomorphism on the algebras $F[X]$ and $M_{n \times n}(F)$.

(12) Suppose V is a vector space over F. Let $S: V \cong V$ be an isomorphism of V. Show that the map $T \to S^{-1}TS$ is an algebra homomorphism of $\mathscr{E}(V)$ which is one to one and onto.

(13) Let F be a field. Show that the vector space $V = F$ (over F) is not the direct sum of any two proper subspaces.

(14) An algebra A over F is said to be commutative if $\alpha\beta = \beta\alpha$ for all $\alpha, \beta \in A$. Suppose V is a vector space over F such that $\dim_F(V) > 1$. Show that $\mathscr{E}(V)$ is not commutative.

(15) Suppose V is a vector space over F. Let $T \in \mathscr{E}(V)$ be idempotent. Show $V = \ker(T) \oplus \operatorname{Im}(T)$.

(16) Let V be a vector space over F, and let $T \in \mathscr{E}(V)$. If $T^3 = T$, show that $V = V_0 \oplus V_1 \oplus V_2$ where the V_i are subspaces of V with the following properties: $\alpha \in V_0 \Rightarrow T(\alpha) = 0$, $\alpha \in V_1 \Rightarrow T(\alpha) = \alpha$, and $\alpha \in V_2 \Rightarrow T(\alpha) = -\alpha$. In this exercise, assume $2 \neq 0$ in F.

(17) Suppose V is a finite-dimensional vector space over F. If $T \in \mathscr{E}(V)$ is nonzero, show there exists an $S \in \mathscr{E}(V)$ such that ST is a nonzero idempotent of $\mathscr{E}(V)$.

(18) Suppose $T \in \mathscr{E}(V)$ is not zero and not an isomorphism of V. Prove there is an $S \in \mathscr{E}(V)$ such that $ST = 0$, but $TS \neq 0$.

(19) Suppose V is a finite-dimensional vector space over F with subspaces $W_1, \ldots, W_k$. Suppose $V = W_1 + \cdots + W_k$, and $\dim(V) = \sum_{i=1}^{k} \dim(W_i)$. Show that $V = W_1 \oplus \cdots \oplus W_k$.

5. QUOTIENT SPACES AND THE ISOMORPHISM THEOREMS

In this section, we develop the notion of a quotient space of V. In order to do that, we need to consider equivalence relations. Suppose A is a nonempty set and $R \subseteq A \times A$ is a relation on A. The reader will recall from Section 2 that we used the notation $x \sim y$ to mean $(x, y) \in R$. The relation $\sim$ is called an equivalence relation if the following conditions are satisfied:

5.1: (a) $x \sim x$ for all $x \in A$.

(b) If $x \sim y$, then $y \sim x$ for all $x, y \in A$.

(c) If $x \sim y$ and $y \sim z$, then $x \sim z$ for all $x, y, z \in A$.

A relation satisfying 5.1(a) is called *reflexive*. If 5.1(b) is satisfied, the relation is said to be *symmetric*. A relation satisfying 5.1(c) is said to be *transitive*. Thus, an equivalence relation is a reflexive, symmetric relation that is transitive.

Example 5.2: Let $A = \mathbb{Z}$, and suppose p is a positive prime. Define a relation $\equiv$ (congruence mod p) on A by $x \equiv y$ if and only if $p \mid x - y$. The reader can easily check that $\equiv$ is an equivalence relation on $\mathbb{Z}$. ☐

The equivalence relation introduced in Example 5.2 is called a *congruence*, and we shall borrow the symbol $\equiv$ to indicate a general equivalence relation. Thus, if $R \subseteq A \times A$ is an equivalence relation on A and $(x, y) \in R$, then we shall write $x \equiv y$. We shall be careful in the rest of this text to use the symbol $\equiv$ only when dealing with an equivalence relation.

Now suppose $\equiv$ is an equivalence relation on a set A. For each $x \in A$, we set $\bar{x} = \{y \in A \mid y \equiv x\}$. $\bar{x}$ is a subset of A containing x. $\bar{x}$ is called the equivalence class of x. The function from A to $\mathscr{P}(A)$ given by $x \to \bar{x}$ satisfies the following properties:

5.3: (a) $x \in \bar{x}$.

  (b) $\bar{x} = \bar{y}$ if and only if $x \equiv y$.

  (c) For any $x, y \in A$, either $\bar{x} = \bar{y}$ or $\bar{x} \cap \bar{y} = \phi$.

  (d) $A = \bigcup_{x \in A} \bar{x}$.

The proofs of the statements in 5.3 are all easy consequences of the definitions. If we examine Example 5.2 again, we see $\mathbb{Z}$ is the disjoint union of the p equivalence classes $\bar{0}, \bar{1}, \ldots, \overline{p - 1}$. It follows from 5.3(c) and 5.3(d) that any equivalence relation on a set A divides A into a disjoint union of equivalence classes. The reader probably has noted that the equivalence classes $\{\bar{0}, \bar{1}, \ldots, \overline{p - 1}\}$ of $\mathbb{Z}$ inherit an addition and multiplication from $\mathbb{Z}$ and form the field $\mathbb{F}_p$ discussed in Example 1.3. This is a common phenomenon in algebra. The set of equivalence classes on a set A often inherits some algebraic operations from A itself. This type of inheritance of algebraic structure is particularly fruitful in the study of vector spaces.

Let V be a vector space over a field F, and suppose W is a subspace of V. The subspace W determines an equivalence relation $\equiv$ on V defined as follows:

5.4:

$$\alpha \equiv \beta \quad \text{if} \quad \alpha - \beta \in W$$

Let us check that the relation $\equiv$ defined in 5.4 is reflexive, symmetric, and transitive. Clearly, $\alpha \equiv \alpha$. If $\alpha \equiv \beta$, then $\alpha - \beta \in W$. Since W is a subspace, $\beta - \alpha \in W$. Therefore, $\beta \equiv \alpha$. Suppose $\alpha \equiv \beta$ and $\beta \equiv \gamma$. Then $\alpha - \beta, \beta - \gamma \in W$. Again, since W is a subspace, $\alpha - \gamma = (\alpha - \beta) + (\beta - \gamma) \in W$, and, thus $\alpha \equiv \gamma$. So, indeed $\equiv$ is an equivalence relation on V. The reader should realize that the equivalence relation $\equiv$ depends on the subspace W. We have deliberately suppressed any reference to W in the symbol $\equiv$ to simplify notation. This will cause no confusion in the sequel.

Definition 5.5: Let W be a subspace of V, and let $\equiv$ denote the equivalence relation defined in 5.4. If $\alpha \in V$, then the equivalence class of α will be denoted by $\bar{\alpha}$. The set of all equivalence classes $\{\bar{\alpha} \mid \alpha \in V\}$ will be denoted by V/W.

Thus, $\bar{\alpha} = \{\beta \in V \mid \beta \equiv \alpha\}$ and $V/W = \{\bar{\alpha} \mid \alpha \in V\}$. Note that the elements in V/W are subsets of V. Hence V/W consists of a collection of elements from $\mathscr{P}(V)$.

Definition 5.6: If W is a subspace of V and $\alpha \in V$, then the subset $\alpha + W = \{\alpha + \gamma \mid \gamma \in W\}$ is called a *coset* of W.

Clearly, $\beta \in \alpha + W$ if and only if $\alpha - \beta \in W$. Thus, the coset $\alpha + W$ is the same set as the equivalence class $\bar{\alpha}$ of α under $\equiv$. So, V/W is the set of all cosets of W. In particular, the equivalence class $\bar{\alpha}$ of α has a nice geometric interpretation. $\bar{\alpha} = \alpha + W$ is the translate of the subspace W through the vector α.

Let us pause for a second and discuss the other names that some of these objects have. A coset $\alpha + W$ is also called an *affine subspace* or *flat* of V. We shall not use the word "flat" again in this text, but we want to introduce formally the set of affine subspaces of V.

Definition 5.7: The set of all affine subspaces of V will be denoted $\mathscr{A}(V)$.

Thus, $A \in \mathscr{A}(V)$ if and only if $A = \alpha + W$ for some subspace $W \subseteq V$ and some $\alpha \in V$. Note that an affine subspace $A = \alpha + W$ is not a subspace of V unless $\alpha = 0$. Thus, we must be careful to use the word "affine" when considering elements in $\mathscr{A}(V)$. Since $\mathscr{A}(V)$ consists of all cosets of all subspaces of V, $V/W \subseteq \mathscr{A}(V) \subseteq \mathscr{P}(V)$ and these inclusions are usually strict.

The set V/W is called the *quotient* of V by W and is read "V mod W". We shall see shortly that V/W inherits a vector space structure from V. Before discussing this point, we gather together some of the more useful properties of affine subspaces in general.

Theorem 5.8: Let V be a vector space over F, and let $\mathscr{A}(V)$ denote the set of all affine subspaces of V.

(a) If $\{A_i \mid i \in \Delta\}$ is an indexed collection of affine subspaces in $\mathscr{A}(V)$, then either $\bigcap_{i \in \Delta} A_i = \phi$ or $\bigcap_{i \in \Delta} A_i \in \mathscr{A}(V)$.

(b) If $A, B \in \mathscr{A}(V)$, then $A + B \in \mathscr{A}(V)$.

(c) If $A \in \mathscr{A}(V)$ and $x \in F$, then $xA \in \mathscr{A}(V)$.

(d) If $A \in \mathscr{A}(V)$ and $T \in \mathrm{Hom}(V, V')$, then $T(A) \in \mathscr{A}(V')$.

(e) If $A' \in \mathscr{A}(V')$ and $T \in \mathrm{Hom}(V, V')$, then $T^{-1}(A')$ is either empty or an affine subspace of V.

Proof: The proofs of (b)–(e) are all straightforward. In (e), $T^{-1}(A') = \{\alpha \in V \mid T(\alpha) \in A'\}$. We give a proof of (a) only. Suppose $A_i = \alpha_i + W_i$ for each $i \in \Delta$. Here W_i is a subspace of V and α_i a vector in V. Suppose $\bigcap_{i \in \Delta} A_i \neq \phi$. Let $\beta \in \bigcap_{i \in \Delta} A_i$. Then for each $i \in \Delta$, $\beta = \alpha_i + \gamma_i$ with $\gamma_i \in W_i$. But then $\beta + W_i = \alpha_i + W_i$, and $\bigcap_{i \in \Delta} A_i = \bigcap_{i \in \Delta} (\beta + W_i)$.

We claim that $\bigcap_{i \in \Delta} (\beta + W_i) = \beta + (\bigcap_{i \in \Delta} W_i)$. Clearly, $\beta + (\bigcap_{i \in \Delta} W_i) \subseteq$

$\bigcap_{i \in \Delta}(\beta + W_i)$, so let $\alpha \in \bigcap_{i \in \Delta}(\beta + W_i)$. Then, for $i \neq j$, $\alpha = \beta + \delta_i = \beta + \delta_j$ with $\delta_i \in W_i$ and $\delta_j \in W_j$. But then $\delta_i = \delta_j$ and $\alpha \in \beta + (W_i \cap W_j)$. Thus, $\bigcap_{i \in \Delta}(\beta + W_i) \subseteq \beta + (\bigcap_{i \in \Delta} W_i)$. Therefore, $\bigcap_{i \in \Delta}(\beta + W_i) = \beta + (\bigcap_{i \in \Delta} W_i)$. Since $\beta + (\bigcap_{i \in \Delta} W_i) \in \mathscr{A}(V)$, the proof of (a) is complete. $\square$

We can generalize Theorem 5.8(d) one step further by introducing the concept of an affine map between two vector spaces. If $\alpha \in V$, then by translation through α, we shall mean the function $S_\alpha : V \to V$ given by $S_\alpha(\beta) = \alpha + \beta$. Any coset $\alpha + W$ is just $S_\alpha(W)$ for the translation S_α. Note that when $\alpha \neq 0$, S_α is not a linear transformation.

Definition 5.9: Let V and V' be two vector spaces over a field F. A function $f : V \to V'$ is called an *affine transformation* if $f = S_\alpha T$ for some $T \in \mathrm{Hom}_F(V, V')$ and some $\alpha \in V'$. The set of all affine transformations from V to V' will be denoted $\mathrm{Aff}_F(V, V')$.

Clearly, $\mathrm{Hom}_F(V, V') \subseteq \mathrm{Aff}_F(V, V') \subseteq (V')^V$. Theorem 5.8(d) can be restated as follows:

Theorem 5.10: If $A \in \mathscr{A}(V)$ and $f \in \mathrm{Aff}_F(V, V')$, then $f(A) \in \mathscr{A}(V')$. $\square$

Let us now return to the special subset V/W of $\mathscr{A}(V)$. The cosets of W can be given the structure of a vector space. We first define a binary operation $\dot{+}$ on V/W by the following formula:

5.11:

$$\bar{\alpha} \dot{+} \bar{\beta} = \overline{\alpha + \beta}$$

In equation 5.11, α and β are vectors in V and $\bar{\alpha}$ and $\bar{\beta}$ are their corresponding equivalence classes. $\bar{\alpha} \dot{+} \bar{\beta}$ is defined to be the equivalence class that contains $\alpha + \beta$. We note that our definition of $\bar{\alpha} \dot{+} \bar{\beta}$ depends only on the equivalence classes $\bar{\alpha}$ and $\bar{\beta}$ and not on the particular elements $\alpha \in \bar{\alpha}$ and $\beta \in \bar{\beta}$ (used to form the right-hand side of 5.11). To see this, suppose $\alpha_1 \in \bar{\alpha}$ and $\beta_1 \in \bar{\beta}$. Then $\alpha_1 - \alpha$ and $\beta_1 - \beta$ are in W. Therefore, $(\alpha_1 + \beta_1) - (\alpha + \beta) \in W$ and $\overline{\alpha_1 + \beta_1} = \overline{\alpha + \beta}$. Thus, $\dot{+} : V/W \times V/W \to V/W$ is a well-defined function. The reader can easily check that $(V/W, \dot{+})$ satisfies axioms V1–V4 of Definition 1.4. $\bar{0}$ is the zero element of V/W, and $\overline{-\alpha}$ is the inverse of $\bar{\alpha}$ under $\dot{+}$. The function $\dot{+}$ is called addition on V/W, and, henceforth, we shall simply write $+$ for this operation. Thus, $\bar{\alpha} + \bar{\beta} = \overline{\alpha + \beta}$ defines the operation of vector addition on V/W.

We can define scalar multiplication on V/W by the following formula:

5.12:

$$x\bar{\alpha} = \overline{x\alpha}$$

In equation 5.12, $x \in F$ and $\bar{\alpha} \in V/W$. Again we observe that if $\alpha_1 \in \bar{\alpha}$, then $\overline{x\alpha_1} = \overline{x\alpha}$. Thus $(x, \bar{\alpha}) \to \overline{x\alpha}$ is a well-defined function from $F \times V/W$ to V/W. The reader can easily check that scalar multiplication satisfies axioms V5–V8 in Definition 1.4. Thus, $(V/W, (\bar{\alpha}, \bar{\beta}) \to \bar{\alpha} + \bar{\beta}, (x, \bar{\alpha}) \to x\bar{\alpha})$ is a vector space over F. We shall refer to this vector space in the future as simply V/W.

Equations 5.11 and 5.12 imply that the natural map $\Pi : V \to V/W$ given by $\Pi(\alpha) = \bar{\alpha}$ is a linear transformation. Clearly, Π is surjective and has kernel W. Thus, if $i : W \to V$ denotes the inclusion of W into V, then we have the following short exact sequence:

5.13:

$$0 \to W \xrightarrow{\ i\ } V \xrightarrow{\ \Pi\ } V/W \to 0$$

In particular, Theorem 3.33 implies the following theorem:

Theorem 5.14: Suppose V is a finite-dimensional vector space over F and W a subspace of V. Then dim V = dim W + dim V/W.   □

We shall finish this section on quotients with three theorems that are collectively known as the isomorphism theorems. These theorems appear in various forms all over mathematics and are very useful.

Theorem 5.15 (First Isomorphism Theorem): Let $T \in \mathrm{Hom}_F(V, V')$, and suppose W is a subspace of V for which T(W) = 0. Let $\Pi : V \to V/W$ be the natural map. Then there exists a unique $\bar{T} \in \mathrm{Hom}_F(V/W, V')$ such that the following diagram commutes:

5.16:

Proof: We define $\bar{T}$ by $\bar{T}(\bar{\alpha}) = T(\alpha)$. Again, we remaind the reader that $\bar{\alpha}$ is a subset of V containing α. To ensure that our definition of $\bar{T}$ makes sense, we must argue that $T(\alpha_1) = T(\alpha)$ for any $\alpha_1 \in \bar{\alpha}$. If $\alpha_1 \in \bar{\alpha}$, then $\alpha_1 - \alpha \in W$. Since T is zero on W, we get $T(\alpha_1) = T(\alpha)$. Thus, our definition of $\bar{T}(\bar{\alpha})$ depends only on the coset $\bar{\alpha}$ and not on any particular representative of $\bar{\alpha}$. Since $\bar{T}(x\bar{\alpha} + y\bar{\beta}) = \bar{T}(\overline{x\alpha + y\beta}) = T(x\alpha + y\beta) = xT(\alpha) + yT(\beta) = x\bar{T}(\bar{\alpha}) + y\bar{T}(\bar{\beta})$,   we see $\bar{T} \in \text{Hom}(V/W, V')$. $\bar{T}\Pi(\alpha) = \bar{T}(\bar{\alpha}) = T(\alpha)$ and so 5.16 commutes. Only the uniqueness of $\bar{T}$ remains to be proved.

If $T' \in \text{Hom}(V/W, V')$ is another map for which $T'\Pi = T$, then $\bar{T} = T'$ on Im Π. But Π is surjective. Therefore, $\bar{T} = T'$. $\square$

Corollary 5.17: Suppose $T \in \text{Hom}_F(V, V')$. Then Im $T \cong V/\text{ker } T$.

Proof: We can view T as a surjective, linear transformation from V to Im T. Applying Theorem 5.15, we get a unique linear transformation $\bar{T}: V/\text{ker } T \to \text{Im } T$ for which the following diagram is commutative:

5.18:

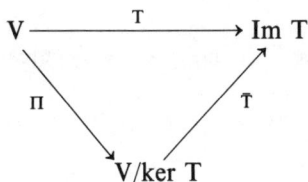

In 5.18, Π is the natural map from V to $V/\text{ker } T$. We claim $\bar{T}$ is an isomorphism. Since $\bar{T}\Pi = T$ and $T: V \to \text{Im } T$ is surjective, $\bar{T}$ is surjective. Suppose $\bar{\alpha} \in \text{ker } \bar{T}$. Then $T(\alpha) = \bar{T}\Pi(\alpha) = \bar{T}(\bar{\alpha}) = 0$. Thus, $\alpha \in \text{ker } T$. But, then $\Pi(\alpha) = 0$. Thus, $\bar{\alpha} = 0$, and $\bar{T}$ is injective. $\square$

The second isomophism theorem deals with multiple quotients. Suppose W is a subspace of V and consider the natural projection $\Pi: V \to V/W$. If W' is a subspace of V containing W, then $\Pi(W')$ is a subspace of V/W. Hence, we can form the quotient space $(V/W)/\Pi(W')$. By Corollary 5.17, $\Pi(W')$ is isomorphic to W'/W. Thus we may rewrite $(V/W)/\Pi(W')$ as $(V/W)/(W'/W)$.

Theorem 5.19 (Second Isomorphism Theorem): Suppose $W \subseteq W'$ are subspaces of V. Then $(V/W)/(W'/W) \cong V/W'$.

Proof: Let $\Pi: V \to V/W$ and $\Pi': V/W \to (V/W)/\Pi(W')$ be the natural projections. Set $T = \Pi'\Pi: V \to (V/W)/\Pi(W')$. Since Π and Π' are both surjective, T is a surjective, linear transformation. Clearly, $W' \subseteq \text{ker } T$. Let $\alpha \in \text{ker } T$. Then $0 = \Pi'\Pi(\alpha)$. Thus, $\bar{\alpha} = \Pi(\alpha) \in \Pi(W')$. Let $\beta \in W'$ such that $\Pi(\beta) = \Pi(\alpha)$. Then $\Pi(\beta - \alpha) = 0$. Thus, $\beta - \alpha \in \text{ker } \Pi = W \subseteq W'$. In particular, $\alpha \in W'$. We have now proved that ker $T = W'$. Applying Corollary 5.17, we have $(V/W)/\Pi(W') = \text{Im } T \cong V/\text{ker } T = V/W'$. $\square$

The third isomorphism theorem deals with sums and quotients.

Theorem 5.20 (Third Isomorphism Theorem): Suppose W and W' are subspaces of V. Then $(W + W')/W \cong W'/(W \cap W')$.

Proof: Let $\Pi: W + W' \to (W + W')/W$ be the natural projection. The inclusion map of W' into $W + W'$ when composed with Π gives us a linear transformation $T: W' \to (W + W')/W$. Since the kernel of Π is W, $\ker T = W \cap W'$. We claim T is surjective. To see this, consider a typical element $\gamma \in (W + W')/W$. γ is a coset of W of the form $\gamma = \delta + W$ with $\delta \in W + W'$. Thus, $\delta = \alpha + \beta$ with $\alpha \in W$ and $\beta \in W'$. But $\alpha + W = W$. So, $\gamma = \delta + W = (\beta + \alpha) + W = \beta + W$. In particular, $T(\beta) = \beta + W = \gamma$, and T is surjective. By Corollary 5.17, $(W + W')/W = \operatorname{Im} T \cong W'/\ker T = W'/W \cap W'$. $\square$

We close this section with a typical application of the isomorphism theorems. Suppose V is an internal direct sum of subspaces $V_1, \ldots, V_n$. Thus, $V = V_1 \oplus \cdots \oplus V_n$. Since $V_i \cap (\sum_{j \neq i} V_j) = (0)$, Theorem 5.20 implies $V/V_i = (V_i + \sum_{j \neq i} V_j)/V_i \cong (\sum_{j \neq i} V_j)/(V_i \cap (\sum_{j \neq i} V_j)) = (\sum_{j \neq i} V_j)/(0) = V_1 \oplus \cdots \oplus \check{V}_i \oplus \cdots \oplus V_n$. Here the little hat (ˇ) above V_i means V_i is not present in this sum.

EXERCISES FOR SECTION 5

(1) Suppose $f \in \operatorname{Hom}_F(V, F)$. If $f \neq 0$, show $V/\ker f \cong F$.

(2) Let $T \in \operatorname{Hom}(V, V)$ and suppose $T(\alpha) = \alpha$ for all $\alpha \in W$, a subspace of V.
   (a) Show that T induces a map $S \in \operatorname{Hom}(V/W, V/W)$.
   (b) If S is the identity map on V/W, show that $R = T - I_V$ has the property that $R^2 = 0$.
   (c) Conversely, suppose $T = I_V + R$ with $R \in \operatorname{Hom}(V, V)$ and $R^2 = 0$. Show that there exists a subspace W of V such that T is the identity on W and the induced map S is the identity on V/W.

(3) A subspace W of V is said to have finite codimension n if $\dim V/W = n$. If W has finite codimension, we write $\operatorname{codim} W < \infty$. Show that if W_1 and W_2 have finite codimension in V, then so does $W_1 \cap W_2$. Show $\operatorname{codim}(W_1 \cap W_2) \leq \operatorname{codim} W_1 + \operatorname{codim} W_2$.

(4) In Exercise 3, suppose V is finite dimensional and $\operatorname{codim} W_1 = \operatorname{codim} W_2$. Show that $\dim(W_1/W_1 \cap W_2) = \dim(W_2/W_1 \cap W_2)$.

(5) Let $T \in \operatorname{Hom}(V, V')$, and suppose T is surjective. Set $K = \ker T$. Show there exists a one-to-one, inclusion-preserving correspondence between the subspaces of V' and the subspaces of V containing K.

(6) Let $T \in \operatorname{Hom}(V, V')$, and let $K = \ker T$. Show that all vectors of V that have the same image under T belong to the same coset of V/K.

(7) Suppose W is a finite-dimensional subspace of V such that V/W is finite dimensional. Show V must be finite dimensional.

(8) Let V be a finite-dimensional vector space. If W is a subspace with $\dim W = \dim V - 1$, then the cosets of W are called hyperplanes in V. Suppose S is an affine subspace of V and $H = \alpha + W$ is a hyperplane. Show that if $H \cap S = \varnothing$, then $S \subseteq \beta + W$ for some $\beta \in V$.

(9) If $S = \alpha + W \in \mathscr{A}(V)$, we define $\dim S = \dim W$. Suppose V is finite dimensional, H a hyperplane in V, and $S \in \mathscr{A}(V)$. Show that $S \cap H \neq \varnothing \Rightarrow \dim(S \cap H) = \dim S - 1$. Assume $S \not\subseteq H$.

(10) Let $S \in \mathscr{A}(V)$ with $\dim S = m - 1$. Show that $S = \{\sum_{i=1}^{m} x_i \alpha_i \mid \sum_{i=1}^{m} x_i = 1\}$ for some choice of m vectors $\alpha_1, \ldots, \alpha_m \in V$.

(11) Suppose $C = \{(V_i, d_i) \mid i \in \mathbb{Z}\}$ is a chain complex. For each $i \in \mathbb{Z}$, set $H_i(C) = \ker d_i / \operatorname{Im} d_{i+1}$. $H_i(C)$ is called the ith homology of C.

(a) Show that C is exact if and only if $H_i(C) = 0$ for all $i \in \mathbb{Z}$.

(b) Let $C = \{(V_i, d_i) \mid i \in \mathbb{Z}\}$ and $C' = \{(V_i', d_i') \mid i \in \mathbb{Z}\}$ be two chain complexes. Show that any chain map $T = \{T_i\}_{i \in \mathbb{Z}} \colon C \to C'$ induces a linear transformation $\bar{T}_i \colon H_i(C) \to H_i(C')$ such that $\bar{T}_i(\alpha + \operatorname{Im} d_{i+1}) = T_i(\alpha) + \operatorname{Im} d_{i+1}'$.

(c) Suppose

$$C: 0 \to V_n \xrightarrow{d_n} V_{n-1} \to \cdots \xrightarrow{d_2} V_1 \to 0$$

is a finite chain complex. Show that $\sum (-1)^i \dim H_i(C) = \sum (-1)^i \dim V_i$. Here each V_i is assumed finite dimensional.

(12) Suppose V is an n-dimensonal vector space over F, and $W_1, \ldots, W_k$ are subspaces of codimension $e_i = n - \dim(W_i)$. Let $S_i = \alpha_i + W_i$ for $i = 1, \ldots, k$. If $S_1 \cap \ldots \cap S_k = \phi$, show $\dim(W_1 \cap \ldots \cap W_k) > n - \sum_{i=1}^{k} e_i$.

(13) Use Exercise 12 to prove the following assertion: Let $S_1 = \alpha_1 + W_1$ and $S_2 = \alpha_2 + W_2$ be two cosets of dimension k [i.e., $\dim(W_i) = k$]. Show that S_1 and S_2 are parallel (i.e., $W_1 = W_2$) if and only if S_1 and S_2 are contained in a coset of dimension $k + 1$, and have empty intersection.

(14) In $\mathbb{R}^3$, show that the intersection of two nonparallel planes (i.e., cosets of dimension 2) is a line (i.e., a coset of dimension 1). The same problem makes sense in any three-dimensional vector space V.

(15) Let S_1, S_2, and S_3 be planes in $\mathbb{R}^3$ such that $S_1 \cap S_2 \cap S_3 = \phi$, but no two S_i are parallel. Show that the lines $S_1 \cap S_2$, $S_1 \cap S_3$ and $S_2 \cap S_3$ are parallel.

(16) Let $f(X) = X^n + a_{n-1}X^{n-1} + \cdots + a_0 \in \mathbb{R}[X]$. Let $W = \{pf \mid p \in \mathbb{R}[X]\}$. Show that W is a subspace of $\mathbb{R}[X]$. Show that $\dim(\mathbb{R}[X]/W) = n$. (*Hint:* Use the division algorithm in $\mathbb{R}[X]$.)

(17) In Theorem 5.15, if T is surjective and $W = \ker T$, then $\bar{T}$ is an isomorphism [prove!]. In particular, $S = (\bar{T})^{-1}$ is a well-defined map from V' to V/W. Show that the process of indefinite integration is an example of such a map S.

6. DUALS AND ADJOINTS

Let V be a vector space over F.

Definition 6.1: $V^* = \mathrm{Hom}_F(V, F)$ is called the *dual* of V.

If V is a finite-dimensional vector space over F, then it follows from Theorem 3.25 that V^* is finite dimensional with dim $V^* =$ dim V. We record this fact with a different proof in 6.2.

Theorem 6.2: Let V be finite dimensional. Then dim $V^* =$ dim V.

Proof: Let $\underline{\alpha} = \{\alpha_1, \ldots, \alpha_n\}$ be a basis of V. For each $i = 1, \ldots, n$, define an element $\alpha_i^* \in V^*$ by $\alpha_i^* = \pi_i(\cdot)_{\underline{\alpha}}$. Here $(\cdot)_{\underline{\alpha}} : V \to F^n$ is the isomorphism determined by the basis $\underline{\alpha}$ and $\pi_i : F^n \to F$ is the natural projection onto the ith coordinate of F^n. Thus, if $\alpha = x_1\alpha_1 + \cdots + x_n\alpha_n \in V$, then α_i^* is given by

6.3:

$$\alpha_i^*(x_1\alpha_1 + \cdots + x_n\alpha_n) = x_i$$

We claim that $\underline{\alpha}^* = \{\alpha_1^*, \ldots, \alpha_n^*\}$ is a basis of V^*. Suppose $\sum_{i=1}^n y_i\alpha_i^* = 0$. Let $j \in \{1, \ldots, n\}$. Then equation 6.3 implies $0 = (\sum_{i=1}^n y_i\alpha_i^*)(\alpha_j) = \sum_{i=1}^n y_i\alpha_i^*(\alpha_j) = y_j$. Thus, $y_1 = \cdots y_n = 0$, and $\underline{\alpha}^*$ is linearly independent over F.

If $T \in V^*$, then $T = \sum_{i=1}^n T(\alpha_i)\alpha_i^*$. This last equation follows immediately from 6.3. Thus, $L(\underline{\alpha}^*) = V^*$, and $\underline{\alpha}^*$ is a basis of V^*. In particular, dim $V^* = |\underline{\alpha}^*| = n =$ dim V. $\square$

The basis $\underline{\alpha}^* = \{\alpha_1^*, \ldots, \alpha_n^*\}$ constructed in 6.3 is called the dual basis of $\underline{\alpha}$. Thus, every basis $\underline{\alpha}$ of a finite-dimensional vector space V has a corresponding dual basis $\underline{\alpha}^*$ of V^*. Furthermore, $V \cong V^*$ under the linear map T, which sends every $\alpha_i \in \underline{\alpha}$ to the corresponding $\alpha_i^* \in \underline{\alpha}^*$.

If V is not finite dimensional over F, then the situation is quite different. Theorem 6.2 is false when dim $V = \infty$. If dim $V = \infty$, then dim $V^* >$ dim V. Instead of proving that fact, we shall content ourselves with an example.

Example 6.4: Let $V = \bigoplus_{i=1}^\infty F$, that is, V is the direct sum of the vector spaces $\{V_i = F \mid i \in \mathbb{N}\}$. It follows from Exercise 2 of Section 4 that $V^* = \mathrm{Hom}_F(\bigoplus_{i=1}^\infty F, F) \cong \prod_{i=1}^\infty \mathrm{Hom}_F(F, F) \cong \prod_{i=1}^\infty F$. From Theorem 4.13, we know that dim $V = |\mathbb{N}|$. A simple counting exercise will convince the reader that dim $V^* = \mathrm{dim}(\prod_{i=1}^\infty F)$ is strictly larger than $|\mathbb{N}|$. $\square$

Before stating our next result, we need the following definition:

Definition 6.5: Let V, V', and W be vector spaces over F, and let $\omega: V \times V' \to W$ be a function. We call ω a bilinear map if for all $\alpha \in V$, $\omega(\alpha, \cdot) \in \text{Hom}_F(V', W)$ and for all $\beta \in V'$, $\omega(\cdot, \beta) \in \text{Hom}_F(V, W)$.

Thus, a function $\omega: V \times V' \to W$ is a bilinear map if and only if $\omega(x\alpha_1 + y\alpha_2, \beta) = x\omega(\alpha_1, \beta) + y\omega(\alpha_2, \beta)$, and $\omega(\alpha, x\beta_1 + y\beta_2) = x\omega(\alpha, \beta_1) + y\omega(\alpha, \beta_2)$ for all x, $y \in F$, $\alpha, \alpha_1, \alpha_2 \in V$ and $\beta, \beta_1, \beta_2 \in V'$. If V is any vector space over F, there is a natural bilinear map $\omega: V \times V^* \to F$ given by

6.6:

$$\omega(\alpha, T) = T(\alpha)$$

In equation 6.6, $\alpha \in V$ and $T \in V^*$. The fact that ω is a bilinear map is obvious. ω determines a natural, injective, linear transformation $\psi: V \to V^{**}$ in the following way. If $\alpha \in V$, set $\psi(\alpha) = \omega(\alpha, \cdot)$. Thus, for any $T \in V^*$, $\psi(\alpha)(T) = \omega(\alpha, T) = T(\alpha)$. If x, $y \in F$, α, $\beta \in V$ and $T \in V^*$, then $\psi(x\alpha + y\beta)(T) = \omega(x\alpha + y\beta, T) = x\omega(\alpha, T) + y\omega(\beta, T) = (x\psi(\alpha) + y\psi(\beta))(T)$. Consequently, $\psi \in \text{Hom}_F(V, V^{**})$. To see that ψ is injective, we need to generalize equation 6.3. Suppose $\underline{\alpha} = \{\alpha_i \mid i \in \Delta\}$ is a basis of V (finite or infinite). Then for every $i \in \Delta$, we can define a dual transformation $\alpha_i^* \in V^*$ as follows: For each nonzero vector $\alpha \in V$, there exists a unique finite subset $\Delta(\alpha) = \{j_1, \ldots, j_n \mid j_k \in \Delta\}$ of Δ such that $\alpha = x_{j_1}\alpha_{j_1} + \cdots + x_{j_n}\alpha_{j_n}$. Here $x_{j_1}, \ldots, x_{j_n}$ are all nonzero scalars in F. We then define $\alpha_i^*(\alpha) = x_{j_k}$ if $i = j_k$ for some $k = 1, \ldots, n$. If $i \notin \Delta(\alpha)$, we set $\alpha_i^*(\alpha) = 0$. If $\alpha = 0$, we of course define $\alpha_i^*(\alpha) = 0$. Clearly $\alpha_i^* \in V^*$, and

$$\alpha_i^*(\alpha_j) = \begin{cases} 1 & \text{if } i = j \\ 0 & \text{if } i \neq j \end{cases}$$

Now if $\alpha \in \ker \psi$, then $T(\alpha) = 0$ for all $T \in V^*$. In particular, $\alpha_i^*(\alpha) = 0$ for all $i \in \Delta$. This clearly implies $\alpha = 0$, and, thus, ψ is injective.

We note in passing that the set $\underline{\alpha}^* = \{\alpha_i^* \mid i \in \Delta\} \subseteq V^*$, which we have just constructed above, is clearly linearly independent over F. If dim $V < \infty$, this is just the dual basis of V^* coming from $\underline{\alpha}$. If dim $V = \infty$, then $\underline{\alpha}^*$ does not span V^*, and, therefore, cannot be called a dual basis. At any rate, we have proved the first part of the following theorem:

Theorem 6.7: Let V be a vector space over F and suppose $\omega: V \times V^* \to F$ is the bilinear map given in equation 6.6. Then the map $\psi: V \to V^{**}$ given by $\psi(\alpha) = \omega(\alpha, \cdot)$ is an injective linear transformation. If dim $V < \infty$, then ψ is a natural isomorphism.

Proof: Only the last sentence in Theorem 6.7 remains to be proved. If dim $V < \infty$, then Theorem 6.2 implies dim $V = $ dim $V^{**} < \infty$. Since ψ is injective, our result follows from Theorem 3.33(b). $\square$

The word "natural" in Theorem 6.7 has a precise meaning in category theory, but here we mean only that the isomorphism $\psi: V \cong V^{**}$ is independent of any choice of bases in V and V^{**}. The word "natural" when applied to an isomorphism $\psi: V \to V^{**}$ also means certain diagrams must be commutative. See Exercise 4 at the end of this section for more details. We had noted previously that when dim $V < \infty$, then $V \cong V^*$. This type of isomorphism is not natural, since it is constructed by first picking a basis $\underline{\alpha} = \{\alpha_1, \ldots, \alpha_n\}$ of V and then mapping α_i to α_i^* in V^*.

The bilinear map $\omega: V \times V^* \to F$ can also be used to set up certain correspondences between $\mathscr{P}(V)$ and $\mathscr{P}(V^*)$.

Definition 6.8: If A is any subset of V, let $A^{\perp} = \{\beta \in V^* \mid \omega(\alpha, \beta) = 0 \text{ for all } \alpha \in A\}$.

Thus, $A^{\perp}$ is precisely the set of all vectors in V^* that vanish on A. It is easy to see that $A^{\perp}$ is in fact a subspace of V^*. We have a similar definition for subsets of V^*.

Definition 6.9: If A is a subset of V^*, Let $A^{\perp} = \{\alpha \in V \mid \omega(\alpha, \beta) = 0 \text{ for all } \beta \in A\}$.

Thus, if $A \subseteq V^*$, then $A^{\perp}$ is the set of all vectors in V that are zero under the maps in A. Clearly, $A^{\perp}$ is a subspace of V for any $A \subseteq V^*$.

Theorem 6.10: Let A and B be subsets of V (or V^*).

(a) $A \subseteq B$ implies $A^{\perp} \supseteq B^{\perp}$.

(b) $L(A)^{\perp} = A^{\perp}$.

(c) $(A \cup B)^{\perp} = A^{\perp} \cap B^{\perp}$.

(d) $A \subseteq A^{\perp\perp}$.

(e) If W is a subspace of a finite-dimensional vector space V, then dim $V = $ dim $W + $ dim $W^{\perp}$.

Proof: (a)–(d) are straightforward, and we leave their proofs as exercises. We prove (e). Let $\{\alpha_1, \ldots, \alpha_m\}$ be a basis of W. We extend this set to a basis $\underline{\alpha} = \{\alpha_1, \ldots, \alpha_m, \alpha_{m+1}, \ldots, \alpha_n\}$ of V. Thus, dim $W = m$ and dim $V = n$. Let $\underline{\alpha}^* = \{\alpha_1^*, \ldots, \alpha_n^*\}$ be a dual basis of $\underline{\alpha}$. We complete the proof of (e) by arguing that $\{\alpha_{m+1}^*, \ldots, \alpha_n^*\}$ is a basis of $W^{\perp}$.

If $m + 1 \leqslant j \leqslant n$, then $\alpha_j^*(\alpha_i) = 0$ for $i = 1, \ldots, m$. In particular, $\alpha_{m+1}^*, \ldots, \alpha_n^* \in W^{\perp}$. Since $\{\alpha_{m+1}^*, \ldots, \alpha_n^*\} \subseteq \underline{\alpha}^*$, $\{\alpha_{m+1}^*, \ldots, \alpha_n^*\}$ is linearly independent over F. We must show $L(\{\alpha_{m+1}^*, \ldots, \alpha_n^*\}) = W^{\perp}$. Let $\beta \in W^{\perp}$. Then $\beta = \sum_{i=1}^{n} c_i \alpha_i^*$. Since $\alpha_1, \ldots, \alpha_m \in W$, we have for any $j = 1, \ldots, m$,

$0 = \beta(\alpha_j) = (\sum_{i=1}^{n} c_i \alpha_i^*)(\alpha_j) = \sum_{i=1}^{n} c_i \alpha_i^*(\alpha_j) = c_j.$ Thus, $\beta = \sum_{i=m+1}^{n} c_i \alpha_i^* \in L(\{\alpha_{m+1}^*, \ldots, \alpha_n^*\}).$ □

If $T \in \mathrm{Hom}_F(V, W)$, then T determines a linear transformation $T^* \in \mathrm{Hom}_F(W^*, V^*)$, which we call the *adjoint* of T.

Definition 6.11: Let $T \in \mathrm{Hom}_F(V, W)$. Then $T^* \in \mathrm{Hom}_F(W^*, V^*)$ is the linear transformation defined by $T^*(f) = fT$ for all $f \in W^*$.

Since the composite $V \to^T W \to^f F$ of the linear transformations f and T is again a linear map from V to F, we see $T^*(f) \in V^*$. If $x, y \in F$ and $f_1, f_2 \in W^*$, then $T^*(xf_1 + yf_2) = (xf_1 + yf_2)T = x(f_1 T) + y(f_2 T) = xT^*(f_1) + yT^*(f_2)$. Thus, T^* is a linear transformation from W^* to V^*.

Theorem 6.12: Let V and W be vector spaces over F. The map $T \to T^*$ from $\mathrm{Hom}_F(V,W) \to \mathrm{Hom}_F(W^*,V^*)$ is an injective transformation. If V and W are finite dimensional, then this map is an isomorphism.

Proof: Let $\chi: \mathrm{Hom}(V, W) \to \mathrm{Hom}(W^*, V^*)$ be defined by $\chi(T) = T^*$. Our comments above imply χ is a well-defined function. Suppose $x,y \in F$, $T_1, T_2 \in \mathrm{Hom}(V, W)$, and $f \in W^*$. Then $\chi(xT_1 + yT_2)(f) = (xT_1 + yT_2)^*(f) = f(xT_1 + yT_2) = x(fT_1) + y(fT_2) = xT_1^*(f) + yT_2^*(f) = (xT_1^* + yT_2^*)(f) = (x\chi(T_1) + y\chi(T_2))(f)$. Thus, $\chi(xT_1 + yT_2) = x\chi(T_1) + y\chi(T_2)$, and χ is a linear transformation.

Suppose $T \in \ker \chi$. Then for every $f \in W^*$, $0 = \chi(T)(f) = T^*(f) = fT$. Now if we follow the same argument given in the proof of Theorem 6.7, we know that if β is a nonzero vector in W, then there exists an $f \in W^*$ such that $f(\beta) \neq 0$. Thus, $fT = 0$ for all $f \in W^*$ implies Im T = (0). Therefore, T = 0, and χ is injective.

Now suppose V and W are finite dimensional. Then Theorems 6.2 and 3.25 imply $\dim\{\mathrm{Hom}_F(V, W)\} = \dim\{\mathrm{Hom}_F(W^*, V^*)\}$. Since χ is injective, Theorem 3.33(b) implies χ is an isomorphism. □

We note in passing that forming the adjoint of a product is the product of the adjoints in the opposite order. More specifically, suppose $T \in \mathrm{Hom}_F(V, W)$ and $S \in \mathrm{Hom}_F(W, Z)$. Then $ST \in \mathrm{Hom}_F(V, Z)$. If $f \in Z^*$, then $(ST)^*(f) = f(ST) = (fS)T = T^*(fS) = T^*(S^*(f)) = T^*S^*(f)$. Thus, we get equation 6.13:

6.13:

$$(ST)^* = T^*S^*$$

The connection between adjoints and Theorem 6.10 is easily described.

Theorem 6.14: Let $T \in \mathrm{Hom}_F(V, W)$. Then

(a) $(\mathrm{Im}\ T^*)^\perp = \ker T$.
(b) $\ker T^* = (\mathrm{Im}\ T)^\perp$.

Proof: (a) Let $\alpha \in (\text{Im } T^*)^{\perp}$, and suppose $\omega: V \times V^* \to F$ is the bilinear map defined in equation 6.6. Then $\omega(\alpha, \text{Im } T^*) = 0$. Thus, for all $f \in W^*$, $0 = \omega(\alpha, T^*(f)) = \omega(\alpha, fT) = fT(\alpha) = f(T(\alpha))$. But we have seen that $f(T(\alpha)) = 0$ for all $f \in W^*$ implies $T(\alpha) = 0$. Thus, $\alpha \in \ker T$. Conversely, if $\alpha \in \ker T$, then $0 = f(T(\alpha)) = \omega(\alpha, T^*(f))$ and $\alpha \in (\text{Im } T^*)^{\perp}$.

(b) Suppose $f \in \ker T^*$. Then $0 = T^*(f) = fT$. In particular, $f(T(\alpha)) = 0$ for all $\alpha \in V$. Therefore, $0 = \omega(T(\alpha), f)$ and $f \in (\text{Im } T)^{\perp}$. Thus, $\ker T^* \subseteq (\text{Im } T)^{\perp}$. The steps in this proof are easily reversed and so $(\text{Im } T)^{\perp} \subseteq \ker T^*$. $\square$

Theorem 6.14 has an interesting corollary. If $T \in \text{Hom}_F(V, W)$, let us define the rank of T, $\text{rk}\{T\}$, to be $\dim(\text{Im } T)$. Thus, $\text{rk}\{T\} = \dim(\text{Im } T)$. Then we have the following:

Corollary 6.15: Let V and W be finite-dimensional vector spaces over F, and let $T \in \text{Hom}_F(V, W)$. Then $\text{rk}\{T\} = \text{rk}\{T^*\}$.

Proof: The following integers are all equal:

$$\text{rk}\{T\} = \dim(\text{Im } T) = \dim V - \dim(\ker T) \qquad \text{[Theorem 3.33(c)]}$$
$$= \dim V - \dim\{(\text{Im } T^*)^{\perp}\} \qquad \text{[Theorem 6.14]}$$
$$= \dim V^* - \dim\{(\text{Im } T^*)^{\perp}\} \qquad \text{[Theorem 6.2]}$$
$$= \dim(\text{Im } T^*) \qquad \text{[Theorem 6.10(e)]}$$
$$= \text{rk}\{T^*\} \quad \square$$

Corollary 6.15 has a familiar interpretation when we switch to matrices. If $\underline{\alpha}$ is any basis of V and $\underline{\beta}$ any basis of W, then Theorem 3.25 implies $\text{rk}\{T\} = \text{rk}(\Gamma(\underline{\alpha}, \underline{\beta})(T))$. Let $A = \Gamma(\underline{\alpha}, \underline{\beta})(T)$. In Theorem 6.16 below, we shall show that the matrix representation of $T^*: W^* \to V^*$ with respect to $\underline{\beta}^*$ and $\underline{\alpha}^*$ is given by the transpose of A. Thus, $\Gamma(\underline{\beta}^*, \underline{\alpha}^*)(T^*) = A^t$. In particular, Corollary 6.15 is the familiar statement that a matrix A and its transpose A^t have the same rank.

Theorem 6.16: Let V and W be finite-dimensional vector spaces over F. Suppose $\underline{\alpha}$ and $\underline{\beta}$ are bases of V and W, respectively. Let $\underline{\alpha}^*$ and $\underline{\beta}^*$ be the corresponding dual bases in V^* and W^*. Then for all $T \in \text{Hom}_F(V, W)$, we have

6.17:

$$\Gamma(\underline{\beta}^*, \underline{\alpha}^*)(T^*) = (\Gamma(\underline{\alpha}, \underline{\beta})(T))^t$$

Proof: Suppose $\underline{\alpha} = \{\alpha_1, \ldots, \alpha_n\}$ and $\underline{\beta} = \{\beta_1, \ldots, \beta_m\}$. Set $A = \Gamma(\underline{\alpha}, \underline{\beta})(T)$. Then $A = (a_{ij}) \in M_{m \times n}(F)$, and from 3.24, we have $T(\alpha_j) = \sum_{i=1}^m a_{ij}\beta_i$ for all $j = 1, \ldots, n$.

$\Gamma(\beta^*,\ \underline{\alpha}^*)(T^*)$ is the $n \times m$ matrix that makes the following diagram commute:

$$
\begin{array}{ccc}
W^* & \xrightarrow{\quad T^* \quad} & V^* \\
\downarrow{\scriptstyle [\cdot]_{\underline{\beta}^*}} & & \downarrow{\scriptstyle [\cdot]_{\underline{\alpha}^*}} \\
M_{m \times 1}(F) & \xrightarrow{\quad \Gamma(\underline{\beta}^*,\ \underline{\alpha}^*)(T^*) \quad} & M_{n \times 1}(F)
\end{array}
$$

The transpose of A is the $n \times m$ matrix $A^t = (b_{pq})$, where $b_{pq} = a_{qp}$ for all $p = 1, \ldots, n$, and $q = 1, \ldots, m$. It follows from 3.24 that $\Gamma(\underline{\beta}^*, \underline{\alpha}^*)(T^*) = A^t$ provided that the following equation is true:

6.18:

$$
T^*(\beta_q^*) = \sum_{p=1}^{n} b_{pq}\alpha_p^* \qquad \text{for all} \quad q = 1, \ldots, m
$$

Fix $q = 1, \ldots, m$. To show that $T^*(\beta_q^*)$ and $\sum_{p=1}^{n} b_{pq}\alpha_p^*$ are the same vector in V^*, it suffices to show that these two maps agree on the basis $\underline{\alpha}$ of V. For any $r = 1, \ldots, n, (T^*(\beta_q^*))(\alpha_r) = \beta_q^*(T(\alpha_r)) = \beta_q^*(\sum_{i=1}^{m} a_{ir}\beta_i) = \sum_{i=1}^{m} a_{ir}\beta_q^*(\beta_i) = a_{qr}$. On the other hand, $(\sum_{p=1}^{n} b_{pq}\alpha_p^*)(\alpha_r) = \sum_{p=1}^{n} b_{pq}\alpha_p^*(\alpha_r) = b_{rq} = a_{qr}$. Thus, equation 6.18 is established, and the proof of Theorem 6.16 is complete. $\square$

EXERCISES FOR SECTION 6

(1) Prove (a)–(d) in Theorem 6.10.

(2) Let V and W be finite-dimensional vector spaces over F with bases $\underline{\alpha}$ and β, respectively. Suppose T $\mathrm{Hom}_F(V, W)$. Show that $\mathrm{rk}\{T\} = \mathrm{rk}(\Gamma(\underline{\alpha}, \underline{\beta})(T))$.

(3) Let $0 \neq \beta \in V$ and $f \in V^* - (0)$. Define T: $V \to V$ by $T(\alpha) = f(\alpha)\beta$. A function defined in this way is called a *dyad*.
   (a) Show $T \in \mathrm{Hom}(V, V)$ such that $\dim(\mathrm{Im}\, T) = 1$.
   (b) If $S \in \mathrm{Hom}(V, V)$ such that $\dim(\mathrm{Im}\, S) = 1$, show that S is a dyad.
   (c) If T is a dyad on V, show that T^* is a dyad on V^*.

(4) Let V and W be finite-dimensional vector spaces over F. Let $\psi_V: V \to V^{**}$ and $\psi_W: W \to W^{**}$ be the isomorphisms given in Theorem 6.7. Show that for every $T \in \mathrm{Hom}(V, W)$ the following diagram is commutative:

$$
\begin{array}{ccc}
V & \xrightarrow{\quad T \quad} & W \\
\downarrow{\scriptstyle \psi_V} & & \downarrow{\scriptstyle \psi_W} \\
V^{**} & \xrightarrow{\quad T^{**} \quad} & W^{**}
\end{array}
$$

(5) Let $A = \{f_1, \ldots, f_n\} \subseteq V^*$. Show $A^\perp = \bigcap_{i=1}^n \ker f_i$.

(6) Let $A = \{f_1, \ldots, f_n\} \subseteq V^*$ and suppose $g \in V^*$ such that g vanishes on $A^\perp$. Show $g \in L(A)$. [*Hint*: First assume $\dim(V) < \infty$; then use Exercise 3 of Section 5 for the general case.]

(7) Let V and W be finite-dimensional vector spaces over F, and let $\omega: V \times W \to F$ be an arbitrary bilinear map. Let $T: V \to W^*$ and $S: W \to V^*$ be defined from ω as follows: $T(\alpha)(\beta) = \omega(\alpha, \beta)$ and $s(\beta)(\alpha) = \omega(\alpha, \beta)$. Show that $S = T^*$ if we identify W with W^{**} via ψ_W.

(8) Show that $(V/W)^* \cong W^\perp$.

(9) Let V be a finite-dimensional vector space over F. Let $W = V \oplus V^*$. Show that the map $(\alpha, \beta) \to (\beta, \alpha)$ is an isomorphism between W and W^*.

(10) If

$$0 \to V \xrightarrow{\ S\ } W \xrightarrow{\ T\ } Z \to 0$$

is a short exact sequence of vector spaces over F, show that

$$0 \to Z^* \xrightarrow{\ T^*\ } W^* \xrightarrow{\ S^*\ } V^* \to 0$$

is exact.

(11) Let $\{W_i \mid i \in \mathbb{Z}\}$ be a sequence of vector spaces over F. Suppose for each $i \in \mathbb{Z}$, we have a linear transformation $e_i \in \mathrm{Hom}_F(W_i, W_{i+1})$. Then $D = \{(W_i, e_i) \mid i \in \mathbb{Z}\}$ is called a cochain complex if $e_{i+1}e_i = 0$ for all $i \in \mathbb{Z}$. D is said to be exact if $\mathrm{Im}\, e_i = \ker e_{i+1}$ for all $i \in \mathbb{Z}$.

    (a) If $C = \{(C_i, d_i) \mid i \in \mathbb{Z}\}$ is a chain complex, show that $C^* = \{(C_i^*, e_i = d_{i+1}^*) \mid i \in \mathbb{Z}\}$ is a cochain complex.

    (b) If C is exact, show that C^* is also exact.

(12) Prove that $(V_1 \oplus \cdots \oplus V_n)^* \cong V_1^* \oplus \cdots \oplus V_n^*$.

(13) Let V be a finite-dimensional vector space over F with basis $\underline{\alpha} = \{\alpha_1, \ldots, \alpha_n\}$. Define $T: F^n \to V$ by $T(x_1, \ldots, x_n) = \sum_{i=1}^n x_i \alpha_i$. Show that $T^*(f) = (f)_{\underline{\alpha}^*}$ for all $f \in V^*$. Here you will need to identify $(F^n)^*$ with F^n in a natural way.

(14) Let $\{z_i\}_{i=0}^\infty$ be a sequence of complex numbers. Define a map $T: \mathbb{C}[X] \to \mathbb{C}$ by $T(\sum_{k=0}^n a_k X^k) = \sum_{k=0}^n a_k z_k$. Show that $T \in (\mathbb{C}[X])^*$. Show that every $T \in (\mathbb{C}[X])^*$ is given by such a sequence.

(15) Let $V = \mathbb{R}[X]$. Which of the following functions on V are elements in V^*:

    (a) $T(p) = \int_0^1 p(X)\, dX$.

    (b) $T(p) = \int_0^1 p(X)^2\, dX$.

(c) $T(p) = \int_0^1 X^2 p(X) \, dX$.

(d) $T(p) = dp/dX$.

(e) $T(p) = dp/dX|_{x=0}$.

(16) Suppose F is a finite field (e.g., $\mathbb{F}_p$). Let V be a vector space over F of dimension n. For every $m \leqslant n$, show the number of subspaces of V of dimension m is precisely the same as the number of subspaces of V of dimension $n - m$.

(17) An important linear functional on $M_{n \times n}(F)$ is the trace map $\text{Tr}: M_{n \times n}(F) \to F$ defined by $\text{Tr}(A) = \sum_{i=1}^n a_{ii}$ where $A = (a_{ij})$. Show that $\text{Tr}(\,) \in (M_{n \times n}(F))^*$.

(18) In Exercise 17, show $\text{Tr}(AB) = \text{Tr}(BA)$ for all $A, B \in M_{n \times n}(F)$.

(19) Let $m, n \in \mathbb{N}$. Let $f_1, \ldots, f_m \in (F^n)^*$. Define $T: F^n \to F^m$ by $T(\alpha) = (f_1(\alpha), \ldots, f_m(\alpha))$. Show that $T \in \text{Hom}_F(F^n, F^m)$. Show that every $T \in \text{Hom}_F(F^n, F^m)$ is given in this way for some $f_1, \ldots, f_m$.

(20) Let V be a finite-dimensional vector space over $\mathbb{C}$. Suppose $\alpha_1, \ldots, \alpha_n$ are distinct, nonzero vectors in V. Show there exists a $T \in V^*$ such that $T(\alpha_k) \neq 0$ for all $k = 1, \ldots, n$.

7. SYMMETRIC BILINEAR FORMS

In this last section of Chapter I, we discuss symmetric bilinear forms on a vector space V. Unlike the first six sections, the nature of the base field F is important here. In our main theorems, we shall assume V is a finite-dimensional vector space over the reals $\mathbb{R}$.

Let V be a vector space over an arbitrary field F.

Definition 7.1: By a bilinear form ω on V, we shall mean any bilinear map $\omega: V \times V \to F$. We say ω is symmetric if $\omega(\alpha, \beta) = \omega(\beta, \alpha)$ for all $\alpha, \beta \in V$.

Example 7.2: The standard example to keep in mind here is the form $\omega((x_1, \ldots, x_n), (y_1, \ldots, y_n)) = \sum_{i=1}^n x_i y_i$. Clearly, ω is a symmetric, bilinear form on F^n. $\square$

Suppose ω is a bilinear form on a finite-dimensional vector space V. Then for every basis $\underline{\alpha} = \{\alpha_1, \ldots, \alpha_n\}$ of V, we can define an $n \times n$ matrix $M(\omega, \underline{\alpha}) \in M_{n \times n}(F)$ whose (i, j)th entry is given by $\{M(\omega, \underline{\alpha})\}_{i,j} = \omega(\alpha_i, \alpha_j)$. In terms of the usual coordinate map $[\cdot]_{\underline{\alpha}}: V \to M_{n \times 1}(F)$, ω is then given by the following equation:

7.3:

$$\omega(\beta, \delta) = [\beta]_{\underline{\alpha}}^t M(\omega, \underline{\alpha}) [\delta]_{\underline{\alpha}}$$

Clearly, ω is symmetric if and only if $M(\omega, \underline{\alpha})$ is a symmetric matrix.

Definition 7.4: Suppose ω is a bilinear form on V. The function q: $V \to F$ defined by $q(\xi) = \omega(\xi, \xi)$ is called the quadratic form associated with ω.

If V is finite dimensional with basis $\underline{\alpha} = \{\alpha_1, \ldots, \alpha_n\}$, then equation 7.3 implies $q(\xi) = [\xi]^t_{\underline{\alpha}} M(\omega, \underline{\alpha})[\xi]_{\underline{\alpha}} = \sum_{i,j=1}^n a_{ij} x_i x_j$. Here $(x_1, \ldots, x_n)^t = [\xi]_{\underline{\alpha}}$ and $(a_{ij}) = M(\omega, \underline{\alpha})$. Thus, $q(\xi)$ is a quadratic homogeneous polynomial in the coordinates $x_1, \ldots, x_n$ of ξ. That fact explains why q is called a quadratic form on V. In Example 7.2, for instance, $q((x_1, \ldots, x_n)) = \sum_{i=1}^n x_i^2$.

At this point, a natural question arises. Suppose ω is a symmetric, bilinear form on a finite-dimensional vector space V. Can we choose a basis $\underline{\alpha}$ of V so that the representation of ω in equation 7.3 is as simple as possible? What would the corresponding quadratic form q look like in this representation? We shall give answers to both of these questions when $F = \mathbb{R}$. For a more general treatment, we refer the reader to [2].

For the rest of this section, we assume V is a finite-dimensional vector space over $\mathbb{R}$. Let ω be a symmetric, bilinear form on V.

Definition 7.5: A basis $\underline{\alpha} = \{\alpha_1, \ldots, \alpha_n\}$ of V is said to be ω-orthonormal if

(a) $\omega(\alpha_i, \alpha_j) = 0$ whenever $i \neq j$, and
(b) $\omega(\alpha_i, \alpha_i) \in \{-1, 0, 1\}$ for all $i = 1, \ldots, n$.

In Example 7.2, for instance, the canonical basis $\underline{\delta} = \{\delta_i = (0, \ldots, 1, \ldots, 0) | i = 1, \ldots, n\}$ is an ω-orthonormal basis of $\mathbb{R}^n$. Our first theorem in this section guarantees ω-orthonormal bases exist.

Theorem 7.6: Let V be a finite-dimensional vector space over $\mathbb{R}$ and suppose ω is a symmetric, bilinear form on V. Then V has an ω-orthonormal basis.

Proof: We proceed via induction on $n = \dim V$. If $V = (0)$, then the result is trivial. So, suppose $n = 1$. Then any nonzero vector of V is a basis of V. If $\omega(\alpha, \alpha) = 0$ for every $\alpha \in V$, then any nonzero vector of V is an ω-orthonormal basis. Suppose there exists a $\beta \in V$ such that $\omega(\beta, \beta) \neq 0$. Then $c = |\omega(\beta, \beta)|^{-1/2}$ is a positive scalar in $\mathbb{R}$, and $\{c\beta\}$ is an ω-orthonormal basis of V. Thus, we have established the result for all vector spaces of dimension 1 over $\mathbb{R}$.

Suppose $n > 1$, and we have proved the theorem for any vector space over $\mathbb{R}$ of dimension less than n. Since ω is symmetric, we have

7.7:

$$\omega(\xi, \eta) = \frac{q(\xi + \eta) - q(\xi - \eta)}{4} \qquad \text{for all} \quad \xi, \eta \in V$$

In equation 7.7, q is the quadratic form associated with ω. Now if $\omega(\alpha, \alpha) = q(\alpha) = 0$ for all $\alpha \in V$, then 7.7 implies ω is identically zero. In this case, any basis of V is an ω-orthonormal basis. Thus, we can assume there exists a nonzero vector $\beta \in V$ such that $\omega(\beta, \beta) \neq 0$. As in the case n = 1, we can then adjust β by a scalar multiple if need be and find an $\alpha_n \neq 0$ in V such that $\omega(\alpha_n, \alpha_n) \in \{-1, 1\}$.

Next define a linear transformation $f \in V^*$ by $f(\xi) = \omega(\alpha_n, \xi)$. Since $f(\alpha_n) = \omega(\alpha_n, \alpha_n) \neq 0$, f is a nonzero map. Set N = ker f. Since $f \neq 0$, and $\dim_{\mathbb{R}} \mathbb{R} = 1$, f is surjective. Thus, Corollary 5.17 implies $V/N \cong \mathbb{R}$. In particular, Theorem 5.14 implies dim N = dim V − 1. ω when restricted to N is clearly a symmetric bilinear form. Hence our induction hypothesis implies N has an ω-orthonormal basis $\{\alpha_1, \ldots, \alpha_{n-1}\}$.

We claim $\underline{\alpha} = \{\alpha_1, \ldots, \alpha_{n-1}, \alpha_n\}$ is an ω-orthonormal basis of V. Since $f(\alpha_n) \neq 0$, $\alpha_n \notin N$. In particular, $\underline{\alpha}$ is linearly independent over $\mathbb{R}$. Since $\dim_{\mathbb{R}}(V) = n$, $\underline{\alpha}$ is a basis of V. Conditions (a) and (b) of Definitions 7.5 are satisfied for $\{\alpha_1, \ldots, \alpha_{n-1}\}$ since this set is an ω-orthonormal basis of N. Since N = ker f, $\omega(\alpha_i, \alpha_n) = 0$ for i = 1, ..., n − 1. Thus, $\underline{\alpha}$ is an ω-orthonormal basis of V and the proof of Theorem 7.6 is complete. $\square$

The existence of ω-orthonormal bases of V answers our first question about representing ω. Suppose $\underline{\alpha} = \{\alpha_1, \ldots, \alpha_n\}$ is an ω-orthonormal basis of V. Then the matrix $M(\omega, \underline{\alpha})$ is just an $n \times n$ diagonal matrix, $\text{diag}(q(\alpha_1), \ldots, q(\alpha_n))$, with $q(\alpha_i) = \omega(\alpha_i, \alpha_i) \in \{-1, 0, 1\}$. If $\xi, \eta \in V$ with $[\xi]_{\underline{\alpha}} = (x_1, \ldots, x_n)^t$ and $[\eta]_{\underline{\alpha}} = (y_1, \ldots, y_n)^t$, then equation 7.3 implies $\omega(\xi, \eta) = \sum_{i=1}^{n} x_i y_i q(\alpha_i)$. By reordering the elements of $\underline{\alpha}$ if need be, we can assume $\underline{\alpha} = \{\alpha_1, \ldots, \alpha_p\} \cup \{\alpha_{p+1}, \ldots, \alpha_{p+m}\} \cup \{\alpha_{p+m+1}, \ldots, \alpha_{p+m+r}\}$, where

7.8:

$$q(\alpha_i) = \begin{cases} 1 & \text{for} \quad i = 1, \ldots, p \\ -1 & \text{for} \quad i = p + 1, \ldots, p + m \\ 0 & \text{for} \quad i = p + m + 1, \ldots, p + m + r \end{cases}$$

The vector space V then decomposes into the direct sum $V = V_{-1} \oplus V_0 \oplus V_1$, where $V_{-1} = L(\{\alpha_{p+1}, \ldots, \alpha_{p+m}\})$, $V_0 = L(\{\alpha_{p+m+1}, \ldots, \alpha_{p+m+r}\})$, and $V_1 = L(\{\alpha_1, \ldots, \alpha_p\})$.

Our quadratic form q is positive on $V_1 - (0)$, zero on V_0, and negative on $V_{-1} - (0)$. For example, suppose $\beta \in V_{-1} - (0)$. Then $\beta = x_1 \alpha_{p+1} + \cdots + x_m \alpha_{p+m}$ for some $x_1, \ldots, x_m \in F$. Thus, $q(\beta) = \omega(\beta, \beta) = \sum_{i=1}^{m} x_i^2 q(\alpha_{p+i})$. Since $\beta \neq 0$, some x_i is nonzero. Since $q(\alpha_{p+i}) = -1$ for all i = 1, ..., m, we see $q(\beta) < 0$.

The subspaces V_{-1}, V_0, and V_1 are pairwise ω-orthogonal in the sense that $\omega(V_i, V_j) = 0$ whenever $i, j \in \{-1, 0, 1\}$ and $i \neq j$. Thus, any ω-orthonormal basis $\underline{\alpha}$ of V decomposes V into a direct sum $V = V_{-1} \oplus V_0 \oplus V_1$ of pairwise ω-orthogonal subspaces V_j. The sign of the associated quadratic form q is constant on each $V_j - (0)$. An important fact here is that the dimensions of these three subspaces, p, m, and r, depend only on ω and not on the particular ω-orthonormal basis $\underline{\alpha}$ chosen.

Lemma 7.9: Suppose $\beta = \{\beta_1, \ldots, \beta_n\}$ is a second ω-orthonormal basis of V, and let $V = W_{-1} \oplus W_0 \oplus W_1$ be the corresponding decomposition of V. Then $\dim W_j = \dim V_j$ for $j = -1, 0, 1$.

Proof: W_{-1} is the subspace of V spanned by those β_i for which $q(\beta_i) = -1$. Let $\alpha \in W_{-1} \cap (V_0 + V_1)$. If $\alpha \neq 0$, then $q(\alpha) < 0$ since $\alpha \in W_{-1}$. But $\alpha \in V_0 + V_1$ implies $q(\alpha) \geqslant 0$, which is impossible. Thus, $\alpha = 0$. So, $W_{-1} \cap (V_0 + V_1) = (0)$. By expanding the basis of W_{-1} if need be, we can then construct a subspace P of V such that $W_{-1} \subseteq P$, and $P \oplus (V_0 + V_1) = V$. Thus, from Theorem 4.9, we have $\dim(W_{-1}) \leqslant \dim P = \dim V - \dim V_0 - \dim V_1 = \dim(V_{-1})$. Therefore, $\dim(W_{-1}) \leqslant \dim V_{-1}$. Reversing the roles of the W_i and V_i in this proof gives $\dim(V_{-1}) \leqslant \dim(W_{-1})$. Thus, $\dim(W_{-1}) = \dim(V_{-1})$. A similar proof shows $\dim(W_1) = \dim(V_1)$. Then $\dim(W_0) = \dim(V_0)$ by Theorem 4.9. This completes the proof of Lemma 7.9. $\square$

Let us agree when discussing ω-orthonormal bases $\underline{\alpha}$ of V always to order the basis elements $\alpha_i \in \underline{\alpha}$ according to equation 7.8. Then Lemma 7.9 implies that the integers p, m, and r do not depend on $\underline{\alpha}$ but only on ω. In particular, the following definition makes sense.

Definition 7.10: $p - m$ is called the *signature* of q. $p + m$ is called the *rank* of q.

We have now proved the following theorem:

Theorem 7.11: Let ω be a symmetric, bilinear form on a finite-dimensional vector space V over $\mathbb{R}$. Then there exists integers m and p such that if $\underline{\alpha} = \{\alpha_1, \ldots, \alpha_n\}$ is any ω-orthonormal basis of V and $[\xi]_{\underline{\alpha}} = (x_1, \ldots, x_n)^t$, then $q(\xi) = \sum_{i=1}^{p} x_i^2 - \sum_{i=p+1}^{p+m} x_i^2$. $\square$

A quadratic form q, associated with some symmetric bilinear form ω on V, is said to be definite if $q(\xi) = 0$ implies $\xi = 0$. For instance, in Example 7.2, $q((x_1, \ldots, x_n)) = \sum_{i=1}^{n} x_i^2$ is definite when $F = \mathbb{R}$. If $F = \mathbb{C}$, then q is not definite since, for example, $q((1, \sqrt{-1}, 0, \ldots, 0)) = 0$.

If q is a definite quadratic form on a finite-dimensional vector space V over $\mathbb{R}$, then Theorem 7.11 implies $q(\xi) > 0$ for all $\xi \in V - (0)$ or $q(\xi) < 0$ for all $\xi \in V - (0)$. In general, we say a quadratic form q is *positive definite* if $q(\xi) > 0$ for all $\xi \in V - (0)$. We say q is *negative definite* if $q(\xi) < 0$ for all $\xi \in V - (0)$.

Definition 7.12: Let V be a vector space over $\mathbb{R}$. A symmetric, bilinear form ω on V whose associated quadratic form is positive definite is called an *inner product* on V.

Note in our definition that we do not require that V be finite dimensional. We finish this section with a few examples of inner products.

Example 7.13: Let $V = \mathbb{R}^n$, and define ω as in Example 7.2. □

Example 7.14: Let $V = \bigoplus_{i=1}^{\infty} \mathbb{R}$, and define ω by $\omega((x_1, x_2, \dots), (y_1, y_2, \dots)) = \sum_{i=1}^{\infty} x_i y_i$. Since both sequences $\{x_i\}$ and $\{y_i\}$ are eventually zero, ω is well defined and is clearly an inner product on V. □

Example 7.15: Let $V = C([a, b])$. Define $\omega(f, g) = \int_a^b f(x)g(x)\, dx$. Clearly, ω is an inner product on V. □

We shall come back to the study of inner products in Chapter V.

EXERCISES FOR SECTION 7

(1) In our proof of Lemma 7.9, we used the following fact: If W and W' are subspaces of V such that $W \cap W' = (0)$, then there exists a complement of W' that contains W. Give a proof of this fact.

(2) Let $V = M_{m \times n}(F)$, and let $C \in M_{m \times m}(F)$. Define a map $\omega: V \times V \to F$ by the formula $\omega(A, B) = \text{Tr}(A^t C B)$. Show that ω is a bilinear form. Is ω symmetric?

(3) Let $V = M_{n \times n}(F)$. Define a map $\omega: V \times V \to F$ by $\omega(A, B) = n\,\text{Tr}(AB) - \text{Tr}(A)\text{Tr}(B)$. Show that ω is a bilinear form. Is ω symmetric?

(4) Exhibit a bilinear form on $\mathbb{R}^n$ that is not symmetric.

(5) Find a symmetric bilinear form on $\mathbb{C}^n$ whose associated quadratic form is positive definite.

(6) Describe explicitly all symmetric bilinear forms on $\mathbb{R}^3$.

(7) Describe explicitly all skew-symmetric bilinear froms on $\mathbb{R}^3$. A bilinear form ω is skew-symmetric if $\omega(\alpha, \beta) = -\omega(\beta, \alpha)$.

(8) Let $\omega: V \times V \to F$ be a bilinear form on a finite dimensional vector space V. Show that the following conditions are equivalent:
   (a) $\{\alpha \in V \mid \omega(\alpha, \beta) = 0 \text{ for all } \beta \in V\} = (0)$.
   (b) $\{\alpha \in V \mid \omega(\beta, \alpha) = 0 \text{ for all } \beta \in V\} = (0)$.
   (c) $M(\omega, \underline{\alpha})$ is nonsingular for any basis $\underline{\alpha}$ of V.
   We say ω is nondegenerate if ω satisfies the conditions listed above.

(9) Suppose $\omega: V \times V \to F$ is a nondegenerate, bilinear form on a finite-dimensional vector space V. Let W be a subspace of V. Set $W^{\perp} = \{\alpha \in V \mid \omega(\alpha, \beta) = 0 \text{ for all } \beta \in W\}$. Show that $V = W \oplus W^{\perp}$.

(10) With the same hypotheses as in Exercise 9, suppose $f \in V^*$. Prove that there exists an $\alpha \in V$ such that $f(\beta) = \omega(\alpha, \beta)$ for all $\beta \in V$.

(11) Suppose $\omega \colon V \times V \to F$ is a bilinear form on V. Let W_1 and W_2 be subspaces of V. Show that $(W_1 + W_2)^\perp = W_1^\perp \cap W_2^\perp$. If ω is nondegenerate, prove that $(W_1 \cap W_2)^\perp = W_1^\perp + W_2^\perp$.

(12) Let ω be a nondegenerate, bilinear form on a finite-dimensional vector space V. Let ω' be any bilinear form on V. Show there exists a unique $T \in \operatorname{Hom}_F(V, V)$ such that $\omega'(\alpha, \beta) = \omega(T(\alpha), \beta)$ for all $\alpha, \beta \in V$. Show that ω' is nondegenerate if and only if T is bijective.

(13) With the same hypotheses as in Exercise 12, show that for every $T \in \operatorname{Hom}_F(V, V)$ there exists a unique $T' \in \operatorname{Hom}_F(V, V)$ such that $\omega(T(\alpha), \beta) = \omega(\alpha, T'(\beta))$ for all $\alpha, \beta \in V$.

(14) Let Bil(V) denote the set of all bilinear forms on the vector space V. Define addition in Bil(V) by $(\omega + \omega')(\alpha, \beta) = \omega(\alpha, \beta) + \omega'(\alpha, \beta)$, and scalar multiplication by $(x\omega)(\alpha, \beta) = x\omega(\alpha, \beta)$. Prove that Bil(V) is a vector space over F with these definitions. What is the dimension of Bil(V) when V is finite dimensional?

(15) Find an ω-orthonormal basis for $\mathbb{R}^2$ when ω is given by $\omega((x_1, y_1), (x_2, y_2)) = x_1 y_2 + x_2 y_1$.

(16) Argue that $\omega(f, g) = \int_a^b f(x)g(x)\,dx$ is an inner product on C([a, b]).

(17) Let $V = \{p(X) \in \mathbb{R}[X] \mid \deg(p) \leqslant 5\}$. Suppose $\omega \colon V \times V \to \mathbb{R}$ is given by $\omega(f, g) = \int_0^1 f(x)g(x)\,dx$. Find an ω-orthonormal basis of V.

(18) Let V be the subspace of $C([-\pi, \pi])$ spanned by the functions 1, sin(x), cos(x), sin(2x), cos(2x), ..., sin(nx), cos(nx). Find an ω-orthonormal basis of V where ω is the inner product given in Exercise 16.

Chapter II

Multilinear Algebra

1. MULTILINEAR MAPS AND TENSOR PRODUCTS

In Chapter I, we dealt mainly with functions of one variable between vector spaces. Those functions were linear in that variable and were called linear transformations. In this chapter, we examine functions of several variables between vector spaces. If such a function is linear in each of its variables, then the function is called a *multilinear mapping*. Along with any theory of multilinear maps comes a sequence of universal mapping problems whose solutions are the fundamental ideas in multilinear algebra. In this and the next few sections, we shall give a careful explanation of the principal constructions of the subject matter. Applications of the ideas discussed here will abound throughout the rest of the book.

Let us first give a careful definition of a multilinear mapping. As usual, F will denote an arbitrary field. Suppose $V_1, \ldots, V_n$ and V are vector spaces over F. Let $\phi: V_1 \times \cdots \times V_n \to V$ be a function from the finite product $V_1 \times \cdots \times V_n$ to V. We had seen in Section 4 of Chapter I that a typical vector in $V_1 \times \cdots \times V_n$ is an n-tuple $(\alpha_1, \ldots, \alpha_n)$ with $\alpha_i \in V_i$. Thus, we can think of ϕ as a function of the n variable vectors $\alpha_1, \ldots, \alpha_n$.

Definition 1.1: A function $\phi: V_1 \times \cdots \times V_n \to V$ is called a multilinear mapping if for each $i = 1, \ldots, n$, we have

(a) $\phi(\alpha_1, \ldots, \alpha_i + \alpha_i', \ldots, \alpha_n) = \phi(\alpha_1, \ldots, \alpha_i, \ldots, \alpha_n) + \phi(\alpha_1, \ldots, \alpha_i', \ldots, \alpha_n)$,
   and

(b) $\phi(\alpha_1, \ldots, x\alpha_i, \ldots, \alpha_n) = x\phi(\alpha_1, \ldots, \alpha_i, \ldots, \alpha_n)$.

Here $\alpha_i, \alpha_i' \in V_i, \alpha_j \in V_j$ for $j \neq i$, and $x \in F$. Thus, a function $\phi: V_1 \times \cdots \times V_n \to V$ is a multilinear mapping if for all $i \in \{1, \ldots, n\}$ and for all vectors $\alpha_1 \in V_1$, $\ldots, \alpha_{i-1} \in V_{i-1}, \alpha_{i+1} \in V_{i+1}, \ldots, \alpha_n \in V_n$, we have $\phi(\alpha_1, \ldots, \alpha_{i-1}, \cdot, \alpha_{i+1}, \ldots, \alpha_n) \in \mathrm{Hom}_F(V_i, V)$. Before proceeding further, let us give a few examples of multilinear maps.

Example 1.2: If $n = 1$, then a function $\phi: V_1 \to V$ is a multilinear mapping if and only if ϕ is a linear transformation. Thus, linear transformations are just special cases of multilinear maps. □

Example 1.3: If $n = 2$, then a multilinear map $\phi: V_1 \times V_2 \to V$ is what we called a bilinear map in Chapter I. For a concrete example, we have $\omega: V \times V^* \to F$ given by $\omega(\alpha, T) = T(\alpha)$ (equation 6.6 of Chapter I). □

Example 1.4: The determinant, det(A), of an $n \times n$ matrix A can be thought of as a multilinear mapping $\phi: F^n \times \cdots \times F^n \to F$ in the following way: If $\alpha_i = (a_{i1}, \ldots, a_{in}) \in F^n$ for $i = 1, \ldots, n$, then set $\phi(\alpha_1, \ldots, \alpha_n) = \det(a_{ij})$. The fact that ϕ is multilinear is an easy computation, which we leave as an exercise at the end of this section. □

Example 1.5: Suppose A is an algebra over F with multiplication denoted by $\alpha\beta$ for $\alpha, \beta \in A$. Let $n \geqslant 2$. We can then define a function $\mu: A^n \to A$ by $\mu(\alpha_1, \ldots, \alpha_n) = \alpha_1 \alpha_2 \cdots \alpha_n$. Clearly μ is a multilinear mapping. □

If $\phi: V_1 \times \cdots \times V_n \to V$ is a multilinear map and T is a linear transformation from V to W, then clearly, $T\phi: V_1 \times \cdots \times V_n \to W$ is again a multilinear map. We can use this idea along with Example 1.5 above to give a few familiar examples from analysis.

Example 1.6: Let I be an open interval in $\mathbb{R}$. Set $C^\infty(I) = \bigcap_{k=1}^\infty C^k(I)$. Thus, $C^\infty(I)$ consists of those $f \in C(I)$ such that f is infinitely differentiable on I. Clearly, $C^\infty(I)$ is an algebra over $\mathbb{R}$ when we define vector addition $[(f + g)(x) = f(x) + g(x)]$, scalar multiplication $[(yf)(x) = yf(x)]$, and algebra multiplication $[(fg)(x) = f(x)g(x)]$ in the usual ways. Let $D: C^\infty(I) \to C^\infty(I)$ be the function that sends a given $f \in C^\infty(I)$ to its derivative f'. Thus, $D(f) = f'$. Clearly, $D \in \mathrm{Hom}_{\mathbb{R}}(C^\infty(I), C^\infty(I))$.

Let $n \in \mathbb{N}$. Define a map $\phi: \{C^\infty(I)\}^n \to C^\infty(I)$ by $\phi(f_1, \ldots, f_n) = D(f_1 \cdots f_n)$. Our comments immediately proceeding this example imply that ϕ is a multilinear mapping. □

Example 1.7: Let [a, b] be a closed interval in $\mathbb{R}$ and consider C([a, b]). Clearly, C([a, b]) is an $\mathbb{R}$-algebra under the same pointwise operations given in Example 1.6. We can define a multilinear, real valued function $\psi: C([a, b])^n \to \mathbb{R}$ by $\psi(f_1, \ldots, f_n) = \int_a^b f_1 \cdots f_n$. □

Let us denote the collection of multilinear mappings from $V_1 \times \cdots \times V_n$ to V by $\text{Mul}_F(V_1 \times \cdots \times V_n, V)$. If $Z = V_1 \times \cdots \times V_n$, then clearly, $\text{Mul}_F(V_1 \times \cdots \times V_n, V)$ is a subset of the vector space V^Z. In particular, if $f, g \in \text{Mul}_F(V_1 \times \cdots \times V_n, V)$ and $x, y \in F$, then $xf + yg$ is a vector in V^Z. A simple computation shows that $xf + yg$ is in fact a multilinear mapping. This proves the first assertion in the following theorem:

Theorem 1.8: Let $V_1, \ldots, V_n$ and V be vector spaces over F. Set $Z = V_1 \times \cdots \times V_n$. Then

(a) $\text{Mul}_F(V_1 \times \cdots \times V_n, V)$ is a subspace of V^Z.

(b) If $n \geqslant 2$, $\{\text{Mul}_F(V_1 \times \cdots \times V_n, V)\} \cap \{\text{Hom}_F(V_1 \times \cdots \times V_n, V)\} = (0)$.

Proof: We need prove only (b). Suppose $\phi: V_1 \times \cdots \times V_n \to V$ is a multilinear mapping that is also a linear transformation. Fix $i = 1, \ldots, n$. Since ϕ is multilinear, we have $\phi(\alpha_1, \ldots, \alpha_i + \alpha_i, \ldots, \alpha_n) = \phi(\alpha_1, \ldots, \alpha_n) + \phi(\alpha_1, \ldots, \alpha_n)$. Since ϕ is linear, we have $\phi(\alpha_1, \ldots, \alpha_i + \alpha_i, \ldots, \alpha_n) = \phi(\alpha_1, \ldots, \alpha_i, \ldots, \alpha_n) + \phi(0, \ldots, \alpha_i, \ldots, 0)$. Comparing the two results gives $\phi(\alpha_1, \ldots, \alpha_n) = \phi(0, \ldots, \alpha_i, \ldots, 0)$. Again since ϕ is linear, we have $\phi(\alpha_1, \ldots, \alpha_n) = \sum_{j=1}^{n} \phi(0, \ldots, \alpha_j, \ldots, 0)$. Thus, $\phi(\alpha_1, \ldots, \alpha_{i-1}, 0, \alpha_{i+1}, \ldots, \alpha_n) = 0$.

Now $n \geqslant 2$, the α_j are arbitrary, and so is the index i. Therefore, for any $(\alpha_1, \ldots, \alpha_n) \in V_1 \times \cdots \times V_n$, we have $\phi(\alpha_1, \ldots, \alpha_n) = \phi(0, \alpha_2, \ldots, \alpha_n) + \phi(\alpha_1, 0, \ldots, 0) = 0 + 0 = 0$. $\square$

Theorem 1.8(b) says that in general (i.e., when $n \geqslant 2$) a nonzero, multilinear mapping $\phi: V_1 \times \cdots \times V_n \to V$ is not a linear transformation from $V_1 \times \cdots \times V_n$ to V, and vice versa. We must always be careful not to confuse these two concepts when dealing with functions from $V_1 \times \cdots \times V_n$ to V.

We have seen in Examples 1.6 and 1.7 that one method for constructing multilinear mappings on $V_1 \times \cdots \times V_n$ is to choose a fixed multilinear map from $V_1 \times \cdots \times V_n$ to V and then compose it with various linear transformations from V to other vector spaces. A natural question arises here. Can we construct (with possibly a judicious choice of V) all possible multilinear maps on $V_1 \times \cdots \times V_n$ by this method? This question has an affirmative answer, which leads us to the construction of the tensor product of $V_1, \ldots, V_n$. Before proceeding further, let us give a precise statement of the problem we wish to consider.

1.9: Let $V_1, \ldots, V_n$ be vector spaces over F. Is there is a vector space V (over F) and a multilinear mapping $\phi: V_1 \times \cdots \times V_n \to V$ such that if $\psi: V_1 \times \cdots \times V_n \to W$ is any multilinear mapping on $V_1 \times \cdots \times V_n$, then there exists a unique linear transformation $T \in \text{Hom}_F(V, W)$ with $T\phi = \psi$?

Question 1.9 is called the universal mapping problem for multilinear mappings on $V_1 \times \cdots \times V_n$. In terms of commutative diagrams, the universal problem can be stated as follows: Can we construct a multilinear map $\phi: V_1 \times \cdots \times V_n \to V$ with the property that for any multilinear map $\psi: V_1 \times \cdots \times V_n \to W$ there exists a unique $T \in \text{Hom}_F(V, W)$ such that the following diagram is commutative:

1.10:

$$
\begin{array}{ccc}
V_1 \times \cdots \times V_n & \xrightarrow{\ \ \phi\ \ } & V \\
& \psi \searrow & \downarrow T \\
& & W
\end{array}
$$

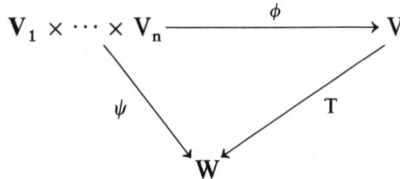

Notice that a solution to 1.9 consists of a vector space V and a multilinear map $\phi: V_1 \times \cdots \times V_n \to V$. The pair (V, ϕ) must satisfy the following property: If W is any vector space over F and $\psi: V_1 \times \cdots \times V_n \to W$ is any multilinear mapping, then there must exist a unique, linear transformation $T: V \to W$ such that 1.10 commutes.

Before constructing a pair (V, ϕ) satisfying the properties in 1.9, let us make the observation that any such pair is essentially unique up to isomorphism. To be more precise, we have the following lemma:

Lemma 1.11: Suppose (V, ϕ) and (V', ϕ') are two solutions to 1.9. Then there exist two isomorphisms $T_1 \in \text{Hom}_F(V, V')$ and $T_2 \in \text{Hom}_F(V', V)$ such that

(a) $T_1 T_2 = I_{V'}, T_2 T_1 = I_V$, and
(b) the following diagram is commutative:

1.12:

$$
\begin{array}{ccc}
V_1 \times \cdots \times V_n & \xrightarrow{\ \ \phi\ \ } & V \\
& \phi' \searrow & \begin{array}{c} T_2 \\ T_1 \end{array} \Big\updownarrow \\
& & V'
\end{array}
$$

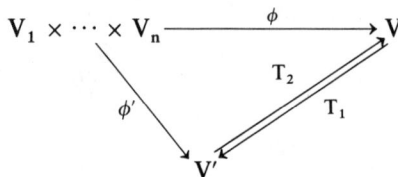

Proof: Since (V, ϕ) is a solution to 1.9 and $\phi': V_1 \times \cdots \times V_n \to V'$ is a multi-linear map, there exists a unique $T_1 \in \text{Hom}_F(V, V')$ such that $T_1 \phi = \phi'$. Similarly, there exists a unique $T_2 \in \text{Hom}_F(V', V)$ such that $T_2 \phi' = \phi$. Putting the two obvious diagrams together, we get

1.13:

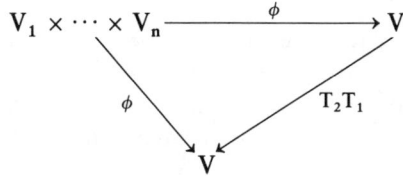

is commutative. Now in diagram 1.13, we can replace $T_2 T_1$ with I_V the identity map on V. Clearly, the diagram stays commutative. Since (V, ϕ) satisfies 1.9, there can be only one linear transformation from V to V making 1.13 commutative. We conclude that $T_2 T_1 = I_V$. Similarly, $T_1 T_2 = I_{V'}$, and the proof of the lemma is complete. $\square$

Thus, if we find any solution to 1.9, then up to isomorphism we have found them all. We now turn to the matter of constructing a solution. We need to recall a few facts about direct sums.

Suppose Δ is a nonempty set. Then we can construct a vector space U over F and a bijective map $\varphi : \Delta \to U$ such that $\varphi(\Delta)$ is a basis of U. To see this, set $U = \bigoplus_{i \in \Delta} F$. Thus, U is the direct sum of $|\Delta|$ copies of F. For each $i \in \Delta$, let δ_i be the vector in U defined by $\delta_i(j) = 0$ if $j \neq i$, and $\delta_i(i) = 1$. It follows from Theorem 4.13 in Chapter I that $B = \{\delta_i | i \in \Delta\}$ is a basis of U. The map $\varphi : \Delta \to B$ given by $\varphi(i) = \delta_i$ is clearly bijective.

Now suppose Δ itself is a vector space over F. Then in $U = \bigoplus_{i \in \Delta} F$, we have vectors of the form $\delta_{(i_1 + \cdots + i_n)} - \delta_{i_1} - \cdots - \delta_{i_n}$ and $\delta_{xi} - x\delta_i$ for $i_1, \ldots, i_n, i \in \Delta$ and $x \in F$. We shall employ these ideas in the construction of a solution to 1.9.

Let $V_1, \ldots, V_n$ be vector spaces over F, and, for notational convenience, set $Z = V_1 \times \cdots \times V_n$. A typical element in the set Z is an n-tuple of the form $(\alpha_1, \ldots, \alpha_n)$ with $\alpha_i \in V_i$. Set $U = \bigoplus_{(\alpha_1, \ldots, \alpha_n) \in Z} F$. Thus, U is the direct sum of $|Z|$ copies of F. As we observed above, U has a basis of the form $\{\delta_{(\alpha_1, \ldots, \alpha_n)} | (\alpha_1, \ldots, \alpha_n) \in V_1 \times \cdots \times V_n\}$. Let U_0 be the subspace of U spanned by all possible vectors of the following two types:

1.14:

$$\delta_{(\alpha_1, \ldots, \alpha_i + \alpha_i', \ldots, \alpha_n)} - \delta_{(\alpha_1, \ldots, \alpha_i, \ldots, \alpha_n)} - \delta_{(\alpha_1, \ldots, \alpha_i', \ldots, \alpha_n)}$$

and

$$\delta_{(\alpha_1, \ldots, x\alpha_i, \ldots, \alpha_n)} - x\delta_{(\alpha_1, \ldots, \alpha_i, \ldots, \alpha_n)}$$

In 1.14, i can be any index between 1 and n, $(\alpha_1, \ldots, \alpha_i, \ldots, \alpha_n)$, $(\alpha_1, \ldots, \alpha_i', \ldots, \alpha_n)$ any elements of Z and x any scalar in F.

Set $V = U/U_0$, the quotient space of U by U_0. There is a natural map $\phi: V_1 \times \cdots \times V_n \to V$ given by $\phi(\alpha_1, \ldots, \alpha_n) = \delta_{(\alpha_1, \ldots, \alpha_n)} + U_0$. Thus, $\phi(\alpha_1, \ldots, \alpha_n)$ is just the coset in U/U_0 containing the vector $\delta_{(\alpha_1, \ldots, \alpha_n)}$. We can now prove the following lemma:

Lemma 1.15: (V, ϕ) is a solution to 1.9.

Proof: Clearly, V is a vector space over F. We must first argue that ϕ is a multilinear mapping. This follows immediately from 1.14 and the definition of U_0. We have

$$\phi(\alpha_1, \ldots, \alpha_i + \alpha_i', \ldots, \alpha_n) = \delta_{(\alpha_1, \ldots, \alpha_i + \alpha_i', \ldots, \alpha_n)} + U_0$$
$$= (\delta_{(\alpha_1, \ldots, \alpha_i, \ldots, \alpha_n)} + \delta_{(\alpha_1, \ldots, \alpha_i', \ldots, \alpha_n)}) + U_0$$
$$= (\delta_{(\alpha_1, \ldots, \alpha_i, \ldots, \alpha_n)} + U_0) + (\delta_{(\alpha_1, \ldots, \alpha_i', \ldots, \alpha_n)} + U_0)$$
$$= \phi(\alpha_1, \ldots, \alpha_i, \ldots, \alpha_n) + \phi(\alpha_1, \ldots, \alpha_i', \ldots, \alpha_n)$$

Also,

$$\phi(\alpha_1, \ldots, x\alpha_i, \ldots, \alpha_n) = \delta_{(\alpha_1, \ldots, x\alpha_i, \ldots, \alpha_n)} + U_0 = x\delta_{(\alpha_1, \ldots, \alpha_i, \ldots, \alpha_n)} + U_0$$
$$= x(\delta_{(\alpha_1, \ldots, \alpha_i, \ldots, \alpha_n)} + U_0)$$
$$= x\phi(\alpha_1, \ldots, \alpha_i, \ldots, \alpha_n)$$

Thus, ϕ is a multilinear mapping.

Now suppose W is another vector space over F and $\psi: V_1 \times \cdots \times V_n \to W$ a multilinear mapping. We must construct a unique linear transformation $T: V \to W$ such that $T\phi = \psi$. To do this, we recall that $B = \{\delta_{(\alpha_1, \ldots, \alpha_n)} | (\alpha_1, \ldots, \alpha_n) \in V_1 \times \cdots \times V_n\}$ is a basis of U. It follows from 3.23 of Chapter I that there exists a unique $T_0 \in \mathrm{Hom}_F(U, W)$ such that $T_0(\delta_{(\alpha_1, \ldots, \alpha_n)}) = \psi(\alpha_1, \ldots, \alpha_n)$ for all $(\alpha_1, \ldots, \alpha_n) \in Z$. Since ψ is multilinear, T_0 vanishes on the generators (1.14) of U_0. Thus, $U_0 \subseteq \ker T_0$. It now follows from the first isomorphism theorem (Theorem 5.15 of Chapter I) that T_0 induces a linear transformation $T: U/U_0 \to W$ such that $T(\delta_{(\alpha_1, \ldots, \alpha_n)} + U_0) = T_0(\delta_{(\alpha_1, \ldots, \alpha_n)}) = \psi(\alpha_1, \ldots, \alpha_n)$ for all $(\alpha_1, \ldots, \alpha_n) \in Z$. Since $\phi(\alpha_1, \ldots, \alpha_n) = \delta_{(\alpha_1, \ldots, \alpha_n)} + U_0$, we have $T\phi = \psi$.

Finally, suppose $T' \in \mathrm{Hom}_F(V, W)$ and $T'\phi = \psi$. We must argue $T' = T$. Since $T'\phi = T\phi$, we see $T = T'$ on $\mathrm{Im}\,\phi$. From our definitions, $L(\mathrm{Im}\,\phi) = V$. Therefore, $T = T'$, and the proof of Lemma 1.15 is complete. $\square$

Definition 1.16: The vector space U/U_0 is called the *tensor product* of $V_1, \ldots, V_n$ (over F) and will henceforth be denoted $V_1 \otimes_F \cdots \otimes_F V_n$.

When the field F is clear from the context, we shall drop it from our notation and simply write $V_1 \otimes \cdots \otimes V_n$ for the tensor product of $V_1, \ldots, V_n$.

Definition 1.17: A coset $\delta_{(\alpha_1,\ldots,\alpha_n)} + U_0$ in the tensor product $U/U_0 = V_1 \otimes \cdots \otimes V_n$ will henceforth be written $\alpha_1 \otimes \cdots \otimes \alpha_n$.

With these changes in notation, our multilinear map $\phi: V_1 \times \cdots \times V_n \to V_1 \otimes \cdots \otimes V_n$ is given by $\phi(\alpha_1, \ldots, \alpha_n) = \alpha_1 \otimes \cdots \otimes \alpha_n$. We shall refer to ϕ as the canonical map of $V_1 \times \cdots \times V_n$ into $V_1 \otimes \cdots \otimes V_n$. Since ϕ is multilinear, we have the following relations in $V_1 \otimes \cdots \otimes V_n$:

1.18:

$$\alpha_1 \otimes \cdots \otimes (\alpha_i + \alpha_i') \otimes \cdots \otimes \alpha_n$$
$$= \alpha_1 \otimes \cdots \otimes \alpha_i \otimes \cdots \otimes \alpha_n + \alpha_1 \otimes \cdots \otimes \alpha_i' \otimes \cdots \otimes \alpha_n$$

and

$$\alpha_1 \otimes \cdots \otimes x\alpha_i \otimes \cdots \otimes \alpha_n = x(\alpha_1 \otimes \cdots \otimes \alpha_i \otimes \cdots \otimes \alpha_n)$$

We also know from our construction of U/U_0 that $V_1 \otimes \cdots \otimes V_n$ is spanned by the image of ϕ. Thus, every vector in $V_1 \otimes \cdots \otimes V_n$ is a finite sum of the form $\sum_{i=1}^{r}(\alpha_{1i} \otimes \cdots \otimes \alpha_{ni})$. Here $\alpha_{1i} \in V_1, \ldots, \alpha_{ni} \in V_n$ for all $i = 1, \ldots, r$.

Finally, let us restate Lemma 1.15 using our new notation.

Theorem 1.19: Let $V_1, \ldots, V_n$ and W be vector spaces over F. Suppose $\psi: V_1 \times \cdots \times V_n \to W$ is a multilinear mapping. Then there exists a unique linear transformation $T \in \mathrm{Hom}_F(V_1 \otimes \cdots \otimes V_n, W)$ such that $T(\alpha_1 \otimes \cdots \otimes \alpha_n) = \psi(\alpha_1, \ldots, \alpha_n)$ for all $(\alpha_1, \ldots, \alpha_n) \in V_1 \times \cdots \times V_n$. $\square$

We shall discuss various functorial properties of $V_1 \otimes \cdots \otimes V_n$ in Section 2. But at this point, having introduced a new vector space $V_1 \otimes \cdots \otimes V_n$, we want to at least give a basis of this space.

Theorem 1.20: Let $V_1, \ldots, V_n$ be vector spaces over F, and suppose $B_1, \ldots, B_n$ are bases of $V_1, \ldots, V_n$, respectively. Then $B = \{ \beta_1 \otimes \cdots \otimes \beta_n \mid \beta_i \in B_i \}$ is a basis of $V_1 \otimes \cdots \otimes V_n$.

Proof: We prove this theorem by using Lemma 1.11. Consider the set $B_1 \times \cdots \times B_n = \{(\beta_1, \ldots, \beta_n) \mid \beta_i \in B_i\}$. Let $V' = \bigoplus_{(\beta_1, \ldots, \beta_n) \in B_1 \times \cdots \times B_n} F$. We have seen from our previous discussion that V' is a vector space over F with basis $\{\delta_{(\beta_1, \ldots, \beta_n)} \mid (\beta_1, \ldots, \beta_n) \in B_1 \times \cdots \times B_n\}$. We define a function $\phi_0: B_1 \times \cdots \times B_n \to V'$ by $\phi_0(\beta_1, \ldots, \beta_n) = \delta_{(\beta_1, \ldots, \beta_n)}$. Now $B_1 \times \cdots \times B_n \subseteq V_1 \times \cdots \times V_n$ and each V_i is the linear span of B_i. It follows that there exists a unique multilinear function $\phi': V_1 \times \cdots \times V_n \to V'$ such that $\phi'(\beta_1, \ldots, \beta_n) = \phi_0(\beta_1, \ldots, \beta_n)$ for all $(\beta_1, \ldots, \beta_n) \in B_1 \times \cdots \times B_n$.

We claim that (V', ϕ') satisfies 1.9. To see this, let $\psi: V_1 \times \cdots \times V_n \to W$ be an arbitrary multilinear mapping. Since $\{\delta_{(\beta_1, \cdots, \beta_n)} | (\beta_1, \ldots, \beta_n) \in B_1 \times \cdots \times B_n\}$ is a basis of V', it follows from 3.23 of Chapter I that there exists a unique linear transformation $T: V' \to W$ such that $T(\delta_{(\beta_1, \ldots, \beta_n)}) = \psi(\beta_1, \ldots, \beta_n)$ for all $(\beta_1, \ldots, \beta_n) \in B_1 \times \cdots \times B_n$. Then $T\phi' = \psi$ and clearly T is the unique linear transformation for which this happens.

We now have two pairs (V', ϕ') and $(V_1 \otimes \cdots \otimes V_n, \phi)$ satisfying 1.9. Hence, Lemma 1.11 implies there exists an isomorphism $S \in \text{Hom}_F(V', V_1 \otimes \cdots \otimes V_n)$ such that

1.21:

$$
\begin{array}{ccc}
V_1 \times \cdots \times V_n & \xrightarrow{\phi'} & V' \\
& \phi \searrow & \downarrow{\scriptstyle S\cong} \\
& & V_1 \otimes \cdots \otimes V_n
\end{array}
$$

is commutative. Now for all $(\beta_1, \ldots, \beta_n) \in B_1 \times \cdots \times B_n$, we have $\beta_1 \otimes \cdots \otimes \beta_n = \phi(\beta_1, \ldots, \beta_n) = S\phi'(\beta_1, \ldots, \beta_n) = S(\delta_{(\beta_1, \ldots, \beta_n)})$. Since S is an isomorphism, it maps any basis in V' to a basis in $V_1 \otimes \cdots \otimes V_n$. We conclude that $B = \{\beta_1 \otimes \cdots \otimes \beta_n | \beta_i \in B_i\}$ is a basis of $V_1 \otimes \cdots \otimes V_n$. $\square$

Corollary 1.22: Suppose $V_1, \ldots, V_n$ are finite-dimensional vector spaces over F. Let $m_i = \dim V_i$. Then $V_1 \otimes \cdots \otimes V_n$ is finite dimensional, and $\dim(V_1 \otimes \cdots \otimes V_n) = m_1 m_2 \cdots m_n$. $\square$

In the exercises at the end of this section, the definition of an algebra homomorphism will be needed. The definition of an associative algebra A (with identity) over the field F was given in Definition 4.19 of Chapter I. Suppose A_1 and A_2 are two algebras over F.

Definition 1.23: A function $\varphi: A_1 \to A_2$ is called an *algebra homomorphism* if $\varphi \in \text{Hom}_F(A_1, A_2)$ and $\varphi(\alpha\beta) = \varphi(\alpha)\varphi(\beta)$ for all $\alpha, \beta \in A_1$.

Example 1.24: Let V be any vector space over F, and consider the two algebras $A_1 = F[X]$ and $A_2 = \mathscr{E}(V)$. Then every $T \in \mathscr{E}(V)$ determines an algebra homomorphism $\varphi_T: A_1 \to A_2$ defined as follows:

1.25:

$$
\varphi_T(a_0 + a_1 X + \cdots + a_n X^n) = a_0 I_V + a_1 T + \cdots + a_n T^n
$$

The fact that φ_T is an algebra homomorphism is easy. Note that

$\varphi_T(1) = T^0 \stackrel{.}{=} I_V$. Thus, φ_T sends the multiplicative identity 1 of F[X] to the multiplicative identity I_V of $\mathscr{E}(V)$. $\square$

If the reader prefers matrices to linear transformations, we can construct a similar algebra homomorphism $\varphi_C: A_1 = F[X] \to A_3 = M_{n \times n}(F)$ for any matrix $C \in A_3$. Set $\varphi_C(a_0 + a_1X + \cdots + a_nX^n) = a_0I + a_1C + \cdots + a_nC^n$. We shall use these two types of algebra homomorphisms extensively in Chapter III. Other examples of algebra maps will be considered in the exercises at the end of this section.

EXERCISES FOR SECTION 1

(1) Complete the details of Example 1.4, that is, argue $\phi(\alpha_1, \ldots, \alpha_n) = \det(a_{ij})$ is a multilinear mapping.

(2) In the proof of Theorem 1.20, we used the following fact: if B_i is a basis of V_i and $\phi_0: B_1 \times \cdots \times B_n \to V'$ is a set map, then ϕ_0 has a unique extension to a multilinear map $\phi': V_1 \times \cdots \times V_n \to V'$. Give a proof of this fact.

(3) Suppose $V_1, \ldots, V_n$ and V are finite-dimensional vector spaces over F with dim $V_i = m_i$ and dim $V = p$. Show that $\dim_F\{Mul_F(V_1 \times \cdots \times V_n, V)\} = pm_1 \cdots m_n$.

(4) Let $\varphi: F^m \times F^n \to M_{m \times n}(F)$ be defined as follows: If $\alpha = (x_1, \ldots, x_m) \in F^m$ and $\beta = (y_1, \ldots, y_n) \in F^n$, let $\varphi(\alpha, \beta)$ be the m $\times$ n matrix whose (i, j)th entry is x_iy_j. Show that φ is a multilinear (i.e., bilinear) mapping.

(5) Let $V_1 = M_{p \times q}(F)$, $V_2 = M_{m \times n}(F)$ and $V_3 = M_{pm \times qn}(F)$. Define a map $\varphi: V_1 \times V_2 \to V_3$ as follows: If $A = (a_{ij}) \in V_1$ and $B = (b_{rs}) \in V_2$, let $\varphi(A, B)$ be the pm $\times$ qn matrix defined by the following block decomposition:

$$\varphi(A, B) = \begin{pmatrix} a_{11}B | \cdots | a_{1q}B \\ \vdots \qquad \vdots \\ a_{p1}B | \cdots | a_{pq}B \end{pmatrix}$$

Show that φ is a bilinear mapping from $V_1 \times V_2$ to V_3. $\varphi(A, B)$ is usually written $A \otimes B$ and is called the *Kronecker product* of A and B.

(6) Show that $Mul_F(V_1 \times \cdots \times V_n, V) \cong Hom_F(V_1 \otimes \cdots \otimes V_n, V)$. Does this give a simple proof of Exercise 3?

(7) Suppose $V_1, \ldots, V_n$ are vector spaces over F and for each $i = 1, \ldots, n$, let $f_i \in V_i^*$. Show that $\psi: V_1 \times \cdots \times V_n \to F$ given by $\psi(\alpha_1, \ldots, \alpha_n) = f_1(\alpha_1)f_2(\alpha_2) \cdots f_n(\alpha_n)$ is a multilinear mapping.

(8) Give an example of a multilinear mapping $\varphi: V_1 \times \cdots \times V_n \to V$ such that Im $\varphi = \{\varphi(\alpha_1, \ldots, \alpha_n) | (\alpha_1, \ldots, \alpha_n) \in V_1 \times \cdots \times V_n\}$ is not a subspace of V.

(9) A vector $\alpha \in V_1 \otimes \cdots \otimes V_n$ is said to be decomposable if $\alpha \in \text{Im } \phi$. Here ϕ is the canonical map from $V_1 \times \cdots \times V_n$ to $V_1 \otimes \cdots \otimes V_n$. Are all vectors in $V_1 \otimes \cdots \otimes V_n$ decomposable? If not, construct an example.

(10) Show that $\alpha = \alpha_1 \otimes \cdots \otimes \alpha_n$ in $V_1 \otimes \cdots \otimes V_n$ is zero if and only if some α_i is zero.

(11) Suppose V is a finite-dimensional vector space over F. Let $\underline{\alpha} = \{\alpha_1, \ldots, \alpha_n\}$ be a basis of V. Show that the isomorphism $\Gamma(\underline{\alpha}, \underline{\alpha}): \mathscr{E}(V) \to M_{n \times n}(F)$ given in equation 3.24 of Chapter I is an algebra homomorphism.

(12) Let V be a vector space over F. For each integer $n \geqslant 0$, we define $V^{\otimes n}$ as follows:

$$V^{\otimes n} = \begin{cases} F & \text{if } n = 0 \\ V & \text{if } n = 1 \\ V \otimes \cdots \otimes V \ (n \text{ times}) & \text{if } n \geqslant 2 \end{cases}$$

Set $\mathscr{T}(V) = \bigoplus_{n \geqslant 0}^{\infty} V^{\otimes n}$. Show that $\mathscr{T}(V)$ is an associative algebra over F when we define multiplication of vectors in $\mathscr{T}(V)$ by $(\alpha_1 \otimes \cdots \otimes \alpha_n)(\beta_1 \otimes \cdots \otimes \beta_m) = \alpha_1 \otimes \cdots \otimes \alpha_n \otimes \beta_1 \otimes \cdots \otimes \beta_m$. $\mathscr{T}(V)$ is called the *tensor algebra* of V.

(13) Show that the tensor algebra of V constructed in Exercise 12 has the following universal mapping property: If A is any (associative) algebra over F and $T \in \text{Hom}_F(V, A)$, then there exists a unique algebra homomorphism $\varphi: \mathscr{T}(V) \to A$ such that $\varphi(\alpha) = T(\alpha)$ for all $\alpha \in V$.

(14) Let F be a field. We regard F as an algebra over F. Show that any algebra homomorphism $\varphi: F \to F$ is either zero or an isomorphism.

(15) Suppose $A \in M_{n \times n}(F)$ is nonsingular. Define a map on $M_{n \times n}(F)$ by the equation $\varphi(B) = A^{-1}BA$. Show that φ is an algebra homomorphism from $M_{n \times n}(F)$ to $M_{n \times n}(F)$ that is bijective.

(16) With the same notation as in Exercise 15, suppose the map $\xi: M_{n \times n}(F) \to M_{n \times n}(F)$ given by $\xi(B) = AB$ is an algebra homomorphism. What can you say about A in this case?

2. FUNCTORIAL PROPERTIES OF TENSOR PRODUCTS

In this section, we present a series of theorems that tell us how to manipulate tensor products and use them in various applications. Our first theorem essentially says that forming tensor products is an associative operation.

Theorem 2.1: Let $V_1, \ldots, V_n$ and $W_1, \ldots, W_m$ be vector spaces over F. Then there exists an isomorphism $T: (V_1 \otimes \cdots \otimes V_n) \otimes (W_1 \otimes \cdots \otimes W_m) \to$

$V_1 \otimes \cdots \otimes V_n \otimes W_1 \otimes \cdots \otimes W_m$ such that $T((\alpha_1 \otimes \cdots \otimes \alpha_n) \otimes (\beta_1 \otimes \cdots \otimes \beta_m))$ $= \alpha_1 \otimes \cdots \otimes \alpha_n \otimes \beta_1 \otimes \cdots \otimes \beta_m$. Here $\alpha_i \in V_i$ and $\beta_j \in W_j$ for all $i = 1, \ldots, n$ and $j = 1, \ldots, m$.

The proofs of the theorems in this section can usually be done in two different ways. We can appeal to Lemma 1.11 or use Theorem 1.20. We shall present a mixture of both types of proof here. Since we are dealing with vector spaces, we could prove every theorem in this section by using Theorem 1.20. The advantage to proceeding via Lemma 1.11 (i.e., a basis-free proof) is that this type of proof is valid in more general situations (e.g., modules over commutative rings).

Proof of 2.1: Let $B_i, i = 1, \ldots, n$, be a basis of V_i. Let $C_j, j = 1, \ldots, m$, be a basis of W_j. Applying Theorem 1.20, we have the following facts:

(a) $\Gamma_1 = \{\alpha_1 \otimes \cdots \otimes \alpha_n | (\alpha_1, \ldots, \alpha_n) \in B_1 \times \cdots \times B_n\}$ is a basis of $V_1 \otimes \cdots \otimes V_n$.

(b) $\Gamma_2 = \{\beta_1 \otimes \cdots \otimes \beta_m | (\beta_1, \ldots, \beta_m) \in C_1 \times \cdots \times C_m\}$ is a basis of $W_1 \otimes \cdots \otimes W_m$.

(c) $\Gamma_3 = \{\alpha_1 \otimes \cdots \otimes \alpha_n \otimes \beta_1 \otimes \cdots \otimes \beta_m | (\alpha_1, \ldots, \alpha_n, \beta_1, \ldots, \beta_m) \in B_1 \times \cdots \times B_n \times C_1 \times \cdots \times C_m\}$ is a basis of $V_1 \otimes \cdots \otimes V_n \otimes W_1 \otimes \cdots \otimes W_m$.

(d) $\{(\alpha_1 \otimes \cdots \otimes \alpha_n) \otimes (\beta_1 \otimes \cdots \otimes \beta_m) | (\alpha_1 \otimes \cdots \otimes \alpha_n, \beta_1 \otimes \cdots \otimes \beta_m) \in \Gamma_1 \times \Gamma_2\}$ is a basis of $(V_1 \otimes \cdots \otimes V_n) \otimes (W_1 \otimes \cdots \otimes W_m)$.

Using 3.23 of Chapter I, we can construct a linear transformation $T:(V_1 \otimes \cdots \otimes V_n) \otimes (W_1 \otimes \cdots \otimes W_m) \to V_1 \otimes \cdots \otimes V_n \otimes W_1 \otimes \cdots \otimes W_m$ such that $T((\alpha_1 \otimes \cdots \otimes \alpha_n) \otimes (\beta_1 \otimes \cdots \otimes \beta_m)) = \alpha_1 \otimes \cdots \otimes \alpha_n \otimes \beta_1 \otimes \cdots \otimes \beta_m$ for all $(\alpha_1 \otimes \cdots \otimes \alpha_n, \beta_1 \otimes \cdots \otimes \beta_m) \in \Gamma_1 \times \Gamma_2$. Clearly (d) and (c) imply T is an isomorphism. The fact that $T((\alpha_1 \otimes \cdots \otimes \alpha_n) \otimes (\beta_1 \otimes \cdots \otimes \beta_m)) = \alpha_1 \otimes \cdots \otimes \alpha_n \otimes \beta_1 \otimes \cdots \otimes \beta_m$ for any $\alpha_i \in V_i$ and $\beta_j \in W_j$ is now a straightforward computation that we leave to the exercises. $\square$

Let us say a word about the proof of Theorem 2.1 via Lemma 1.11. There is a natural multilinear mapping $\psi: V_1 \times \cdots \times V_n \times W_1 \times \cdots \times W_m \to (V_1 \otimes \cdots \otimes V_n) \otimes (W_1 \otimes \cdots \otimes W_m)$ given by $\psi(\alpha_1, \ldots, \alpha_n, \beta_1, \ldots, \beta_m) = (\alpha_1 \otimes \cdots \otimes \alpha_n) \otimes (\beta_1 \otimes \cdots \otimes \beta_m)$. We could then argue that the pair $((V_1 \otimes \cdots \otimes V_n) \otimes (W_1 \otimes \cdots \otimes W_m), \psi)$ satisfies the universal mapping property given in 1.9. Lemma 1.11 would then imply $(V_1 \otimes \cdots \otimes V_n) \otimes (W_1 \otimes \cdots \otimes W_m) \cong V_1 \otimes \cdots \otimes V_n \otimes W_1 \otimes \cdots \otimes W_m$ via a linear transformation T for which $T((\alpha_1 \otimes \cdots \otimes \alpha_n) \otimes (\beta_1 \otimes \cdots \otimes \beta_m)) = \alpha_1 \otimes \cdots \otimes \alpha_n \otimes \beta_1 \otimes \cdots \otimes \beta_m$. We ask the reader to provide the details of this proof in the exercises at the end of this section.

There is a special case of Theorem 2.1 that is worth noting explicitly.

Corollary 2.2: $(V_1 \otimes V_2) \otimes V_3 \cong V_1 \otimes (V_2 \otimes V_3)$.

Proof: By Theorem 2.1, both of these vector spaces are isomorphic to $V_1 \otimes V_2 \otimes V_3$. $\square$

The point of Theorem 2.1 and Corollary 2.2 is that we can drop all parentheses when forming successive tensor products of vector spaces. Our next theorem says that forming tensor products is essentially a commutative operation as well.

Theorem 2.3: Let $(i_1, \ldots, i_n)$ be a permutation of $(1, \ldots, n)$. Then there exists an isomorphism $T: V_1 \otimes \cdots \otimes V_n \cong V_{i_1} \otimes \cdots \otimes V_{i_n}$ such that $T(\alpha_1 \otimes \cdots \otimes \alpha_n) = \alpha_{i_1} \otimes \cdots \otimes \alpha_{i_n}$.

Proof: The map $\psi_1: V_1 \times \cdots \times V_n \to V_{i_1} \otimes \cdots \otimes V_{i_n}$ given by $\psi_1(\alpha_1, \ldots, \alpha_n) = \alpha_{i_1} \otimes \cdots \otimes \alpha_{i_n}$ is clearly multilinear. Hence using the universal mapping property of $(V_1 \otimes \cdots \otimes V_n, \phi)$, we have a unique linear transformation $T: V_1 \otimes \cdots \otimes V_n \to V_{i_1} \otimes \cdots \otimes V_{i_n}$ such that $T\phi = \psi_1$. Thus, $\alpha_{i_1} \otimes \cdots \otimes \alpha_{i_n} = \psi_1(\alpha_1, \ldots, \alpha_n) = T\phi(\alpha_1, \ldots, \alpha_n) = T(\alpha_1 \otimes \cdots \otimes \alpha_n)$.

Now let $\phi': V_{i_1} \times \cdots \times V_{i_n} \to V_{i_1} \otimes \cdots \otimes V_{i_n}$ be the canonical multilinear map. The map $\psi_2: V_{i_1} \times \cdots \times V_{i_n} \to V_1 \otimes \cdots \otimes V_n$ given by $\psi_2(\alpha_{i_1}, \ldots, \alpha_{i_n}) = \alpha_1 \otimes \cdots \otimes \alpha_n$ is clearly multilinear. Hence there exists a unique linear transformation $T': V_{i_1} \otimes \cdots \otimes V_{i_n} \to V_1 \otimes \cdots \otimes V_n$ such that $T'\phi' = \psi_2$. Thus, $\alpha_1 \otimes \cdots \otimes \alpha_n = \psi_2(\alpha_{i_1}, \ldots, \alpha_{i_n}) = T'\phi'(\alpha_{i_1}, \ldots, \alpha_{i_n}) = T'(\alpha_{i_1} \otimes \cdots \otimes \alpha_{i_n})$. Clearly, T and T' are inverses. $\square$

We may view F itself as a vector space over F. We can then consider the tensor product $V \otimes_F F$. Here V is an arbitrary vector space over F. Since $\{1\}$ is a basis of F as a vector space over itself, Theorem 1.20 implies $V \otimes_F F \cong V$ under the map sending $\alpha \otimes x$ to αx. Similar remarks can be made for $F \otimes_F V$. Thus, we have the following theorem:

Theorem 2.4: $V \otimes_F F \cong V \cong F \otimes_F V$. $\square$

We next turn our attention to the relations between tensor products and homomorphisms. Suppose that for each $i = 1, \ldots, n$, we have a linear transformation $T_i \in \mathrm{Hom}_F(V_i, V_i')$. Thus, V_i and V_i' are vector spaces over F, and $T_i: V_i \to V_i'$ is a linear transformation. We then have a natural multilinear mapping $\varphi: V_1 \times \cdots \times V_n \to V_1' \otimes \cdots \otimes V_n'$ given by $\varphi(\alpha_1, \ldots, \alpha_n) = T_1(\alpha_1) \otimes \cdots \otimes T_n(\alpha_n)$. Since each T_i is linear, φ is clearly multilinear. It follows from Theorem 1.19 that there exists a unique linear transformation $S: V_1 \otimes \cdots \otimes V_n \to V_1' \otimes \cdots \otimes V_n'$ such that $S\phi = \varphi$. Here $\phi: V_1 \times \cdots \times V_n \to V_1 \otimes \cdots \otimes V_n$ is the canonical multilinear mapping given by $\phi(\alpha_1, \ldots, \alpha_n) = \alpha_1 \otimes \cdots \otimes \alpha_n$. The map S is called the *tensor product* of the T_i and is usually written $S = T_1 \otimes \cdots \otimes T_n$. Since $S\phi = \varphi$, we have $(T_1 \otimes \cdots \otimes T_n)(\alpha_1 \otimes \cdots \otimes \alpha_n) = T_1(\alpha_1) \otimes \cdots \otimes T_n(\alpha_n)$ for all $(\alpha_1, \ldots, \alpha_n) \in V_1 \times \cdots \times V_n$. Let us summarize this discussion with the following definition:

Definition 2.5: Suppose $T_i \in \text{Hom}_F(V_i, V_i')$ for $i = 1, \ldots, n$. $T_1 \otimes \cdots \otimes T_n$ is the linear transformation from $V_1 \otimes \cdots \otimes V_n$ to $V_1' \otimes \cdots \otimes V_n'$ defined by the following equation:

$$(T_1 \otimes \cdots \otimes T_n)(\alpha_1 \otimes \cdots \otimes \alpha_n) = T_1(\alpha_1) \otimes \cdots \otimes T_n(\alpha_n)$$

In our notation, we have deliberately suppressed the field F. When dealing with more than one field, we shall write $T_1 \otimes_F \cdots \otimes_F T_n$: $V_1 \otimes_F \cdots \otimes_F V_n \to V_1' \otimes_F \cdots \otimes_F V_n'$ instead of the simpler notation used in 2.5.

In our next theorem, we gather together some of the more obvious facts about tensor products of linear transformations.

Theorem 2.6: Let V_i, V_i', and V_i'', $i = 1, \ldots, n$, be vector spaces over F. Suppose T_i, $T_i' \in \text{Hom}_F(V_i, V_i')$ and $S_i \in \text{Hom}_F(V_i'', V_i)$. Then the following assertions are true:

(a) If each T_i is surjective, so is $T_1 \otimes \cdots \otimes T_n$.

(b) If each T_i is injective, so is $T_1 \otimes \cdots \otimes T_n$.

(c) If each T_i is an isomorphism, so is $T_1 \otimes \cdots \otimes T_n$.

(d) $(T_1 \otimes \cdots \otimes T_n)(S_1 \otimes \cdots \otimes S_n) = (T_1 S_1) \otimes \cdots \otimes (T_n S_n)$.

(e) If each T_i is an isomorphism, $(T_1 \otimes \cdots \otimes T_n)^{-1} = T_1^{-1} \otimes \cdots \otimes T_n^{-1}$.

(f) $T_1 \otimes \cdots \otimes (T_i + T_i') \otimes \cdots \otimes T_n = T_1 \otimes \cdots \otimes T_i \otimes \cdots \otimes T_n + T_1 \otimes \cdots \otimes T_i' \otimes \cdots \otimes T_n$.

(g) $T_1 \otimes \cdots \otimes x T_i \otimes \cdots \otimes T_n = x(T_1 \otimes \cdots \otimes T_i \otimes \cdots \otimes T_n)$.

(h) $I_{V_1} \otimes \cdots \otimes I_{V_n} = I_{(V_1 \otimes \cdots \otimes V_n)}$

Proof: (c)–(h) are all straightforward and are left to the reader. We prove (a) and (b).

(a) It follows from our construction of the tensor product that $V_1' \otimes \cdots \otimes V_n'$ is spanned as a vector space over F by all vectors of the form $\alpha_1' \otimes \cdots \otimes \alpha_n'$ with $(\alpha_1', \ldots, \alpha_n') \in V_1' \times \cdots \times V_n'$. Let $(\alpha_1', \ldots, \alpha_n') \in V_1' \times \cdots \times V_n'$. Since each T_i is surjective, there exists an $\alpha_i \in V_i$ such that $T_i(\alpha_i) = \alpha_i'$. Thus, $\alpha_1' \otimes \cdots \otimes \alpha_n' = T_1(\alpha_1) \otimes \cdots \otimes T_n(\alpha_n) = (T_1 \otimes \cdots \otimes T_n)\ (\alpha_1 \otimes \cdots \otimes \alpha_n)$. Thus, $V_1' \otimes \cdots \otimes V_n' = L(\text{Im}(T_1 \otimes \cdots \otimes T_n)) = \text{Im}(T_1 \otimes \cdots \otimes T_n)$. Hence, $T_1 \otimes \cdots \otimes T_n$ is surjective.

(b) Suppose T_i is injective for each $i = 1, \ldots, n$. Let $B_i = \{\alpha_{ik} \mid k \in \Delta_i\}$ be a basis of V_i. Since T_i is injective, $T_i(B_i) = \{T_i(\alpha_{ik}) \mid k \in \Delta_i\}$ is a linearly independent subset of V_i'. In particular, $T_i(B_i)$ is part of a basis of V_i' (Theorem 2.6, Chapter I). It now follows from Theorem 1.20 that the set

$$B = \{T_1(\alpha_{1k_1}) \otimes \cdots \otimes T_n(\alpha_{nk_n}) \mid (k_1, \ldots, k_n) \in \Delta_1 \times \cdots \times \Delta_n\}$$

is a linearly independent subset of $V_1' \otimes \cdots \otimes V_n'$.

Now let $\alpha \in \ker(T_1 \otimes \cdots \otimes T_n)$. Again using Theorem 1.20, α can be written uniquely in the following form:

2.7:

$$\alpha = \sum_{(k_1,\ldots,k_n) \in \Delta_1 \times \cdots \times \Delta_n} c_{k_1,\ldots,k_n} \alpha_{1k_1} \otimes \cdots \otimes \alpha_{nk_n}$$

In equation 2.7, every $c_{k_1,\ldots,k_n} \in F$ and all but possibly finitely many of these scalars are zero. If we now apply $T_1 \otimes \cdots \otimes T_n$ to α and use the fact the B is linearly independent over F, we see $c_{k_1,\ldots,k_n} = 0$ for every $(k_1,\ldots,k_n) \in \Delta_1 \times \cdots \times \Delta_n$. Thus, $\alpha = 0$ and $T_1 \otimes \cdots \otimes T_n$ is injective. $\square$

Recall that a complex of vector spaces (over F),

$$V'' \xrightarrow{\ S\ } V \xrightarrow{\ T\ } V' \longrightarrow 0$$

is said to be exact if T is surjective, and $\operatorname{Im} S = \ker T$. Suppose for each $i = 1,\ldots,n$, we have an exact complex of the following form:

2.8:

$$V_i'' \xrightarrow{\ S_i\ } V_i \xrightarrow{\ T_i\ } V_i' \longrightarrow 0$$

Theorem 2.6(a) implies $T_1 \otimes \cdots \otimes T_n : V_1 \otimes \cdots \otimes V_n \to V_1' \otimes \cdots \otimes V_n'$ is a surjective linear transformation. We want to identify the kernel of $T_1 \otimes \cdots \otimes T_n$.

For each $i = 1,\ldots,n$, we can consider the linear transformation $I_{V_1} \otimes \cdots \otimes S_i \otimes \cdots \otimes I_{V_n} : V_1 \otimes \cdots \otimes V_i'' \otimes \cdots \otimes V_n \to V_1 \otimes \cdots \otimes V_i \otimes \cdots \otimes V_n$. Let $W_i = \operatorname{Im}(I_{V_1} \otimes \cdots \otimes S_i \otimes \cdots \otimes I_{V_n})$. Then W_i is the subspace of $V_1 \otimes \cdots \otimes V_n$ spanned by all vectors of the form $\{\alpha_1 \otimes \cdots \otimes S_i(\alpha_i'') \otimes \cdots \otimes \alpha_n \mid (\alpha_1,\ldots,\alpha_i'',\ldots,\alpha_n) \in V_1 \times \cdots \times V_i'' \times \cdots \times V_n\}$. We can then form the subspace $W = W_1 + \cdots + W_n \subseteq V_1 \otimes \cdots \otimes V_n$. We can now prove the following lemma:

Lemma 2.9: $W = \ker(T_1 \otimes \cdots \otimes T_n)$.

Proof: Fix $i = 1,\ldots,n$, and consider a typical generator $\beta = \alpha_1 \otimes \cdots \otimes S_i(\alpha_i'') \otimes \cdots \otimes \alpha_n$ of W_i. $(T_1 \otimes \cdots \otimes T_n)(\beta) = T_1(\alpha_1) \otimes \cdots \otimes T_i S_i(\alpha_i'') \otimes \cdots \otimes T_n(\alpha_n)$. Since 2.8 is exact, $T_i S_i(\alpha_i'') = 0$. Thus, $(T_1 \otimes \cdots \otimes T_n)(\beta) = 0$. Since $\ker(T_1 \otimes \cdots \otimes T_n)$ is a subspace of $V_1 \otimes \cdots \otimes V_n$, we conclude that $W = W_1 + \cdots + W_n \subseteq \ker(T_1 \otimes \cdots \otimes T_n)$.

The opposite inclusion, $\ker(T_1 \otimes \cdots \otimes T_n) \subseteq W$, is a bit more difficult to establish. We begin by defining a multilinear mapping $\psi: V'_1 \times \cdots \times V'_n \to (V_1 \otimes \cdots \otimes V_n)/W$ as follows:

If $(\alpha'_1, \ldots, \alpha'_n) \in V'_1 \times \cdots \times V'_n$, then there exists a vector $(\alpha_1, \ldots, \alpha_n) \in V_1 \times \cdots \times V_n$ such that $T_i(\alpha_i) = \alpha'_i$ for all $i = 1, \ldots, n$. This follows from the fact that each T_i is surjective. We then define $\psi(\alpha'_1, \ldots, \alpha'_n)$ to be the following coset:

2.10:

$$\psi(\alpha'_1, \ldots, \alpha'_n) = (\alpha_1 \otimes \cdots \otimes \alpha_n) + W$$

Now it is not obvious that ψ is well defined. We must check that if $(\beta_1, \ldots, \beta_n)$ is a second vector in $V_1 \times \cdots \times V_n$ with the property that $T_i(\beta_i) = \alpha'_i$ for all $i = 1, \ldots, n$, then $(\beta_1 \otimes \cdots \otimes \beta_n) + W = (\alpha_1 \otimes \cdots \otimes \alpha_n) + W$.

Since $T_i(\beta_i) = T_i(\alpha_i) = \alpha'_i$ for $i = 1, \ldots, n$ and each sequence in 2.8 is exact, there exists a $\mu_i \in V''_i$ such that $S_i(\mu_i) = \alpha_i - \beta_i$. In particular, we have the following relations in $V_1 \otimes \cdots \otimes V_n$:

2.11:

$$\alpha_1 \otimes \cdots \otimes \alpha_n - \beta_1 \otimes \alpha_2 \otimes \cdots \otimes \alpha_n \in W_1$$

$$\beta_1 \otimes \alpha_2 \otimes \cdots \otimes \alpha_n - \beta_1 \otimes \beta_2 \otimes \alpha_3 \otimes \cdots \otimes \alpha_n \in W_2$$

$$\vdots$$

$$\beta_1 \otimes \cdots \otimes \beta_{n-1} \otimes \alpha_n - \beta_1 \otimes \cdots \otimes \beta_n \in W_n$$

Adding the relations in 2.11 gives $\alpha_1 \otimes \cdots \otimes \alpha_n - \beta_1 \otimes \cdots \otimes \beta_n \in W_1 + \cdots + W_n = W$. Thus, $(\alpha_1 \otimes \cdots \otimes \alpha_n) + W = (\beta_1 \otimes \cdots \otimes \beta_n) + W$, and equation 2.10 gives us a well-defined function $\psi: V'_1 \times \cdots \times V'_n \to (V_1 \otimes \cdots \otimes V_n)/W$. The fact that ψ is multilinear is obvious.

Let $\phi: V'_1 \times \cdots \times V'_n \to V'_1 \otimes \cdots \otimes V'_n$ be the canonical multilinear map. Using the universal mapping property of $(V'_1 \otimes \cdots \otimes V'_n, \phi)$, we conclude that there exists a unique linear transformation $T: V'_1 \otimes \cdots \otimes V'_n \to (V_1 \otimes \cdots \otimes V_n)/W$ such that $T\phi = \psi$. Thus, for all $(\alpha'_1, \ldots, \alpha'_n) \in V'_1 \times \cdots \times V'_n$, and any $(\alpha_1, \ldots, \alpha_n) \in V_1 \times \cdots \times V_n$ such that $T_i(\alpha_i) = \alpha'_i$, $i = 1, \ldots, n$, we have $(\alpha_1 \otimes \cdots \otimes \alpha_n) + W = \psi(\alpha'_1, \ldots, \alpha'_n) = T\phi(\alpha'_1, \ldots, \alpha'_n) = T(\alpha'_1 \otimes \cdots \otimes \alpha'_n)$.

Now consider the composite linear transformation given by

$$V_1 \otimes \cdots \otimes V_n \xrightarrow{\ T_1 \otimes \cdots \otimes T_n\ } V'_1 \otimes \cdots \otimes V'_n \xrightarrow{\qquad T \qquad} (V_1 \otimes \cdots \otimes V_n)/W.$$

Set $S = T(T_1 \otimes \cdots \otimes T_n)$. Then for all $(\alpha_1, \ldots, \alpha_n) \in V_1 \times \cdots \times V_n$, we have $S(\alpha_1 \otimes \cdots \otimes \alpha_n) = T(T_1(\alpha_1) \otimes \cdots \otimes T_n(\alpha_n)) = (\alpha_1 \otimes \cdots \otimes \alpha_n) + W$. Thus, S is nothing but the natural map from $V_1 \otimes \cdots \otimes V_n$ to $(V_1 \otimes \cdots \otimes V_n)/W$. In particular, it follows from 5.13 of Chapter I that $\ker S = W$. Since $\ker(T_1 \otimes \cdots \otimes T_n) \subseteq \ker S$, we conclude that $\ker(T_1 \otimes \cdots \otimes T_n) \subseteq W$. This completes the proof of the lemma. $\square$

We have now proved the following theorem:

Theorem 2.12: Let

$$V_i'' \xrightarrow{S_i} V_i \xrightarrow{T_i} V_i' \longrightarrow 0$$

be an exact complex of vector spaces over F for each $i = 1, \ldots, n$. Let $W_i = \mathrm{Im}(I_{V_1} \otimes \cdots \otimes S_i \otimes \cdots \otimes I_{V_n})$ and set $W = W_1 + \cdots + W_n$. Then

$$0 \to W \to V_1 \otimes \cdots \otimes V_n \xrightarrow{T_1 \otimes \cdots \otimes T_n} V_1' \otimes \cdots \otimes V_n' \to 0$$

is a short exact sequence. Thus, $T_1 \otimes \cdots \otimes T_n$ is surjective and $W = \ker(T_1 \otimes \cdots \otimes T_n)$. $\square$

There is a special case of Theorem 2.12 that we present as a separate theorem.

Theorem 2.13: Suppose

$$0 \to V'' \xrightarrow{S} V \xrightarrow{T} V' \to 0$$

is a short exact sequence of vector spaces over F. Then for any vector space W,

2.14:

$$0 \to V'' \otimes W \xrightarrow{S \otimes I_W} V \otimes W \xrightarrow{T \otimes I_W} V' \otimes W \to 0$$

is a short exact sequence.

Proof: $T \otimes I_W$ is surjective by Theorem 2.6(a). $S \otimes I_W$ is injective by Theorem 2.6(b). If we apply Theorem 2.12 to the two exact complexes:

$$V'' \xrightarrow{S} V \xrightarrow{T} V' \longrightarrow 0$$

$$0 \longrightarrow W \xrightarrow{I_W} W \longrightarrow 0$$

we see that $\ker(T \otimes I_W) = \text{Im}(S \otimes I_W)$. Thus 2.14 is exact and the proof of the theorem is complete. $\square$

The next natural question to ask about tensor products is how they behave with respect to direct sums. We answer this question in our next theorem, but leave most of the technical details as exercises at the end of this section.

Theorem 2.15: Suppose $\{V_i \,|\, i \in \Delta\}$ is a collection of vector spaces over F. Then for any vector space V we have

$$\{\bigoplus_{i \in \Delta} V_i\} \otimes V \cong \bigoplus_{i \in \Delta} \{V_i \otimes V\}$$

Proof: Let $\theta_j \colon V_j \to \bigoplus_{i \in \Delta} V_i$ and $\pi_j \colon \bigoplus_{i \in \Delta} V_i \to V_j$ be the canonical injections and surjections introduced in Definition 4.2 of Chapter I. Then we have the following facts:

(a) $\pi_j \theta_j = I_{V_j}$ for all $j \in \Delta$.
(b) $\pi_j \theta_i = 0$ if $i \neq j$.
(c) For any $\xi \in \bigoplus_{i \in \Delta} V_i$, $\pi_j(\xi) = 0$ except possibly for finitely many $j \in \Delta$.
(d) $\sum_{i \in \Delta} \theta_i \pi_i = I$, the identity map on $\bigoplus_{i \in \Delta} V_i$.

Perhaps we should make a few comments about (d). If $\xi \in \bigoplus_{i \in \Delta} V_i$, then ξ is a Δ-tuple with at most finitely many nonzero components. Thus, $\sum_{i \in \Delta} \theta_i \pi_i(\xi)$ is a finite sum whose value is clearly ξ. This is what the statement $\sum_{i \in \Delta} \theta_i \pi_i = I$ means in (d).

Now for each $j \in \Delta$, we can consider the linear transformation $\theta_j \otimes I_V \colon V_j \otimes V \to \{\bigoplus_{i \in \Delta} V_i\} \otimes V$. An easy computation shows that $\{\bigoplus_{i \in \Delta} V_i\} \otimes V$ is the internal direct sum of the subspaces $\{\text{Im}(\theta_i \otimes I_V) \,|\, i \in \Delta\}$. Thus, $\{\bigoplus_{i \in \Delta} V_i\} \otimes V = \bigoplus_{i \in \Delta} \text{Im}(\theta_i \otimes I_V)$. Since each θ_i is injective, Theorem 2.6 implies $\theta_i \otimes I_V$ is injective. Hence $V_i \otimes V \cong \text{Im}(\theta_i \otimes I_V)$. It now follows that $\bigoplus_{i \in \Delta}(V_i \otimes V) \cong \bigoplus_{i \in \Delta} \text{Im}(\theta_i \otimes I_V) = \{\bigoplus_{i \in \Delta} V_i\} \otimes V$. $\square$

We next study a construction using tensor products that is very useful in linear algebra. Suppose V is a vector space over F, and let K be a second field containing F. For example, $F = \mathbb{R}$ and $K = \mathbb{C}$. We have seen in Chapter I that K is a vector space (even an algebra) over F. Thus, we can form the tensor product $V \otimes_F K$ of the vector spaces V and K over F.

$V \otimes_F K$ is a vector space over F. We want to point out that there is a natural K-vector space structure on $V \otimes_F K$ as well. Vector addition in $V \otimes_F K$ as a K-vector space is the same as before. Namely, if $\xi = \sum_{i=1}^{n}(\alpha_i \otimes_F x_i)$, and $\eta = \sum_{j=1}^{m}(\beta_j \otimes_F y_j)$ are two vectors in $V \otimes_F K$ (thus $\alpha_i, \beta_j \in V$, and $x_i, y_j \in K$), then

$$\xi + \eta = \alpha_1 \otimes_F x_1 + \cdots + \alpha_n \otimes_F x_n + \beta_1 \otimes_F y_1 + \cdots + \beta_m \otimes_F y_m$$

We need to define scalar multiplication of vectors in $V \otimes_F K$ with scalars in K. Let $x \in K$, and consider the linear map $\mu_x \in \text{Hom}_F(K, K)$ defined by $\mu_x(y) = xy$. Clearly, μ_x is an F-linear transformation on K. In particular, $I_V \otimes_F \mu_x$ is a well-defined F-linear transformation on $V \otimes_F K$. Now if $\xi = \sum_{i=1}^n (\alpha_i \otimes_F x_i)$ is a typical vector in $V \otimes_F K$, we define scalar multiplication $x\xi$ by the following formula:

2.16: $x\xi = (I_V \otimes_F \mu_x)(\xi)$.

Thus, $x\xi = \sum_{i=1}^n (\alpha_i \otimes_F x x_i)$. Our previous discussion in this section implies that equation 2.16 gives us a well-defined function from $K \times (V \otimes_F K) \to V \otimes_F K$. The fact that this scalar multiplication satisfies axioms V5–V8 in Definition 1.4 of Chapter I is straightforward. Thus, via the operations defined above, the F-vector space $V \otimes_F K$ becomes a vector space over K.

Throughout the rest of this book, whenever we view $V \otimes_F K$ as a vector space over K, then addition and scalar multiplication will be as defined above. The process whereby we pass from a vector space V over F to the vector space $V \otimes_F K$ over K is called extending the scalars to K.

Since $F \subseteq K$, Theorem 2.6 implies that the natural map $I_V \otimes_F i$: $V \otimes_F F \to V \otimes_F K$ is injective. Here $i: F \to K$ is the inclusion map. Now $V \cong V \otimes_F F$ by Theorem 2.4. Putting these two maps together gives us a natural, injective map $\hat{i} \in \text{Hom}_F(V, V \otimes_F K)$ given by $\hat{i}(\alpha) = \alpha \otimes_F 1$. By 2.16, $\text{Im}\,\hat{i}$ generates $V \otimes_F K$ as a K-vector space. We shall often identify V with $\text{Im}\,\hat{i} = V \otimes_F 1$ in $V \otimes_F K$. We note that $V \otimes_F 1$ is an F-subspace of $V \otimes_F K$. This follows immediately from 2.16. For if $x \in F$, then $x(\alpha \otimes_F 1) = \alpha \otimes_F x = x\alpha \otimes_F 1 \in V \otimes_F 1$. Thus, when we extend the scalars from F to K, we produce a K-vector space, $V \otimes_F K$, which contains V, that is, $\text{Im}\,\hat{i}$, as an F-subspace, and such that $V \otimes_F K$ is the K-linear span of V. We can now construct a K-basis of $V \otimes_F K$.

Theorem 2.17: Let V be a vector space over F, and suppose K is a field containing F. If B is a basis of V, then $\{\alpha \otimes_F 1 \mid \alpha \in B\}$ is a basis of the K-vector space $V \otimes_F K$.

Proof: Let $\Gamma = \{\alpha \otimes_F 1 \mid \alpha \in B\}$. Since $\{1\}$ is subset of K that is linearly independent over F, Theorem 1.20 implies the vectors in Γ are linearly independent over F. In particular, $|\Gamma| = |B|$, and no element of Γ is zero. We must argue Γ is linearly independent over K, and $L_K(\Gamma) = V \otimes_F K$. Here $L_K(\Gamma)$ is all K-linear combinations of the vectors in Γ.

Let us first argue that Γ is linearly independent over K. Let $\alpha_1, \ldots, \alpha_n \in B$, $k_1, \ldots, k_n \in K$, and suppose $\sum_{i=1}^n k_i(\alpha_i \otimes_F 1) = 0$. Let $C = \{z_j \mid j \in \Delta\}$ be a basis of K over F. Then each k_i can be written uniquely in the following form:

2.18:

$$k_i = \sum_{j \in \Delta} x_{ij} z_j, \qquad i = 1, \ldots, n$$

In Equation 2.18, the x_{ij} are scalars in F and each sum on the right-hand side is finite. Thus, for each $i = 1, \ldots, n$, $x_{ij} = 0$ except possibly for finitely many $j \in \Delta$. We now have

$$\sum_{i=1}^{n} k_i(\alpha_i \otimes_F 1) = \sum_{i=1}^{n} (\alpha_i \otimes_F k_i) = \sum_{i=1}^{n} \left\{ \alpha_i \otimes_F \left(\sum_{j \in \Delta} x_{ij} z_j \right) \right\} = \sum_{i=1}^{n} \sum_{j \in \Delta} x_{ij}(\alpha_i \otimes_F z_j)$$

Since the vectors $\{ \alpha_i \otimes_F z_j \,|\, \alpha_i \in B, \, z_j \in C \}$ are linearly independent over F by Theorem 1.20, we conclude that $x_{ij} = 0$ for all i and j. In particular, $k_1 = \cdots = k_n = 0$, and Γ is linearly independent over K.

To complete the proof, we must show $L_K(\Gamma) = V \otimes_F K$. Since $V \otimes_F K$ is spanned as a vector space over F by vectors of the form $\alpha \otimes_F k$ ($\alpha \in V, k \in K$) and $F \subseteq K$, it suffices to show that $\alpha \otimes_F k \in L_K(\Gamma)$. This last inclusion is easy. Write $\alpha = \sum_{i=1}^{n} x_i \alpha_i$ with $\alpha_1, \ldots, \alpha_n \in B$ and $x_1, \ldots, x_n \in F$. Then

$$\alpha \otimes_F k = \left(\sum_{i=1}^{n} x_i \alpha_i \right) \otimes_F k = \sum_{i=1}^{n} (x_i \alpha_i \otimes_F k) = \sum_{i=1}^{n} (\alpha_i \otimes_F x_i k)$$
$$= \sum_{i=1}^{n} x_i k(\alpha_i \otimes_F 1) \in L_K(\Gamma) \quad \square$$

There are two important corollaries to Theorem 2.17 that are worth noting here.

Corollary 2.19: Suppose V is a finite-dimensional vector space over F and K is a field containing F. Then $\dim_F(V) = \dim_K(V \otimes_F K)$.

Proof: If $\{ \alpha_1, \ldots, \alpha_n \}$ is a basis of V over F, then $\{ \alpha_1 \otimes_F 1, \ldots, \alpha_n \otimes_F 1 \}$ is a basis of $V \otimes_F K$ over K by Theorem 2.17 $\quad \square$

Corollary 2.20: Suppose V is a finite-dimensional vector space over F and K is a field containing F. Then $\text{Hom}_F(V, V) \otimes_F K \cong \text{Hom}_K(V \otimes_F K, V \otimes_F K)$ as vector spaces over K.

Proof: If $\dim_F V = n$, then $\dim_F(\text{Hom}_F(V, V)) = n^2$ by Theorem 3.25 of Chapter I. Thus $\dim_K(\text{Hom}_F(V, V) \otimes_F K) = n^2$ by Corollary 2.19. On the other hand, the same corollary implies $\dim_K(V \otimes_F K) = n$. Consequently, $\dim_K(\text{Hom}_K(V \otimes_F K, V \otimes_F K)) = n^2$ by Theorem 3.25 again. Since the K-vector spaces $\text{Hom}_F(V, V) \otimes_F K$ and $\text{Hom}_K(V \otimes_F K, V \otimes_F K)$ have the same dimension, they are isomorphic by Theorem 3.15 of Chapter I. $\quad \square$

A word about Corollary 2.20 is in order here. We proved this result by counting dimensions. This type of argument gives us a quick proof of the corollary but tends to obscure the nature of the isomorphism between the two vector spaces. It is worthwhile to construct an explicit K-linear isomorphism $\psi \colon \text{Hom}_F(V, V) \otimes_F K \to \text{Hom}_K(V \otimes_F K, V \otimes_F K)$. We proceed as follows: Consider the map $\chi \colon \text{Hom}_F(V, V) \times K \to \text{Hom}_K(V \otimes_F K, V \otimes_F K)$ defined by

$\chi(T, k) = k(T \otimes_F I_K)$. Here $T \in \text{Hom}_F(V, V)$, and $k \in K$. From the discussion preceding Definition 2.5, we know that $T \otimes_F I_K \in \text{Hom}_F(V \otimes_F K, V \otimes_F K)$. We claim $T \otimes_F I_K$ is in fact a K-linear map on $V \otimes_F K$. To see this, we use equation 2.16. We have $(T \otimes_F I_K)(k(\alpha \otimes_F k')) = (T \otimes_F I_K)(\alpha \otimes_F kk') = T(\alpha) \otimes_F kk' = k[T(\alpha) \otimes_F k'] = k[(T \otimes_F I_K)(\alpha \otimes_F k')]$. Thus, $T \otimes_F I_K \in \text{Hom}_K(V \otimes_F K, V \otimes_F K)$. Again by 2.16, $k(T \otimes_F I_K)$ is the K-linear transformation on $V \otimes_F K$ given by $[k(T \otimes_F I_K)](\alpha \otimes_F k') = k(T(\alpha) \otimes_F k') = T(\alpha) \otimes_F kk'$. In particular, $\text{Im} \chi \subseteq \text{Hom}_K(V \otimes_F K, V \otimes_F K)$. χ is clearly an F-bilinear mapping, and, thus, factors through the tensor product $\text{Hom}_F(V, V) \otimes_F K$. So, we have the following commutative diagram:

2.21:

$$\begin{array}{ccc}
\text{Hom}_F(V, V) \times K & \xrightarrow{\phi} & \text{Hom}_F(V, V) \otimes_F K \\
& & \\
\chi \searrow & & \swarrow \psi \\
& & \\
& \text{Hom}_K(V \otimes_F K, V \otimes_F K) &
\end{array}$$

In 2.21, $\phi(T, k) = T \otimes_F k$, and ψ is the unique, F-linear transformation making 2.21 commute. Thus, $\psi(T \otimes_F k) = \psi\phi(T, k) = \chi(T, k) = k(T \otimes_F I_K)$. Using equation 2.16, we can verify that ψ is in fact a K-linear transformation. We have $\psi(k_2(T \otimes_F k_1)) = \psi(T \otimes_F k_2 k_1) = k_2 k_1(T \otimes_F I_K) = k_2 \psi(T \otimes_F k_1)$. Thus, ψ is K-linear.

Finally, we must argue that ψ is an isomorphism. We do this by repeated applications of Theorem 2.17. Let $\underline{\alpha} = \{\alpha_1, \ldots, \alpha_n\}$ be basis of V. Define $T_{ij} \in \text{Hom}_F(V, V)$ by $T_{ij}(\alpha_p) = \alpha_i$ if $p = j$ and zero otherwise. It follows from Theorem 3.25 of Chapter I that $\{T_{ij} \mid i, j = 1, \ldots, n\}$ is a basis of $\text{Hom}_F(V, V)$. Theorem 2.17 then implies $\{T_{ij} \otimes_F 1 \mid i, j = 1, \ldots, n\}$ is a K-basis of $\text{Hom}_F(V, V) \otimes_F K$. On the other hand, $\{\alpha_i \otimes_F 1 \mid i = 1, \ldots, n\}$ is a basis of $V \otimes_F K$. Thus, $\{S_{ij} \mid i, j = 1, \ldots, n\}$ (where $S_{ij}(\alpha_p \otimes_F 1) = \alpha_i \otimes_F 1$ if $p = j$ and zero otherwise) is a K-basis of $\text{Hom}_K(V \otimes_F K, V \otimes_F K)$. Now one easily checks that $\psi(T_{ij} \otimes_F 1) = S_{ij}$. Thus, ψ is an isomorphism of K-vector spaces.

Let us rephrase some of our last remarks in terms of matrices. Suppose V is a finite-dimensional vector space over F. Let $T \in \text{Hom}_F(V, V)$. Let $\underline{\alpha} = \{\alpha_1, \ldots, \alpha_n\}$ be a basis of V over F. Set $A = \Gamma(\underline{\alpha}, \underline{\alpha})(T)$. Thus, A is the matrix representation of T relative to $\underline{\alpha}$. Now suppose we extend scalars to a field $K \supseteq F$ by passing to $V \otimes_F K$. Theorem 2.17 implies $\underline{\alpha} \otimes 1 = \{\alpha_1 \otimes_F 1, \ldots, \alpha_n \otimes_F 1\}$ is a K-basis of $V \otimes_F K$. The F-linear map $T: V \to V$ has a natural extension $\psi(T \otimes_F 1) = T \otimes_F I_K$ to a K-linear map on $V \otimes_F K$. Here ψ is the isomorphism in diagram 2.21. We have seen that V is imbedded in the extension $V \otimes_F K$ as the subspace $V \otimes_F 1$. If we identify V with $V \otimes_F 1$, then $T \otimes_F I_K$ restricted to V is just T. Thus, we may think of $T \otimes_F I_K$ as an extension of T. Clearly, $\Gamma(\underline{\alpha} \otimes 1,$

$\underline{\alpha} \otimes 1)(T \otimes_F I_\kappa) = A$. Thus, the matrix representation of the extension of T relative to the extended basis is the same as the matrix representation of T on V.

One of the most important examples of extending scalars is the complexification of a real vector space. We finish this section with a brief discussion of that notion.

Definition 2.22: Let V be a vector space over $\mathbb{R}$. The tensor product $V \otimes_F \mathbb{C}$ is called the complexification of V.

We shall shorten our notation here and let $V^\mathbb{C}$ denote the complexification of V. Thus, $V^\mathbb{C} = \{\Sigma(\alpha_i \otimes_\mathbb{R} z_i) | \alpha_i \in V, \ z_i \in \mathbb{C}\}$. Our previous discussion implies that $V^\mathbb{C}$ is a vector space over $\mathbb{C}$ with scalar multiplication given by $z'(\alpha \otimes_\mathbb{R} z) = \alpha \otimes_\mathbb{R} z'z$. If B is an $\mathbb{R}$-basis of V, then $B \otimes 1 = \{\alpha \otimes_\mathbb{R} 1 | \alpha \in B\}$ is a basis of $V^\mathbb{C}$ over $\mathbb{C}$.

There is an important $\mathbb{R}$-linear map on $V^\mathbb{C}$ that comes from complex conjugation on $\mathbb{C}$. Recall that if $z = x + iy$ ($x, y \in \mathbb{R}$, $i = \sqrt{-1}$) is a complex number, then $\bar{z} = x - iy$ is called the conjugate of z. Clearly the map $\sigma: \mathbb{C} \to \mathbb{C}$ given by $\sigma(z) = \bar{z}$ is an $\mathbb{R}$-linear transformation. Thus, $I_V \otimes_\mathbb{R} \sigma \in \text{Hom}_\mathbb{R}(V^\mathbb{C}, V^\mathbb{C})$. Recall that $I_V \otimes_\mathbb{R} \sigma$ is given by the following equation:

2.23:

$$(I_V \otimes_\mathbb{R} \sigma)\left(\sum_{k=1}^n (\alpha_k \otimes_\mathbb{R} z_k)\right) = \sum_{k=1}^n (\alpha_k \otimes_\mathbb{R} \bar{z}_k)$$

Since σ is an $\mathbb{R}$-isomorphism of $\mathbb{C}$, Theorem 2.6 implies $I_V \otimes_\mathbb{R} \sigma$ is an $\mathbb{R}$-isomorphism of $V^\mathbb{C}$. Note that $I_V \otimes_\mathbb{R} \sigma$ is not a $\mathbb{C}$-linear transformation of $V^\mathbb{C}$.

Definition 2.24: Let V be a vector space over $\mathbb{R}$, and let $T \in \text{Hom}_\mathbb{R}(V, V)$. The extension $T \otimes_\mathbb{R} I_\mathbb{C}$ will be called the *complexification* of T and written $T^\mathbb{C}$.

Thus, $T^\mathbb{C}$ is the $\mathbb{C}$-linear transformation on $V^\mathbb{C}$ given by

2.25:

$$T^\mathbb{C}\left(\sum_{k=1}^n (\alpha_k \otimes_\mathbb{R} z_k)\right) = \sum_{k=1}^n (T(\alpha_k) \otimes_\mathbb{R} z_k)$$

Clearly, $T^\mathbb{C} = \psi(T \otimes_\mathbb{R} 1)$ where $\psi: \text{Hom}_\mathbb{R}(V, V) \otimes_\mathbb{R} \mathbb{C} \to \text{Hom}_\mathbb{C}(V^\mathbb{C}, V^\mathbb{C})$ is the $\mathbb{C}$-linear isomorphism given in 2.21. If $\dim_\mathbb{R}(V) < \infty$, and $\underline{\alpha}$ is a basis of V, then $\Gamma(\underline{\alpha}, \underline{\alpha})(T) = \Gamma(\underline{\alpha} \otimes 1, \underline{\alpha} \otimes 1)(T^\mathbb{C})$. Thus, the matrix representation of the complexification of T is the same as that of T (provided we make these statements relative to $\underline{\alpha}$ and $\underline{\alpha} \otimes 1$).

It is often important to decide when an $S \in \mathrm{Hom}_{\mathbb{C}}(V^{\mathbb{C}}, V^{\mathbb{C}})$ is the complexification of some $T \in \mathrm{Hom}_{\mathbb{R}}(V, V)$.

Theorem 2.26: Let V be a finite-dimensional vector space over $\mathbb{R}$, and let $S \in \mathrm{Hom}_{\mathbb{C}}(V^{\mathbb{C}}, V^{\mathbb{C}})$. Then $S = T^{\mathbb{C}}$ for some $T \in \mathrm{Hom}_{\mathbb{R}}(V, V)$ if and only if the following equation is satisfied:

2.27:

$$S(I_V \otimes_{\mathbb{R}} \sigma) = (I_V \otimes_{\mathbb{R}} \sigma)S$$

Proof: If S is a $\mathbb{C}$-linear transformation on $V^{\mathbb{C}}$, then clearly S is an $\mathbb{R}$-linear transformation on $V^{\mathbb{C}}$. $I_V \otimes_{\mathbb{R}} \sigma$ is also an $\mathbb{R}$-linear transformation on $V^{\mathbb{C}}$. Thus, the statement in equation 2.27 is that these two endomorphisms commute as maps in $\mathrm{Hom}_{\mathbb{R}}(V^{\mathbb{C}}, V^{\mathbb{C}})$.

Let us first suppose that S is the complexification of some $T \in \mathrm{Hom}_{\mathbb{R}}(V, V)$. Thus, $S = T \otimes_{\mathbb{R}} I_{\mathbb{C}}$. If $\sum_{k=1}^{n} (\alpha_k \otimes_{\mathbb{R}} z_k) \in V^{\mathbb{C}}$, then

$$[S(I_V \otimes_{\mathbb{R}} \sigma)] \left(\sum_{k=1}^{n} (\alpha_k \otimes_{\mathbb{R}} z_k) \right) = S \left(\sum_{k=1}^{n} (\alpha_k \otimes_{\mathbb{R}} \bar{z}_k) \right) = \sum_{k=1}^{n} (T(\alpha_k) \otimes_{\mathbb{R}} \bar{z}_k)$$

On the other hand,

$$[(I_V \otimes_{\mathbb{R}} \sigma)S] \left(\sum_{k=1}^{n} (\alpha_k \otimes_{\mathbb{R}} z_k) \right) = (I_V \otimes_{\mathbb{R}} \sigma) \left(\sum_{k=1}^{n} (T(\alpha_k) \otimes_{\mathbb{R}} z_k) \right)$$

$$= \sum_{k=1}^{n} (T(\alpha_k) \otimes_{\mathbb{R}} \bar{z}_k)$$

Thus, S satisfies equation 2.27.

Conversely, suppose $S \in \mathrm{Hom}_{\mathbb{C}}(V^{\mathbb{C}}, V^{\mathbb{C}})$ and satisfies equation 2.27. The discussion after Corollary 2.20 implies that $S = \sum_{j=1}^{n} (T_j \otimes_{\mathbb{R}} w_j)$, where $T_j \in \mathrm{Hom}_{\mathbb{R}}(V, V)$ and $w_j \in \mathbb{C}$. To be more precise, $S = \psi(\sum_{j=1}^{n} T_j \otimes_{\mathbb{R}} w_j)$, where ψ is the isomorphism in 2.21, $F = \mathbb{R}$, and $K = \mathbb{C}$. We shall suppress ψ here and write $S = \sum_{j=1}^{n} (T_j \otimes_{\mathbb{R}} w_j)$. Thus, S is given by the following equation:

2.28:

$$S \left(\sum_{k=1}^{m} \alpha_k \otimes_{\mathbb{R}} z_k \right) = \sum_{j=1}^{n} \sum_{k=1}^{m} T_j(\alpha_k) \otimes_{\mathbb{R}} w_j z_k$$

Let $\alpha \otimes_{\mathbb{R}} z \in V^{\mathbb{C}}$. Then

$$[S(I_V \otimes_{\mathbb{R}} \sigma)](\alpha \otimes_{\mathbb{R}} z) = S(\alpha \otimes_{\mathbb{R}} \bar{z}) = \sum_{j=1}^{n} (T_j(\alpha) \otimes_{\mathbb{R}} w_j \bar{z})$$

On the other hand,

$$[(I_V \otimes_\mathbb{R} \sigma)S](\alpha \otimes_\mathbb{R} z) = (I_V \otimes_\mathbb{R} \sigma)\left(\sum_{j=1}^n T_j(\alpha) \otimes_\mathbb{R} w_j z \right) = \sum_{j=1}^n T_j(\alpha) \otimes_\mathbb{R} \bar{w}_j \bar{z}$$

Since S satisfies 2.27, we have $\bar{z}(\sum_{j=1}^n T_j(\alpha) \otimes_\mathbb{R} (w_j - \bar{w}_j)) = 0$. In particular, when $z \neq 0$, $\sum_{j=1}^n (T_j(\alpha) \otimes_\mathbb{R} (w_j - \bar{w}_j)) = 0$ for all $\alpha \in V$.

Now suppose the real and imaginary parts of w_j are x_j and y_j, respectively. Thus, x_j, $y_j \in \mathbb{R}$ and $w_j = x_j + iy_j$. Then $w_j - \bar{w}_j = 2iy_j$, and $\sum_{j=1}^n (T_j(\alpha) \otimes_\mathbb{R} (w_j - \bar{w}_j)) = 0$ implies $\sum_{j=1}^n (T_j(\alpha) \otimes_\mathbb{R} iy_j) = 0$ for all $\alpha \in V$. Since $\{\alpha \otimes_\mathbb{R} 1 \,|\, \alpha \in V\}$ spans $V^\mathbb{C}$ as a vector space over $\mathbb{C}$, we can now conclude that $\sum_{j=1}^n (T_j \otimes_\mathbb{R} iy_j) = 0$ on $V^\mathbb{C}$. But then

$$S = \sum_{j=1}^n (T_j \otimes_\mathbb{R} w_j) = \sum_{j=1}^n (T_j \otimes_\mathbb{R} x_j) + \sum_{j=1}^n (T_j \otimes_\mathbb{R} iy_j)$$

$$= \sum_{j=1}^n (T_j \otimes_\mathbb{R} x_j) = \left(\sum_{j=1}^n x_j T_j \right) \otimes_\mathbb{R} I_\mathbb{C} = \left(\sum_{j=1}^n x_j T_j \right)^\mathbb{C}$$

Thus, S is the complexification of $\sum_{j=1}^n x_j T_j \in \mathrm{Hom}_\mathbb{R}(V, V)$ and the proof of Theorem 2.26 is complete. $\square$

We shall have more to say about the complexification of a real operator T in Chapter III.

EXERCISES FOR SECTION 2

(1) Complete the details of the proof of Theorem 2.1 by showing $T((\alpha_1 \otimes \cdots \otimes \alpha_n) \otimes (\beta_1 \otimes \cdots \otimes \beta_m)) = \alpha_1 \otimes \cdots \otimes \alpha_n \otimes \beta_1 \otimes \cdots \otimes \beta_m$ for any $\alpha_i \in V_i$ and $\beta_j \in W_j$.

(2) Give a basis free proof of Theorem 2.1 by showing that the pair $((V_1 \otimes \cdots \otimes V_n) \otimes (W_1 \otimes \cdots \otimes W_m), \psi)$ satisfies 1.9. Recall $\psi: V_1 \times \cdots \times V_n \times W_1 \times \cdots \times W_m \to (V_1 \otimes \cdots \otimes V_n) \otimes (W_1 \otimes \cdots \otimes W_m)$ is given by $\psi(\alpha_1, \ldots, \alpha_n, \beta_1, \ldots, \beta_m) = (\alpha_1 \otimes \cdots \otimes \alpha_n) \otimes (\beta_1 \otimes \cdots \otimes \beta_m)$.

(3) Generalize Theorem 2.13 as follows: Suppose $C: \cdots \to V_{i+1} \to^{d_{i+1}} V_i \to^{d_i} V_{i-1} \to \cdots$ is an exact chain complex of vector spaces over F. If V is any vector space (over F), show that $C \otimes_F V: \cdots \to V_{i+1} \otimes_F V \to^{d_{i+1} \otimes I_V} V_i \otimes_F V \to^{d_i \otimes I_V} V_{i-1} \otimes_F V \to \cdots$ is an exact chain complex.

(4) Show by example that if $0 \to V_1'' \to^{S_1} V_1 \to^{T_1} V_1' \to 0$ and $0 \to V_2'' \to^{S_2} V_2 \to^{T_2} V_2' \to 0$ are two short exact sequences of vector spaces,

then

$$0 \to V_1'' \otimes V_2'' \xrightarrow{S_1 \otimes S_2} V_1 \otimes V_2 \xrightarrow{T_1 \otimes T_2} V_1' \otimes V_2' \to 0$$

is not necessarily exact.

(5) Complete the details of the proof of Theorem 2.15. Namely, show $\{\bigoplus_{i \in \Delta} V_i\} \otimes V$ is the internal direct sum of the subspaces $\{\text{Im}(\theta_i \otimes I_V) \mid i \in \Delta\}$.

(6) Generalize Corollary 2.20 as follows: Suppose V and W are finite-dimensional vector spaces over F. Let K be a field containing F. Show that $\text{Hom}_F(V, W) \otimes_F K \cong \text{Hom}_K(V \otimes_F K, W \otimes_F K)$ as K-vector spaces.

(7) Is Corollary 2.20 true for infinite-dimensional vector spaces V? If so, give a proof. If not, give an example.

(8) Verify axioms V5–V8 from Definition 1.4 of Chapter I for the scalar multiplication being defined in equation 2.16.

(9) Show that $(V_1 \otimes_F K) \otimes_K \cdots \otimes_K (V_n \otimes_F K) \cong (V_1 \otimes_F \cdots \otimes_F V_n) \otimes_F K$ as K-vector spaces under the map that sends $(\alpha_1 \otimes_F k_1) \otimes_K \cdots \otimes_K (\alpha_n \otimes_F k_n) \to (\alpha_1 \otimes_F \cdots \otimes_F \alpha_n) \otimes_F (k_1 k_2 \cdots k_n)$.

(10) Show that $\psi: \text{Hom}_F(V_1 \otimes_F V_2, V_3) \to \text{Hom}_F(V_1, \text{Hom}_F(V_2, V_3))$ is an isomorphism. Here ψ is defined by $[\psi(f)(\alpha_1)](\alpha_2) = f(\alpha_1 \otimes \alpha_2)$.

(11) Show that $\eta: \text{Hom}_F(V_1, V_2) \otimes_F V_3 \to \text{Hom}_F(V_1, V_2 \otimes V_3)$ is an isomorphism. Here η is defined by $\eta(f \otimes \alpha_3)(\alpha_1) = f(\alpha_1) \otimes \alpha_3$. We assume $\dim V_3 < \infty$.

(12) Suppose V and W are vector spaces over F and $\sum_{i=1}^{n} (\alpha_i \otimes \beta_i) = 0$ in $V \otimes_F W$ ($\alpha_i \in V$, $\beta_i \in W$). Show that there exist finite-dimensional subspaces $V_1 \subseteq V$ and $W_1 \subseteq W$ such that

(a) $\alpha_1, \ldots, \alpha_n \in V_1$, $\beta_1, \ldots, \beta_n \in W_1$, and

(b) $\sum_{i=1}^{n} (\alpha_i \otimes \beta_i) = 0$ in $V_1 \otimes W_1$.

(13) Show that $V \otimes W = 0$ if and only if V or W is zero.

Let us return to problems about the Kronecker product $A \otimes B$ of two matrices (see Exercise 5 of Section 1).

(14) Suppose V and W are finite-dimensional vector spaces over a field F. Let $T \in \text{Hom}_F(V, V)$ and $S \in \text{Hom}_F(W, W)$. Suppose A and B are matrix representations of T and S, respectively. Show that $A \otimes B$ is a matrix representation of $T \otimes S$ on $V \otimes_F W$.

(15) If $A \in M_{n \times n}(F)$ and $B \in M_{m \times m}(F)$, show that $\text{rk}(A \otimes B) = \text{rk}(A)\text{rk}(B)$.

(16) In Exercise 15, show that $\det(A \otimes B) = (\det(A))^m(\det(B))^n$.

(17) Let $V = F[X]$. Show that $V \otimes_F V \cong F[X, Y]$ under the map that sends $f(X) \otimes g(X)$ to $f(X)g(Y)$.

(18) Let $D: F[X] \to F[X]$ be the formal derivative. Thus, $D(\sum_{i=0}^n a_i X^i) = \sum_{i=1}^n i a_i X^{i-1}$. Show that D is a linear transformation on $F[X]$ such that $D(V_n) \subseteq V_n$ for all $n \in \mathbb{N}$. Here V_n is the vector space defined in Exercise 1 (Section 2 of Chapter I).

(19) Interpret the map $D \otimes D$ on $F[X] \otimes F[X]$ using the isomorphism given in Exercise 17. Restrict $D \otimes D$ to $V_n \otimes V_m$ and compute a Kronecker product that represents $D \otimes D$.

(20) Generalize Exercise 17 to $F[X_1, \ldots, X_n]$.

3. ALTERNATING MAPS AND EXTERIOR POWERS

In this section, we study a special class of multilinear maps that are called *alternating*. Before we can present the main definitions, we need to discuss permutations. Suppose $\Delta = \{1, \ldots, n\}$. A permutation of Δ is a bijective map of Δ onto itself. Suppose σ is a permutation of Δ. If $\sigma(1) = j_1, \sigma(2) = j_2, \ldots,$ and $\sigma(n) = j_n$ then $\Delta = \{j_1, \ldots, j_n\}$. We can represent the action of σ on Δ by the following $2 \times n$ array:

3.1:

$$\sigma = \begin{bmatrix} 1 & 2 & \cdots & n \\ j_1 & j_2 & \cdots & j_n \end{bmatrix}$$

Example 3.2: Let $n = 5$, and

$$\sigma = \begin{bmatrix} 1 & 2 & 3 & 4 & 5 \\ 2 & 3 & 4 & 1 & 5 \end{bmatrix}$$

Then σ is the bijection of $\Delta = \{1, 2, 3, 4, 5\}$ given by $\sigma(1) = 2, \sigma(2) = 3, \sigma(3) = 4, \sigma(4) = 1$, and $\sigma(5) = 5$. $\square$

Clearly, the number of distinct permutations of Δ is $n!$. We shall let S_n denote the set of all permutations on $\Delta = \{1, \ldots, n\}$. Thus, $|S_n| = n!$.

Since the elements of S_n are functions on Δ, we can compose any two elements $\sigma, \tau \in S_n$, getting a third permutation $\sigma\tau$ of Δ. Thus, we have a function $S_n \times S_n \to S_n$ given by $(\sigma, \tau) \to \sigma\tau$. The action of $\sigma\tau$ on Δ is computed from σ and τ by using equation 3.1 in the obvious way.

Example 3.3: Let n = 5. Suppose $\sigma, \tau \in S_5$ are given by

$$\sigma = \begin{bmatrix} 1 & 2 & 3 & 4 & 5 \\ 2 & 3 & 4 & 1 & 5 \end{bmatrix}, \quad \tau = \begin{bmatrix} 1 & 2 & 3 & 4 & 5 \\ 5 & 4 & 2 & 3 & 1 \end{bmatrix}$$

Then

$$\sigma\tau = \begin{bmatrix} 1 & 2 & 3 & 4 & 5 \\ 5 & 1 & 3 & 4 & 2 \end{bmatrix}, \quad \tau\sigma = \begin{bmatrix} 1 & 2 & 3 & 4 & 5 \\ 4 & 2 & 3 & 5 & 1 \end{bmatrix}$$

Note that $\sigma\tau \neq \tau\sigma$.   □

The map $(\sigma, \tau) \to \sigma\tau$ on S_n satisfies the following properties:

3.4: (a) $\sigma(\tau\gamma) = (\sigma\tau)\gamma$ for all $\sigma, \tau, \gamma \in S_n$.
   (b) There exists an element $1 \in S_n$ such that $\sigma 1 = 1\sigma = \sigma$ for all $\sigma \in S_n$.
   (c) For every $\sigma \in S_n$, there exists a $\tau \in S_n$ such that $\sigma\tau = \tau\sigma = 1$.

In (b) of 3.4, 1 is just the identity map on Δ. Any set S together with a binary operation $(\sigma, \tau) \to \sigma\tau$ from $S \times S$ to S that satisfies the three conditions in 3.4 is called a *group*. For this reason, the set S_n is often called the *permutation group* on n letters. With this notation, some of our previous theorems can be worded more succinctly. For example, Theorem 2.3 becomes: For all $\sigma \in S_n$, $V_1 \otimes \cdots \otimes V_n \cong V_{\sigma(1)} \otimes \cdots \otimes V_{\sigma(n)}$.
In this section, we shall need the definition of the *sign*, sgn(σ), of a permutation $\sigma \in S_n$. We first define cycles and transpositions. A permutation $\sigma \in S_n$ is called a *cycle* (or more accurately an r-cycle) if σ permutes a sequence of elements $i_1, \ldots, i_r$, $r > 1$, cyclically in the sense that $\sigma(i_1) = i_2$, $\sigma(i_2) = i_3, \ldots, \sigma(i_{r-1}) = i_r, \sigma(i_r) = i_1$, and $\sigma(j) = j$ for all $j \in \Delta - \{i_1, \ldots, i_r\}$.

Example 3.5: If n = 5, then

$$\sigma_1 = \begin{bmatrix} 1 & 2 & 3 & 4 & 5 \\ 5 & 3 & 4 & 1 & 2 \end{bmatrix}$$

is 5-cycle.

$$\sigma_2 = \begin{bmatrix} 1 & 2 & 3 & 4 & 5 \\ 2 & 3 & 1 & 4 & 5 \end{bmatrix}$$

is a 3-cycle.

$$\sigma = \begin{bmatrix} 1 & 2 & 3 & 4 & 5 \\ 2 & 3 & 1 & 5 & 4 \end{bmatrix}$$

is not a cycle. However, σ is the product of two cycles:

$$\sigma = \begin{bmatrix} 1 & 2 & 3 & 4 & 5 \\ 2 & 3 & 1 & 4 & 5 \end{bmatrix} \begin{bmatrix} 1 & 2 & 3 & 4 & 5 \\ 1 & 2 & 3 & 5 & 4 \end{bmatrix} \quad \square$$

When dealing with an r-cycle, σ, which permutes $i_1, \ldots, i_r$ and leaves fixed all other elements of Δ, we can shorten our representation of σ and write $\sigma = (i_1, \ldots, i_r)$. Thus, in Example 3.5, $\sigma_1 = (1, 5, 2, 3, 4)$, $\sigma_2 = (1, 2, 3)$, and $\sigma = (1, 2, 3)(4, 5)$.

We say two cycles (of S_n) are disjoint if they have no common symbol in their representations. Thus, in Example 3.5, σ_1 and σ_2 are not disjoint, but $(1, 2, 3)$ and $(4, 5)$ are disjoint. It is convenient to extend the definition of cycles to the case $r = 1$. We adopt the convention that for any $i \in \Delta$, the 1-cycle (i) is the identity map. Then it should be clear that any $\sigma \in S_n$ is a product of disjoint cycles.

Example 3.6: Let $n = 9$ and

$$\sigma = \begin{bmatrix} 1 & 2 & 3 & 4 & 5 & 6 & 7 & 8 & 9 \\ 2 & 3 & 4 & 1 & 6 & 5 & 8 & 9 & 7 \end{bmatrix}$$

Then $\sigma = (1, 2, 3, 4)(5, 6)(7, 8, 9)$. $\square$

Any 2-cycle $(a, b) \in S_n$ is called a *transposition*. The reader can easily check that any cycle $(i_1, \ldots, i_r)$ is a product of transpositions, namely, $(i_1, \ldots, i_r) = (i_1, i_r)(i_1, i_{r-1}) \cdots (i_1, i_3)(i_1, i_2)$. The factorization of a given cycle as a product of transpositions is not unique. Consider the following example:

Example 3.7: Let $n = 4$. Then $(1, 2, 4, 3) = (1, 3)(1, 4)(1, 2)$. Also $(1, 2, 4, 3) = (4, 3, 1, 2) = (4, 2)(4, 1)(4, 3)$. $\square$

Since every permutation is a product of disjoint cycles and every cycle is a product of transpositions, we get every permutation is a product of transpositions. We know from Example 3.7 that such a factorization is not unique, but we do have the following fact:

Lemma 3.8: Let $\sigma \in S_n$. If σ can be written as a product of an even number of transpositions, then any factorization of σ into a product of transpositions must contain an even number of terms. Similarly, if σ can be written as a product of an odd number of transpositions, then any factorization of σ into a product of transpositions must contain an odd number of terms.

Proof: Let $X_1, \ldots, X_n$ denote indeterminates over the field $\mathbb{R}$, and consider the polynomial $P(X_1, \ldots, X_n) = \prod_{i < j}(X_i - X_j)$. Here the product is taken over all i and j such that $1 \leqslant i < j \leqslant n$. If $\sigma \in S_n$, then we define a new polynomial $\sigma(P)$ by $\sigma(P) = P(X_{\sigma(1)}, \ldots, X_{\sigma(n)}) = \prod_{i < j}(X_{\sigma(i)} - X_{\sigma(j)})$. Clearly, $\sigma(P) = \pm P$. A single transposition applied to P changes the sign of P. Thus, $\sigma(P) = P$ if and only if σ is a product of an even number of transpositions. $\sigma(P) = -P$ if and only if σ is a product of an odd number of transpositions. The proof of the lemma is now clear.   □

Definition 3.9: A permutation $\sigma \in S_n$ is even if σ can be written as a product of an even number of transpositions. If σ is even, we define the sign of σ, $\text{sgn}(\sigma)$, by $\text{sgn}(\sigma) = 1$. The permutation σ is odd if σ can be written as a product of an odd number of transpositions. In this case, we set $\text{sgn}(\sigma) = -1$.

Clearly, a product of two even permutations is again even. A product of two odd permutations is also even. The product of an even and odd permutation is odd. Note that our definition implies that 1, the identity map on Δ, is an even permutation (a product of zero transpositions).

Example 3.10: If $n = 4$, then $\sigma = (1,\ 2,\ 4,\ 3) = (1,\ 3)(1,\ 4)(1,\ 2)$ is odd. $\tau = (1, 2)(3, 4)$ is even. Thus, $\text{sgn}(\sigma) = -1$ and $\text{sgn}(\tau) = 1$.   □

We can now return to our study of multilinear mappings and introduce the concept of an alternating map. Suppose V and W are vector spaces over a field F. Recall that $V^n = \{(\alpha_1, \ldots, \alpha_n) \mid \alpha_i \in V\}$. We shall keep n fixed throughout this discussion.

Definition 3.11: A multilinear mapping $\eta: V^n \to W$ is called alternating if $\eta(\alpha_1, \ldots, \alpha_n) = 0$ whenever some $\alpha_i = \alpha_j$ for $i \neq j$.

Thus, a multilinear mapping η from V^n to W is alternating if η vanishes on all n-tuples $(\alpha_1, \ldots, \alpha_n)$ that contain a repetition. We shall clarify the situation when $n = 1$ by adopting the convention that all linear transformations from V to W are alternating.

Example 3.12: Let us return to the example in 1.4. The map $\phi: F^n \times \cdots \times F^n \to F$ given by $\phi(\alpha_1, \ldots, \alpha_n) = \det(a_{ij})$, where $\alpha_i = (a_{i1}, \ldots, a_{in})$ is multilinear. If any two rows of $A \in M_{n \times n}(F)$ are equal, then $\det A = 0$. Thus, ϕ is an alternating multilinear map.   □

Example 3.13: Suppose $\phi: V^n \to W$ is an arbitrary multilinear mapping. We can construct an alternating map $\text{Alt}(\phi)$ from ϕ with the following definition:

$$\text{Alt}(\phi)(\alpha_1, \ldots, \alpha_n) = \sum_{\sigma \in S_n} \text{sgn}(\sigma)\phi(\alpha_{\sigma(1)}, \ldots, \alpha_{\sigma(n)})$$

A simple counting exercise shows $\text{Alt}(\phi)$ is alternating.   □

We shall denote the set of all alternating multilinear mappings from V^n to W by $\text{Alt}_F(V^n, W)$. Clearly, $\text{Alt}_F(V^n, W)$ is a subspace of $\text{Mul}_F(V \times \cdots \times V, W)$. For if $\eta, \eta_1 \in \text{Alt}_F(V^n, W)$, and $x \in F$, then $\eta + \eta_1$ and $x\eta$ are clearly alternating.

Suppose $\sigma = (i, j)$ is a transposition in S_n with $i < j$. Let $\eta \in \text{Alt}_F(V^n, W)$, and $\alpha = (\alpha_1, \ldots, \alpha_n) \in V^n$. Then

$$(\alpha_{\sigma(1)}, \ldots, \alpha_{\sigma(n)}) = (\alpha_1, \ldots, \overset{(i)}{\alpha_j}, \ldots, \overset{(j)}{\alpha_i}, \ldots, \alpha_n)$$

Thus $(\alpha_{\sigma(1)}, \ldots, \alpha_{\sigma(n)})$ is just the vector α with its ith and jth components interchanged. Suppose we consider the n-tuple $(\alpha_1, \ldots, \overset{(i)}{\alpha_i + \alpha_j}, \ldots, \overset{(j)}{\alpha_i + \alpha_j}, \ldots, \alpha_n)$ with $\alpha_i + \alpha_j$ in both the ith and jth positions. Since η is alternating, we have $0 = \eta(\alpha_1, \ldots, \overset{(i)}{\alpha_i + \alpha_j}, \ldots, \overset{(j)}{\alpha_i + \alpha_j}, \ldots, \alpha_n) = \eta(\alpha_1, \ldots, \overset{(i)}{\alpha_i}, \ldots, \overset{(j)}{\alpha_j}, \ldots, \alpha_n) + \eta(\alpha_1, \ldots, \overset{(i)}{\alpha_j}, \ldots, \overset{(j)}{\alpha_i}, \ldots, \alpha_n)$. Thus, $\eta(\alpha_1, \ldots, \overset{(i)}{\alpha_i}, \ldots, \overset{(j)}{\alpha_j}, \ldots, \alpha_n) = -\eta(\alpha_1, \ldots, \overset{(i)}{\alpha_j}, \ldots, \overset{(j)}{\alpha_i}, \ldots, \alpha_n)$. Since $\text{sgn}(\sigma) = -1$, we can rewrite this last equation as follows:

3.14:

$$\eta(\alpha_{\sigma(1)}, \ldots, \alpha_{\sigma(n)}) = \text{sgn}(\sigma)\eta(\alpha_1, \ldots, \alpha_n)$$

Thus, when we interchange two terms in the sequence $\alpha_1, \ldots, \alpha_n$ the sign of $\eta(\alpha_1, \ldots, \alpha_n)$ changes. Since every permutation is a product of transpositions, equation 3.14 immediately implies the following theorem:

Theorem 3.15: Let $\eta: V^n \to W$ be an alternating multilinear mapping. Let $\sigma \in S_n$. Then for all $(\alpha_1, \ldots, \alpha_n) \in V^n$, $\eta(\alpha_{\sigma(1)}, \ldots, \alpha_{\sigma(n)}) = \text{sgn}(\sigma)\eta(\alpha_1, \ldots, \alpha_n)$. $\square$

Another useful observation concerning alternating maps is given in the following theorem:

Theorem 3.16: Let $\eta: V^n \to W$ be a multilinear mapping. Then η is alternating if and only if $\eta(\alpha_1, \ldots, \alpha_n) = 0$ whenever $\alpha_i = \alpha_{i+1}$ for some i.

Proof: This is a straightforward computation which we leave to the exercises. $\square$

If $\eta: V^n \to W$ is an alternating multilinear map and $T \in \text{Hom}_F(W, W')$, then clearly $T\eta \in \text{Alt}_F(V^n, W')$. This suggests the following analog of the universal mapping problem posed in 1.9.

3.17: Let V be a vector space over F, and fix $n \in \mathbb{N}$. Is there a vector space Z over F and an alternating multilinear map $\eta: V^n \to Z$ such that the pair (Z, η) has the following property? If W is any vector space over F, and $\psi \in \text{Alt}_F(V^n, W)$. then there exists a unique $T \in \text{Hom}_F(Z, W)$ such that $T\eta = \psi$.

The question posed in 3.17 is called the universal mapping problem for alternating multilinear maps. This problem has an obvious solution, which we shall construct shortly. First, let us point out that any solution to 3.17 is essentially unique.

Lemma 3.18: Suppose (Z, η) and (Z', η') are two solutions to 3.17. Then there exist isomorphisms $T_1: Z \cong Z'$ and $T_2: Z' \cong Z$ such that

(a) $T_1 T_2 = I_{Z'}$, $T_2 T_1 = I_Z$, and
(b) the following diagram is commutative:

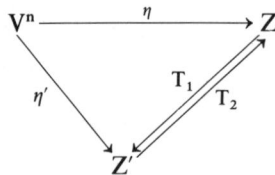

Proof: This proof is identical to that of Lemma 1.11.   □

The next order of business is to construct a solution to 3.17. This is easy using what we already know about tensor products. Consider $V^{\otimes n} = V \otimes_F \cdots \otimes_F V$. Let U be the subspace of $V^{\otimes n}$ generated by all vectors of the form $\alpha_1 \otimes \cdots \otimes \alpha_n$, where the n-tuple $(\alpha_1, \ldots, \alpha_n)$ contains a repetition. Set $Z = V^{\otimes n}/U$. Let $\phi(\alpha_1, \ldots, \alpha_n) = \alpha_1 \otimes \cdots \otimes \alpha_n$ be the canonical multilinear map from V^n to $V^{\otimes n}$. Then define $\eta: V^n \to Z$ by $\eta(\alpha_1, \ldots, \alpha_n) = (\alpha_1 \otimes \cdots \otimes \alpha_n) + U$. Thus, η is just the composite of ϕ with the natural map of $V^{\otimes n}$ onto its quotient space $V^{\otimes n}/U$. The definition of U immediately implies that η is an alternating multilinear map.

Lemma 3.19: The pair (Z, η) constructed above is a solution to 3.17.

Proof: Suppose $\psi: V^n \to W$ is any alternating multilinear map. Then because ψ is multilinear, there exists a unique linear transformation $T_0: V^{\otimes n} \to W$ such that

3.20:

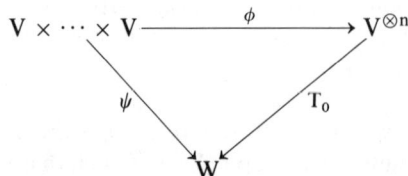

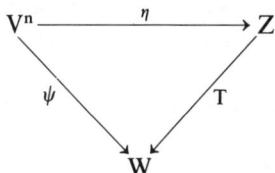

is commutative. If $(\alpha_1, \ldots, \alpha_n) \in V^n$ contains a repetition, then $\psi(\alpha_1, \ldots, \alpha_n) = 0$ since ψ is alternating. Thus, 3.20 implies $T_0(\alpha_1 \otimes \cdots \otimes \alpha_n) = 0$. Since U is generated by all vectors $\alpha_1 \otimes \cdots \otimes \alpha_n$ with $(\alpha_1, \ldots, \alpha_n)$ containing a repetition, we conclude that $T_0(U) = 0$. It now follows from the first isomorphism theorem (Theorem 5.15 of Chapter I) that T_0 induces a linear transformation $T: V^{\otimes n}/U = Z \to W$ given by $T((\alpha_1 \otimes \cdots \otimes \alpha_n) + U) = T_0(\alpha_1 \otimes \cdots \otimes \alpha_n)$. If $(\alpha_1, \ldots, \alpha_n) \in V^n$, then $T\eta(\alpha_1, \ldots, \alpha_n) = T((\alpha_1 \otimes \cdots \otimes \alpha_n) + U) = T_0(\alpha_1 \otimes \cdots \otimes \alpha_n) = \psi(\alpha_1, \ldots, \alpha_n)$. Thus,

3.21:

$$
\begin{array}{ccc}
V^n & \xrightarrow{\;\;\eta\;\;} & Z \\
 & \psi \searrow \quad \swarrow T & \\
 & W &
\end{array}
$$

is commutative.

Finally, we must argue that T is unique. Suppose $T' \in \mathrm{Hom}_F(Z, W)$ makes 3.21 commute. Then $T = T'$ on $\mathrm{Im}\,\eta$. But $Z = L(\mathrm{Im}\,\eta)$. Thus, $T = T'$, and the proof of Lemma 3.19 is complete. $\square$

Definition 3.22: The vector space $V^{\otimes n}/U$ is called the nth exterior power of V (over F) and will henceforth be denoted by $\Lambda_F^n(V)$.

When the base field F is clear from the context, we shall simplify our notation and write $\Lambda^n(V)$. Note that $\Lambda^1(V) = V$. We define $\Lambda^0(V) = F$.

Definition 3.23: The coset $(\alpha_1 \otimes \cdots \otimes \alpha_n) + U$ in $\Lambda^n(V)$ will henceforth be denoted by $\alpha_1 \wedge \cdots \wedge \alpha_n$ and called a wedge product.

Thus, the alternating multilinear map $\eta:\ V^n \to \Lambda^n(V)$ is given by $\eta(\alpha_1, \ldots, \alpha_n) = \alpha_1 \wedge \cdots \wedge \alpha_n$. We have already noted that $\mathrm{Im}\,\eta$ spans $\Lambda^n(V)$. Thus, every vector in $\Lambda^n(V)$ is a finite linear combination of wedge products $\alpha_1 \wedge \cdots \wedge \alpha_n$. We also have the following relations when dealing with wedge products:

3.24: (a) $\alpha_1 \wedge \cdots \wedge \alpha_n = \mathrm{sgn}(\sigma)\alpha_{\sigma(1)} \wedge \cdots \wedge \alpha_{\sigma(n)}$ for all $\sigma \in S_n$.

(b) $\alpha_1 \wedge \cdots \wedge (x\alpha_i) \wedge \cdots \wedge \alpha_n = x(\alpha_1 \wedge \cdots \wedge \alpha_n)$ for all $x \in F$.

(c) $\alpha_1 \wedge \cdots \wedge (\alpha_i + \alpha_i') \wedge \cdots \wedge \alpha_n$
$$= \alpha_1 \wedge \cdots \wedge \alpha_i \wedge \cdots \wedge \alpha_n + \alpha_1 \wedge \cdots \wedge \alpha_i' \wedge \cdots \wedge \alpha_n.$$

Let us restate Lemma 3.19 using our new notation.

Theorem 3.25: Let V and W be vector spaces over F and suppose $\psi\colon V^n \to W$ is an alternating multilinear map. Then there exists a unique linear transformation $T \in \operatorname{Hom}_F(\Lambda^n(V), W)$ such that for all $(\alpha_1, \ldots, \alpha_n) \in V^n$, $\psi(\alpha_1, \ldots, \alpha_n) = T(\alpha_1 \wedge \cdots \wedge \alpha_n)$. $\square$

Having constructed the nth exterior power $\Lambda_F^n(V)$ of V, the next order of business is to find a basis for this space. Suppose B is a basis of V. As usual, set $B^n = \{(\alpha_1, \ldots, \alpha_n) \mid \alpha_i \in B\}$. Let B(n) denote the subset of B^n consisting of those n-tuples which have distinct entries. Thus, $B(n) = \{(\alpha_1, \ldots, \alpha_n) \in B^n \mid \alpha_i \neq \alpha_j$ whenever $i \neq j\}$. It is possible that $B(n) = \phi$. In this case, $n > |B|$. But then every wedge product $\beta_1 \wedge \cdots \wedge \beta_n$ in $\Lambda_F^n(V)$ is zero. Thus, $\Lambda_F^n(V) = 0$, and the empty set is a basis of $\Lambda_F^n(V)$. So, we can assume with no loss of generality that $n \leqslant |B|$.

We define an equivalence relation $\equiv$ on the set B(n) by the following formula:

3.26: $(\alpha_1, \ldots, \alpha_n) \equiv (\alpha_1', \ldots, \alpha_n')$ if and only if there exists a $\sigma \in S_n$ such that $\alpha_{\sigma(1)} = \alpha_1', \ldots, \alpha_{\sigma(n)} = \alpha_n'$.

Thus, two n-tuples in B(n) are equivalent if some permutation of the entries in the first n-tuple gives the second n-tuple. The fact that $\equiv$ is indeed an equivalence relation, that is, that $\equiv$ satisfies the axioms in 5.1 of Chapter I is obvious. We shall let $\overline{B(n)}$ denote the set of equivalence classes of B(n). Recall that the elements of $\overline{B(n)}$ are subsets of B(n). B(n) is the disjoint union of the distinct elements of $\overline{B(n)}$. If $(\alpha_1, \ldots, \alpha_n) \in B(n)$, we shall let $\langle \alpha_1, \ldots, \alpha_n \rangle$ denote the equivalence class in $\overline{B(n)}$ which contains $(\alpha_1, \ldots, \alpha_n)$. Thus, $\overline{B(n)} = \{\langle \alpha_1, \ldots, \alpha_n \rangle \mid (\alpha_1, \ldots, \alpha_n) \in B(n)\}$.

Now for each element $x \in \overline{B(n)}$, we can choose an n-tuple $(\alpha_1, \ldots, \alpha_n) \in B(n)$ such that $\langle \alpha_1, \ldots, \alpha_n \rangle = x$. For a given x, there may be many such n-tuples, but they are all representatives of the same equivalence class x. For each $x \in \overline{B(n)}$, pick a representative of x, say λ_x, in B(n). Thus, λ_x is an n-tuple $(\alpha_1, \ldots, \alpha_n) \in B(n)$ such that $\langle \alpha_1, \ldots, \alpha_n \rangle = x$. We have now defined a set mapping $(x \to \lambda_x)$ of $\overline{B(n)}$ to B(n). There are of course many choices for such a map. Choose any such map and set $C(n) = \{\lambda_x \mid x \in \overline{B(n)}\}$. Then C(n) is a collection of n-tuples in B(n), one n-tuple for every equivalence class $x \in \overline{B(n)}$. We can now state the following theorem:

Theorem 3.27: Let V be a vector space over F. Suppose B is a basis of V. Construct the set C(n) as above. Then $\Delta = \{\alpha_1 \wedge \cdots \wedge \alpha_n \mid (\alpha_1, \ldots, \alpha_n) \in C(n)\}$ is a basis of $\Lambda_F^n(V)$. In particular, $\dim_F(\Lambda_F^n(V)) = |\overline{B(n)}|$.

Proof: The set C(n) consists of one representative in B(n) for each equivalence class $x \in \overline{B(n)}$. Thus, $|C(n)| = |\overline{B(n)}|$. In particular, if Δ is a basis of $\Lambda_F^n(V)$, then $\dim_F(\Lambda_F^n(V)) = |\Delta| = |C(n)| = |\overline{B(n)}|$. So, we need only argue Δ is a basis of $\Lambda_F^n(V)$. The proof we give here is analogous to that of Theorem 1.20.

Let $\hat{V} = \bigoplus_{(\alpha_1,\ldots,\alpha_n)\in C(n)} F$. As we have seen in Section 1, $\hat{V}$ is a vector space over F with basis $\{\delta_{(\alpha_1,\ldots,\alpha_n)} \mid (\alpha_1,\ldots,\alpha_n)\in C(n)\}$. Here $\delta_{(\alpha_1,\ldots,\alpha_n)}$ is the function from C(n) into F given by $\delta_{(\alpha_1,\ldots,\alpha_n)}((\beta_1,\ldots,\beta_n)) = 1$ if $(\beta_1,\ldots,\beta_n) = (\alpha_1,\ldots,\alpha_n)$ and zero otherwise. We shall show that $\hat{V} \cong \Lambda_F^n(V)$.

We define a map $\mu_0 \colon B^n \to \hat{V}$ as follows: If $(\beta_1,\ldots,\beta_n)\in B^n$ contains a repetition, set $\underline{\mu_0(\beta_1,\ldots,\beta_n) = 0}$. Suppose $(\beta_1,\ldots,\beta_n)\in B(n)$. Then $\langle\beta_1,\ldots,\beta_n\rangle = x\in \overline{B(n)}$. If $\lambda_x = (\alpha_1,\ldots,\alpha_n)\in C(n)$, then $\langle\alpha_1,\ldots,\alpha_n\rangle = \langle\beta_1,\ldots,\beta_n\rangle$. Thus, $(\beta_1,\ldots,\beta_n)\equiv(\alpha_1,\ldots,\alpha_n)$. So, there exists a unique $\sigma\in S_n$ such that $(\beta_{\sigma(1)},\ldots,\beta_{\sigma(n)}) = (\alpha_1,\ldots,\alpha_n)$. In this case, define $\mu_0(\beta_1,\ldots,\beta_n) = \mathrm{sgn}(\sigma)\delta_{(\alpha_1,\ldots,\alpha_n)}$. We can extend the map μ_0 in the usual way (see Exercise 2 at the end of Section 1 in this chapter) to a multilinear map $\mu\colon V^n \to \hat{V}$. An easy computation using the definition of μ_0 shows μ is alternating.

We now claim the pair $(\hat{V}, \mu)$ satisfies the universal mapping property in 3.17. To see this, suppose W is a vector space over F, and $\psi\colon V^n \to W$ an alternating multilinear map. Using 3.23 of Chapter I, we can define a linear transformation $T\colon \hat{V} \to W$ by $T(\delta_{(\alpha_1,\ldots,\alpha_n)}) = \psi(\alpha_1,\ldots,\alpha_n)$ for all $(\alpha_1,\ldots,\alpha_n)\in C(n)$. If $(\beta_1,\ldots,\beta_n)\in B^n$ and contains a repetition, then $T\mu(\beta_1,\ldots,\beta_n) = T(0) = 0$. Since ψ is alternating, $\psi(\beta_1,\ldots,\beta_n) = 0$. Suppose $(\beta_1,\ldots,\beta_n)\in B(n)$. Then $(\beta_{\sigma(1)},\ldots,\beta_{\sigma(n)}) = (\alpha_1,\ldots,\alpha_n)\in C(n)$ for some $\sigma\in S_n$. Thus, $T\mu(\beta_1,\ldots,\beta_n) = T(\mathrm{sgn}(\sigma)\delta_{(\alpha_1,\ldots,\alpha_n)}) = \mathrm{sgn}(\sigma)T(\delta_{(\alpha_1,\ldots,\alpha_n)}) = \mathrm{sgn}(\sigma)\psi(\alpha_1,\ldots,\alpha_n)$. On the other hand, since ψ is alternating, Theorem 3.15 implies $\psi(\beta_1,\ldots,\beta_n) = \mathrm{sgn}(\sigma)\psi(\beta_{\sigma(1)},\ldots,\beta_{\sigma(n)}) = \mathrm{sgn}(\sigma)\psi(\alpha_1,\ldots,\alpha_n)$. Thus, $T\mu$ and ψ are two alternating multilinear maps on V^n that agree on B^n. It follows that $T\mu = \psi$.

Finally, we must argue T is unique. If $T'\colon \hat{V} \to W$ is a second linear transformation such that $T'\mu = \psi$, then $T'(\delta_{(\alpha_1,\ldots,\alpha_n)}) = T'\mu(\alpha_1,\ldots,\alpha_n) = \psi(\alpha_1,\ldots,\alpha_n) = T\mu(\alpha_1,\ldots,\alpha_n) = T(\delta_{(\alpha_1,\ldots,\alpha_n)})$ for all $(\alpha_1,\ldots,\alpha_n)\in C(n)$. Thus, T and T' agree on a basis of $\hat{V}$ and, consequently, must be equal.

We can now apply Lemma 3.18 to the pairs $(\Lambda_F^n(V),\eta)$ and $(\hat{V},\mu)$. In particular, there exists an isomorphism $S\colon \hat{V} \cong \Lambda_F^n(V)$ such that $S\mu = \eta$. If $(\alpha_1,\ldots,\alpha_n)\in C(n)$, then $\alpha_1 \wedge \cdots \wedge \alpha_n = \eta(\alpha_1,\ldots,\alpha_n) = S\mu(\alpha_1,\ldots,\alpha_n) = S(\delta_{(\alpha_1,\ldots,\alpha_n)})$. Thus, S is an isomorphism taking the basis $\{\delta_{(\alpha_1,\ldots,\alpha_n)} \mid (\alpha_1,\ldots,\alpha_n)\in C(n)\}$ in $\hat{V}$ to the set Δ in $\Lambda_F^n(V)$. It follows that Δ is a basis of $\Lambda_F^n(F)$. □

Corollary 3.28: Suppose V is a finite-dimensional vector space over F with $\dim_F(V) = N$. Then $\dim_F(\Lambda_F^n(V)) = \binom{N}{n}$.

Before giving a proof of 3.28, let us discuss its meaning. If $0 \leqslant n \leqslant N$, then $\binom{N}{n}$ is the binomial coefficient $N!/n!(N - n)!$. If $n > N$, then $\binom{N}{n} = 0$.

Proof of Corollary 3.28: If $n = 0$, the result is trivial. Suppose $1 \leqslant n \leqslant N$. Let $B = \{\alpha_1,\ldots,\alpha_N\}$ be a basis of V. Theorem 3.27 implies that $\Delta = \{\alpha_{i_1} \wedge \cdots \wedge \alpha_{i_n} \mid 1 \leqslant i_1 < \cdots < i_n \leqslant N\}$ is a basis of $\Lambda_F^n(V)$. The cardinality of Δ is clearly the number of ways of picking an n-element subset from the set $\{1, 2, \ldots, N\}$. Therefore, $|\Delta| = \binom{N}{n}$

Suppose $n > N$. Then we had noted previously that $B(n) = \phi$, and $\Lambda_F^n(V) = 0$. Thus, $\dim_F(\Lambda_F^n(V)) = \binom{N}{n} = 0$. $\square$

There is another corollary that can be derived from Theorem 3.27. Suppose V is an n-dimensional vector space over F. Let $\underline{\alpha} = \{\alpha_1, \ldots, \alpha_n\}$ be a basis of V. Then we have a nonzero, alternating map $\phi: V^n \to F$ given by $\phi(\alpha_1, \ldots, \alpha_n) = \det(a_{ij})$, where $(\alpha_i)_{\underline{\alpha}} = (a_{i1}, \ldots, a_{in})$. Our next corollary says that ϕ is essentially the only alternating map from V^n to F.

Corollary 3.29: Let V be an n-dimensional vector space over F. Then $\dim_F(\text{Alt}_F(V^n, F)) = 1$.

Proof: The map ϕ constructed above is alternating. Hence $\phi \in \text{Alt}_F(V^n, F)$. Suppose $\psi \in \text{Alt}_F(V^n, F)$. If η denotes the canonical map given in Lemma 3.19, then there exists a unique linear transformation $T \in \text{Hom}_F(\Lambda_F^n(V), F)$ such that $T\eta = \psi$. Similarly, there exists a $T_1 \in \text{Hom}_F(\Lambda_F^n(V), F)$ such that $T_1\eta = \phi$.

Now Corollary 3.28 implies $\dim_F(\Lambda^n V) = 1$. Consequently, $\dim_F\{\text{Hom}_F(\Lambda_F^n(V), F)\} = 1$. Since $\phi \neq 0$, we conclude $T_1 \neq 0$. In particular, $\{T_1\}$ is a basis for $\text{Hom}_F(\Lambda_F^n(V), F)$. Therefore $T = xT_1$ for some $x \in F$. We then have $\psi = T\eta = xT_1\eta = x\phi$. Thus, $\{\phi\}$ is a basis of $\text{Alt}_F(V^n, F)$ and the proof of Corollary 3.29 is complete. $\square$

At this point, we could begin to discuss the functorial properties of exterior powers. Almost all the results in Section 2 have analogs in our present situation. Since this is not a text in multilinear algebra per se, we shall leave most of these types of results to the exercises at the end of this section. The reader who wishes to read further in this subject matter should consult [5] or [4].

We shall finish this section with a description of the induced map on exterior powers derived from a given $T \in \text{Hom}_F(V, W)$. Suppose V and W are vector spaces over F and let T be a linear transformation from V to W. Let $\eta: V^n \to \Lambda_F^n(V)$ be the natural alternating map given by $\eta(\alpha_1, \ldots, \alpha_n) = \alpha_1 \wedge \cdots \wedge \alpha_n$. The linear transformation T induces an alternating, multilinear mapping $\psi_T: V^n \to \Lambda_F^n(W)$ given by $\psi_T(\alpha_1, \ldots, \alpha_n) = T(\alpha_1) \wedge \cdots \wedge T(\alpha_n)$. The fact that ψ_T is indeed alternating is clear. It now follows from Theorem 3.25 that there exists a unique linear transformation $S \in \text{Hom}_F(\Lambda_F^n(V), \Lambda_F^n(W))$ such that $S\eta = \psi_T$.

Definition 3.30: The unique linear transformation S for which $S\eta = \psi_T$ will henceforth be denoted $\Lambda^n(T)$.

Thus, $\Lambda^n(T) \in \text{Hom}_F(\Lambda_F^n(V), \Lambda_F^n(W))$, and for all $(\alpha_1, \ldots, \alpha_n) \in V^n$, $\Lambda^n(T)(\alpha_1 \wedge \cdots \wedge \alpha_n) = T(\alpha_1) \wedge \cdots \wedge T(\alpha_n)$.

Clearly, $\Lambda^n(T_1 T_2) = \Lambda^n(T_1)\Lambda^n(T_2)$ for $T_2 \in \text{Hom}_F(V, W)$ and $T_1 \in \text{Hom}_F(W, Z)$. We also have the important analogs of Theorem 2.6.

Theorem 3.31: Let V and W be vector spaces over F, and let $T \in \text{Hom}_F(V, W)$. Then the following assertions are true:

    (a) If T is injective, so is $\Lambda^n(T)$.

    (b) If T is surjective, so is $\Lambda^n(T)$.

    (c) If T is an isomorphism, so is $\Lambda^n(T)$.

Proof: Consider bases of V and W and apply Theorem 3.27. $\square$

    $\Lambda^n(T)$ is usually called the nth exterior power of T.

EXERCISES FOR SECTION 3

  (1) Show that every permutation $\sigma \in S_n$ is a product of disjoint cycles.

  (2) Elaborate on the details of Example 3.13. Specifically, show $\text{Alt}(\cdot)$: $\text{Mul}_F(V^n, W) \to \text{Alt}_F(V^n, W)$ is a well-defined linear transformation.

  (3) Prove Theorem 3.16.

  (4) The exterior algebra $\Lambda(V)$ of V is defined to be the following direct sum: $\Lambda(V) = \bigoplus_{n=0}^{\infty} \Lambda^n(V)$.

    (a) Show that $\Lambda(V)$ is an algebra over F when we define the product of two elements $\alpha = \alpha_1 \wedge \cdots \wedge \alpha_p \in \Lambda^p(V)$ and $\beta = \beta_1 \wedge \cdots \wedge \beta_m \in \Lambda^m(V)$ by $\alpha\beta = \alpha_1 \wedge \cdots \wedge \alpha_p \wedge \beta_1 \wedge \cdots \wedge \beta_m$.

    (b) Show that $\Lambda(V)$ is an anticommutative algebra. This means for all $\alpha \in \Lambda^p(V)$ and $\beta \in \Lambda^m(V)$, $\alpha\beta = (-1)^{pm}\beta\alpha$.

    (c) Show that there exists an injective linear transformation $T: V \to \Lambda(V)$ such that $(T(\alpha))^2 = 0$ for all $\alpha \in V$.

  (5) Show that the exterior algebra $\Lambda(V)$ of V has the following universal mapping property: Given any algebra A over F and a linear transformation $T \in \text{Hom}_F(V, A)$ such that $(T(\alpha))^2 = 0$ for all $\alpha \in V$, then there exists a unique F-algebra homomorphism $\varphi: \Lambda(V) \to A$ such that $\varphi(\alpha) = T(\alpha)$ for all $\alpha \in V$.

  (6) If $\dim_F(V) = n$, what is $\dim_F(\Lambda(V))$?

  (7) Suppose V is a finite-dimensional vector space over F. Show that $\{\Lambda_F^n(V)\}^* \cong \Lambda_F^n(V^*)$. Is this true if V is infinite dimensional?

  (8) Give an example of a short exact sequence $0 \to V \to W \to Z \to 0$ of vector spaces over F such that the corresponding complex $0 \to \Lambda^n V \to \Lambda^n W \to \Lambda^n Z \to 0$ is not exact.

  (9) Suppose V is a vector space over F, and let K be a field containing F. Show $\Lambda_F^n(V) \otimes_F K \cong \Lambda_K^n(V \otimes_F K)$ as K-vector spaces.

(10) Let V and W be vector spaces over F. Show that $\Lambda_F^n(V \oplus W) \cong \bigoplus_{i+j=n} (\Lambda^i(V) \otimes_F \Lambda^j(W))$.

4. SYMMETRIC MAPS AND SYMMETRIC POWERS

In this last section on multilinear algebra, we study another special class of multilinear maps. Let V and W be vector spaces over F and let $n \in \mathbb{N}$.

Definition 4.1: A multilinear mapping $\phi: V^n \to W$ is said to be symmetric if $\phi(\alpha_1, \ldots, \alpha_n) = \phi(\alpha_{\sigma(1)}, \ldots, \alpha_{\sigma(n)})$ for all $(\alpha_1, \ldots, \alpha_n) \in V^n$ and all $\sigma \in S_n$.

We shall denote the set of all symmetric multilinear mappings from V^n to W by $\text{Sym}_F(V^n, W)$. Clearly, $\text{Sym}_F(V^n, W)$ is a subspace of $\text{Mul}_F(V^n, W)$. Note that $\text{Sym}_F(V^n, W) \supseteq \text{Alt}_F(V^n, W)$ whenever $F \supseteq \mathbb{F}_2$. Let us consider some examples:

Example 4.2: If $n = 2$, then $\text{Sym}_F(V^2, W)$ is just the set of all symmetric bilinear maps from $V \times V$ to W. In particular, any inner product is a symmetric multilinear map. □

Example 4.3: If $\phi: V^n \to W$ is any multilinear map, we can construct a symmetric map $S(\phi) \in \text{Sym}_F(V^n, W)$ from ϕ with the following definition:

$$S(\phi)(\alpha_1, \ldots, \alpha_n) = \sum_{\sigma \in S_n} \phi(\alpha_{\sigma(1)}, \ldots, \alpha_{\sigma(n)}) \quad \square$$

We note in passing that a multilinear map $\phi \in \text{Mul}_F(V^n, W)$ is in fact symmetric if $\phi(\alpha_1, \ldots, \alpha_n)$ remains unaltered whenever two adjacent terms in $(\alpha_1, \ldots, \alpha_n)$ are interchanged. If $\phi \in \text{Sym}_F(V^n, W)$, and $T \in \text{Hom}_F(W, Z)$, then clearly, $T\phi \in \text{Sym}_F(V^n, Z)$. As with alternating maps, this suggest the following universal mapping problem for symmetric multilinear maps:

4.4: Let V be a vector space over F and $n \in \mathbb{N}$. Is there a vector space Z over F and a symmetric multilinear map $\phi: V^n \to Z$ such that the pair (Z, ϕ) has the following property: If W is any vector space over F, and $\psi \in \text{Sym}_F(V^n, W)$, then there exists a unique $T \in \text{Hom}_F(Z, W)$ such that $T\phi = \psi$?

As with alternating maps, it is an easy matter to argue that any solution to 4.4 is essentially unique.

Lemma 4.5: Suppose (Z, ϕ) and (Z', ϕ') are two solutions to 4.4. Then there exist isomorphisms $T_1: Z \cong Z'$ and $T_2: Z' \cong Z$ such that

(a) $T_1 T_2 = I_{Z'}$ and $T_2 T_1 = I_Z$, and

(b) the following diagram is commutative:

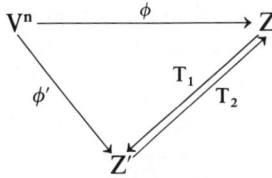

Proof: This proof is identical to that of Lemma 1.11. □

As the reader can see, much of what we do here is completely analogous to the case of alternating maps. For this reason, our treatment of the results for symmetric maps will be much abbreviated.

We now construct a solution to 4.4. Set $Z = V^{\otimes n}/N$ where N is the subspace of $V^{\otimes n}$ generated by all vectors of the form $\alpha_1 \otimes \cdots \otimes \alpha_n - \alpha_{\sigma(1)} \otimes \cdots \otimes \alpha_{\sigma(n)}$ $(\sigma \in S_n)$. Let $\phi : V^n \to Z$ be given by $\phi(\alpha_1, \ldots, \alpha_n) = \alpha_1 \otimes \cdots \otimes \alpha_n + N$. The definition of N implies ϕ is a symmetric multilinear map.

Lemma 4.6: (Z, ϕ) solves 4.4.

Proof: Suppose $\psi \in \mathrm{Sym}_F(V^n, W)$. By Theorem 1.19, there exists a $T_0 \in \mathrm{Hom}_F(V^{\otimes n}, W)$ such that $T_0(\alpha_1 \otimes \cdots \otimes \alpha_n) = \psi(\alpha_1, \ldots, \alpha_n)$. Since ψ is symmetric, T_0 vanishes on N. Consequently, T_0 induces a linear transformation $T : Z \to W$ given by $T((\alpha_1 \otimes \cdots \otimes \alpha_n) + N) = T_0(\alpha_1 \otimes \cdots \otimes \alpha_n)$. Clearly $T\phi = \psi$. The fact that T is unique is the same argument as for multilinear or alternating maps. □

Definition 4.7: The vector space $V^{\otimes n}/N$ is called the nth symmetric power of V and will henceforth be denoted by $S_F^n(V)$. The cosets $\alpha_1 \otimes \cdots \otimes \alpha_n + N$ will henceforth be written $[\alpha_1] \cdots [\alpha_n]$.

With this notation, we have the following theorem:

Theorem 4.8: Let V be a vector space over F, and suppose $\psi \in \mathrm{Sym}_F(V^n, W)$. Then there exists a unique linear transformation $T \in \mathrm{Hom}_F(S_F^n(V), W)$ such that $\psi(\alpha_1, \ldots, \alpha_n) = T([\alpha_1] \cdots [\alpha_n])$ for all $(\alpha_1, \ldots, \alpha_n) \in V^n$. □

From our construction of $S_F^n(V)$, we see that the set $\{[\alpha_1] \cdots [\alpha_n] \mid (\alpha_1, \ldots, \alpha_n) \in V^n\}$ spans $S_F^n(V)$ as a vector space over F. The following relations are all obvious.

4.9: (a) $[\alpha_1] \cdots [\alpha_i + \alpha_i'] \cdots [\alpha_n] = [\alpha_1] \cdots [\alpha_i] \cdots [\alpha_n] + [\alpha_1] \cdots [\alpha_i'] \cdots [\alpha_n]$.

(b) $[\alpha_1] \cdots [x\alpha_i] \cdots [\alpha_n] = x([\alpha_1] \cdots [\alpha_i] \cdots [\alpha_n])$.

(c) $[\alpha_1] \cdots [\alpha_n] = [\alpha_{\sigma(1)}] \cdots [\alpha_{\sigma(n)}]$ for all $\sigma \in S_n$.

To construct a basis of $S_F^n(V)$, we proceed in the same spirit as Theorem 3.27. Let B be any basis of V. Define an equivalence relation $\equiv$ on B^n by $(\alpha_1, \ldots, \alpha_n) \equiv (\alpha_1', \ldots, \alpha_n')$ if and only if $(\alpha_{\sigma(1)}, \ldots, \alpha_{\sigma(n)}) = (\alpha_1', \ldots, \alpha_n')$ for some $\sigma \in S_n$. From each equivalence class of $\equiv$, pick one representative and call this set of representatives B^*. An argument completely analogous to that of Theorem 3.27 gives us $\{[\alpha_1] \cdots [\alpha_n] \mid (\alpha_1, \ldots, \alpha_n) \in B^*\}$ is a basis of $S_F^n(V)$.

Theorem 4.10: Suppose B is a basis of V. Form B^* as described above. Then $\Delta = \{[\alpha_1] \cdots [\alpha_n] \mid (\alpha_1, \ldots, \alpha_n) \in B^*\}$ is a basis of $S_F^n(V)$. $\square$

Now suppose $\dim_F(V) = N < \infty$. Let $B = \{\alpha_1, \ldots, \alpha_N\}$ be a basis of V. The number of elements in B^* is the number of distinct monomials of degree n in the symbols $[\alpha_1], \ldots, [\alpha_N]$. Thus, $|B^*|$ is equal to the number of distinct products of the form $[\alpha_1]^{e_1} \cdots [\alpha_N]^{e_N}$ with $e_1 + \cdots + e_N = n$. An easy counting argument gives us precisely $\binom{N-1+n}{n}$ of these monomials. Thus, we have proven the following corollary:

Corollary 4.11: If $\dim_F(V) = N$, then $\dim_F(S_F^n(V)) = \binom{N-1+n}{n}$. $\square$

Finally, a linear transformation $T: V \to W$ induces a map on symmetric powers as follows: Let $\phi: V^n \to S_F^n(V)$ be the natural symmetric map given by $\phi(\alpha_1, \ldots, \alpha_n) = [\alpha_1] \cdots [\alpha_n]$. T induces a symmetric multilinear mapping ψ_T from V^n to $S_F^n(W)$ given by $\psi_T(\alpha_1, \ldots, \alpha_n) = [T(\alpha_1)] \cdots [T(\alpha_n)]$. By Theorem 4.8, there exists a unique linear transformation $S_F^n(T) \in \mathrm{Hom}_F(S_F^n(V), S_F^n(W))$ such that $S_F^n(T)\phi = \psi_T$. Thus, for all $(\alpha, \ldots, \alpha_n) \in V^n$, $S_F^n(T)$ $([\alpha_1] \cdots [\alpha_n]) = [T(\alpha_1)] \cdots [T(\alpha_n)]$.

If $T_1 \in \mathrm{Hom}_F(V, W)$ and $T_2 \in \mathrm{Hom}_F(W, Z)$, then clearly, $S_F^n(T_2 T_1) = S_F^n(T_2)S_F^n(T_1)$. We also have the analog of Theorem 3.31.

Theorem 4.12: Let V and W be vector spaces over F and suppose $T \in \mathrm{Hom}_F(V, W)$.

(a) If T is injective, then $S_F^n(T)$ is injective.

(b) If T is surjective, then $S_F^n(T)$ is surjective.

(c) If T is an isomorphism, then so is $S_F^n(T)$. $\square$

The linear transformation $S_F^n(T)$ is called the nth symmetric power of T.

EXERCISES FOR SECTION 4

(1) Complete the details of Example 4.3, that is, argue $S(\phi)$ is indeed a symmetric multilinear mapping.

(2) Suppose V is a vector space over F. Show that the following complex is a short exact sequence:

$$0 \longrightarrow \Lambda_F^2(V) \overset{T}{\longrightarrow} V \otimes_F V \overset{T'}{\longrightarrow} S_F^2(V) \longrightarrow 0.$$

Here T is given by $T(\alpha \wedge \beta) = \alpha \otimes \beta - \beta \otimes \alpha$, and T' is given by $T'(\beta \otimes \delta) = [\beta][\delta]$.

(3) Is Exercise 2 still true if 2 is replaced by n?

(4) Let $A = F[X_1, \ldots, X_n]$ denote the set of all polynomials in the variables $X_1, \ldots, X_n$ with coefficients in the field F. Thus, $F[X_1, \ldots, X_n]$ is the set consisting of all finite sums of the following form $\sum_{(i_1, \ldots, i_n)} c_{(i_1 \ldots i_n)} X_1^{i_1} \cdots X_n^{i_n}$ with $c_{(i_1, \ldots, i_n)} \in F$ and $(i_1, \ldots, i_n) \in (\mathbb{N} \cup \{0\})^n$.

  (a) Show that A is an infinite-dimensional vector space over F with basis $\Delta = \{X_1^{i_1} \cdots X_n^{i_n} \mid (i_1, \ldots, i_n) \in (\mathbb{N} \cup \{0\})^n\}$ when we define addition and scalar multiplication as follows:

$$\left(\sum c_{(i_1 \cdots i_n)} X_1^{i_1} \cdots X_n^{i_n}\right) + \left(\sum d_{(i_1 \cdots i_n)} X_1^{i_1} \cdots X_n^{i_n}\right)$$
$$= \sum (c_{(i_1 \cdots i_n)} + d_{(i_1 \cdots i_n)}) X_1^{i_1} \cdots X_n^{i_n}$$

  and

$$x\left(\sum c_{(i_1 \cdots i_n)} X_1^{i_1} \cdots X_n^{i_n}\right) = \sum x c_{(i_1 \cdots i_n)} X_1^{i_1} \cdots X_n^{i_n}.$$

  (b) Suppose we define the product of two monomials $X_1^{e_1} \cdots X_n^{e_n}$, and $X_1^{f_1} \cdots X_n^{f_n}$ in A by the formula $(X_1^{e_1} \cdots X_n^{e_n})(X_1^{f_1} \cdots X_n^{f_n}) = X_1^{e_1 + f_1} \cdots X_n^{e_n + f_n}$. Show that we can extend this definition of multiplication in a natural way to a product on A such that A becomes a commutative algebra over F, that is, $fg = gf$ for all $f, g \in A$.

  (c) Let $A_p = L(\{X_1^{e_1} \cdots X_n^{e_n} \mid e_1 + \cdots + e_n = p\})$. Show that $\dim_F(A_p) = \binom{n-1+p}{p}$.

  (d) Show that A is a graded F-algebra, that is, $A = \bigoplus_{p=0}^{\infty} A_p$ and $A_p A_q \subseteq A_{p+q}$ for all $p, q \geqslant 0$.

(5) Let V be a vector space over F. The symmetric algebra $S(V)$ is defined to be the following direct sum: $S(V) = \bigoplus_{p=0}^{\infty} S_F^p(V)$. Here as usual, $S_F^0(V) = F$.

  (a) Show $S(V)$ is a commutative graded algebra over F when we define products by the formula $([\alpha_1] \cdots [\alpha_p])([\beta_1] \cdots [\beta_q]) = [\alpha_1] \cdots [\alpha_p][\beta_1] \cdots [\beta_q]$.

  (b) Show that there exists a natural, injective linear transformation $T \in \operatorname{Hom}_F(V, S(V))$ such that $T(\alpha)T(\beta) = T(\beta)T(\alpha)$ for all $\alpha, \beta \in V$.

(6) Show that the pair $(S(V), T)$ constructed in Exercise 5 has the following universal mapping property: If A is any F-algebra, and $\psi \in \operatorname{Hom}_F(V, A)$ such that $\psi(\alpha)\psi(\beta) = \psi(\beta)\psi(\alpha)$ for all $\alpha, \beta \in V$, then there exists a unique algebra homomorphism $\varphi: S(V) \to A$ such that $\varphi T = \psi$.

(7) If $\dim_F(V) = n$, show $S(V) \cong F[X_1, \ldots, X_n]$ as F-algebras.

(8) Let V and W be vector spaces over F. Show that $S_F^n(V \oplus W) \cong \bigoplus_{i+j=n} \{S_F^i(V) \otimes_F S_F^j(W)\}$.

(9) If V is a vector space over F and K a field containing F, show $S_F^n(V) \otimes_F K \cong S_K^n(V \otimes_F K)$ as K-vector spaces.

Chapter III

Canonical Forms of Matrices

1. PRELIMINARIES ON FIELDS

In this chapter, we return to the fundamental problem posed in Section 3 of Chapter I. Suppose V is a finite-dimensional vector space over a field F, and let $T \in \mathscr{E}(V) = \mathrm{Hom}_F(V, V)$. What is the simplest matrix representation of T? If $\underline{\alpha} = \{\alpha_1, \ldots, \alpha_n\}$ is any basis of V and $A = \Gamma(\underline{\alpha}, \underline{\alpha})(T)$, then we are asking, What is the simplest matrix $B \in M_{n \times n}(F)$ that is similar to A? Of course, that problem is a bit ambiguous since no attempt has been made to define what the word "simplest" means here. Intuitively, one feels that a given matrix representation A of T is simple if A contains a large number of zeros as entries. Most of the canonical form theorems that appear in this chapter present various matrix representations of T that contain large numbers of zeros strategically placed in the matrix. As one might expect, we get different canonical forms depending on what we are willing to assume about T.

Let us first set up the notation that we shall use for the rest of this chapter. V will denote a finite-dimensional vector space of dimension n over a field F. $\mathscr{E}(V)$ will denote the set of endomorphisms of V. Thus, $\mathscr{E}(V) = \mathrm{Hom}_F(V, V)$. We have noted in previous chapters that $\mathscr{E}(V)$ is an algebra over F. If $\underline{\alpha}$ is any basis of V, then $\Gamma(\underline{\alpha}, \underline{\alpha})(\cdot): \mathscr{E}(V) \to M_{n \times n}(F)$ is an F-algebra isomorphism. Let $T \in \mathscr{E}(V)$. Our goal in this chapter is to study various matrix representations of T.

Recall from Chapter II that T determines an algebra homomorphism $\varphi: F[X] \to \mathscr{E}(V)$ given by $\varphi(f(X)) = f(T)$. Here if $f(X) = c_n X^n + \cdots + c_1 X + c_0 \in F[X]$, then $f(T) = c_n T^n + \cdots + c_1 T + c_0 I_V$. In particular, $\varphi(f(X) + g(X)) = f(T) + g(T)$, $\varphi(cf(X)) = cf(T)$, and $\varphi(f(X)g(X)) = f(T)g(T)$. Note also that $\varphi(1) = I_V$, that is, φ takes the multiplicative identity element 1 in the algebra F[X] to the multiplicative identity I_V in $\mathscr{E}(V)$. Another important point to note here is that any identity $f(X) = g(X)$ in F[X] is mapped by φ into the corresponding identity $f(T) = g(T)$ in $\mathscr{E}(V)$.

We shall constantly use this map φ to study the behavior of T on V. In order to facilitate such a study, we need to know some basic algebraic facts about the polynomial algebra F[X]. We present these facts in the rest of this section.

The algebra F[X] is often called the *ring of polynomials* in the indeterminate X over F. We have seen that F[X] is an infinite-dimensional vector space over F with basis the monomials $\{1 = X^0, X, X^2, \ldots\}$. In particular, two polynomials $f(X) = a_n X^n + \cdots + a_1 X + a_0$ and $g(X) = b_m X^m + \cdots + b_1 X + b_0$ in F[X] with $a_n \neq 0 \neq b_m$ are equal if and only if $n = m$ and $a_n = b_n$, $a_{n-1} = b_{n-1}, \ldots, a_0 = b_0$. If $f(X)$ is a nonzero polynomial in F[X], then $f(X)$ can be written uniquely in the following form: $f(X) = a_n X^n + \cdots + a_1 X + a_0$ with $n \geqslant 0, a_n, a_{n-1}, \ldots, a_0 \in F$, and $a_n \neq 0$. The integer n here is called the *degree of f*. We shall use the notation $\partial(f)$ to indicate the degree of f. Thus, $\partial(\cdot)$ is a function from $F[X] - \{0\}$ to $\mathbb{N} \cup \{0\}$. Notice that we do not give a degree to the zero polynomial 0. The degree function $\partial(\cdot)$ has all the same familiar properties that the reader is acquainted with from studying polynomials with coefficients in $\mathbb{R}$. Thus, we have the following facts:

1.1: (a) $\partial(f) \geqslant 0$ for all $f \in F[X] - \{0\}$.

    (b) $\partial(f) = 0$ if and only if $f = a_0 \in F - \{0\}$.

    (c) $\partial(fg) = \partial(f) + \partial(g)$ for nonzero f, $g \in F[X]$.

    (d) $\partial(f + g) \leqslant \max\{\partial(f), \partial(g)\}$ for nonzero f, g and f + g.

We also have the division algorithm:

1.2: Let $f(X), g(X) \in F[X]$ with $g \neq 0$. Then there exist unique polynomials $h(X)$ and $r(X)$ in F[X] such that $f(X) = h(X)g(X) + r(X)$, where $r(X) = 0$ or $\partial(r) < \partial(g)$.

The proof of 1.2 is nothing more than the long division process you learned in grade school. We leave it as an exercise at the end of this section.

Let f, $g \in F[X]$. We say f divides g if there exists an $h \in F[X]$ such that $fh = g$. If f divides g, we shall write $f \mid g$. We say f and g are associates if $f \mid g$ and $g \mid f$. It follows easily from 1.1(c) that f and g are associates if and only if $f = cg$ for some nonzero constant $c \in F$. For example, $2X + 2$ and $X + 1$ are associates in $\mathbb{Q}[X]$, whereas $X + 1$ and X are not associates.

The notion of a greatest common divisor of a set of polynomials $f_1, \ldots, f_n$ is the same as in ordinary arithmetic. We say $d(X) \in F[X]$ is a greatest common divisor of $f_1, \ldots, f_n$ if d satisfies the following two properties:

1.3: (a) $d \mid f_i$ for $i = 1, \ldots, n$.

    (b) If $g \mid f_i$ for $i = 1, \ldots, n$, then $g \mid d$.

If d is a greatest common divisor of $f_1, \ldots, f_n$, then clearly cd is also a greatest common divisor of $f_1, \ldots, f_n$ for any nonzero constant $c \in F$. On the other hand, if e is a second greatest common divisor of $f_1, \ldots, f_n$, then 1.3(b) implies $d \mid e$ and $e \mid d$. Hence, $e = cd$ for some $c \in F - \{0\}$. Thus, a greatest common divisor of $f_1, \ldots, f_n$ is unique up to associates in $F[X]$. One of the most important properties of the algebra $F[X]$ is the fact that any finite set of polynomials has a greatest common divisor.

Lemma 1.4: Let $f_1, \ldots, f_n \in F[X]$. Then $f_1, \ldots, f_n$ have a greatest common divisor d. Furthermore, $d = a_1 f_1 + \cdots + a_n f_n$ for some $a_1, \ldots, a_n \in F[X]$.

Proof: We shall sketch a proof of 1.4 and leave the details to the reader. We first note that a greatest common divisor of $f_1, \ldots, f_n$ is nothing but a greatest common divisor of f_n and l where l is a greatest common divisor of $f_1, \ldots, f_{n-1}$. Hence by induction, it suffices to prove the lemma for two polynomials f and g. We can also assume $g \neq 0$. We now apply the division algorithm over and over until the remainder becomes zero. More specifically, we have

1.5:

$$
\begin{aligned}
f &= a_1 g + f_2 &&\text{with} \quad \partial(f_2) < \partial(g) \\
g &= a_2 f_2 + f_3 &&\text{with} \quad \partial(f_3) < \partial(f_2) \\
&\;\;\vdots &&\qquad\qquad \vdots \\
f_{r-2} &= a_{r-1} f_{r-1} + f_r &&\text{with} \quad \partial(f_r) < \partial(f_{r-1})
\end{aligned}
$$

and

$$
f_{r-1} = a_r f_r
$$

An easy argument shows f_r is a greatest common divisor of f and g. Back substituting in 1.5 gives $f_r = Af + Bg$ for some $A, B \in F[X]$. $\square$

We shall let g.c.d.$(f_1, \ldots, f_n)$ denote a greatest common divisor of $f_1, \ldots, f_n$. Although a g.c.d.$(f_1, \ldots, f_n)$ is unique only up to some nonzero constant in F, this will cause no confusion in the sequel. Note that g.c.d.$(f_1, \ldots, f_n) = 1$ whenever some f_i is a nonzero constant in F.

In the sequel, we shall also need the dual concept of a least common multiple l.c.m.$(f_1, \ldots, f_n)$ of $f_1, \ldots, f_n$. We shall discuss this notion in the exercises at the end of this section.

A polynomial $f(X) \in F[X]$ is said to be constant if f or $\partial(f)$ is zero. Thus, f is a constant if and only if $f \in F$. We say a polynomial $f(X)$ in $F[X]$ is *irreducible* (over F) if f is not constant, and whenever $f = gh$ with $g, h \in F[X]$, then one of g or h is a constant. This notion of irreducibility definitely depends on the field F.

Example 1.6: $f(X) = X^2 + 1$ is irreducible over $\mathbb{R}$, but factors nontrivially, $f(X) = (X - i)(X + i)$, over $\mathbb{C}$. $\square$

Note that linear polynomials, that is, $aX + b$ with $a \neq 0$, are always irreducible in $F[X]$.

Now suppose $f(X)$ is not constant. If f is not irreducible, then $\partial(f) > 0$ and $f = gh$ with $0 < \partial(g), \partial(h)$ and $\partial(g), \partial(h) < \partial(f)$. If g and h are not irreducible, they in turn factor. Since the degree of the resulting factors keeps dropping, we eventually arrive at a factorization of f in the form $f = g_1, \ldots, g_n$, with each g_i irreducible. Thus, we have proved the first part of the following theorem:

Theorem 1.7: Every nonconstant polynomial $f(X) \in F[X]$ can be factored in an essentially unique way as a product of irreducible polynomials.

A factorization $f = g_1 \cdots g_n$ with each g_i irreducible is said to be essentially unique if given any other factorization $f = g'_1 \cdots g'_m$ with each g'_i irreducible, then $n = m$, and there exists a permutation $\sigma \in S_n$ such that g_i and $g'_{\sigma(i)}$ are associates for all $i = 1, \ldots, n$. Clearly, a given factorization can be only essentially unique. For example, $(X + 1)(X + 2) = (2X + 2)(\frac{1}{2}X + 1)$ are two factorizations of $X^2 + 3X + 2$ in $\mathbb{Q}[X]$.

Proof of 1.7: We have already shown that $f = g_1 \cdots g_n$ for some irreducible polynomials $g_1, \ldots, g_n \in F[X]$. We must argue that this factorization is essentially unique. We give a brief sketch of how to do this. The essential point here is to make the following observation about irreducible polynomials:

1.8: If $g(X)$ is irreducible, and $g \,|\, hk$, then $g \,|\, h$ or $g \,|\, k$.

1.8 follows easily from Lemma 1.4 and the observation that if g is irreducible, then $\text{g.c.d.}(g, p) = 1$ or g for any $p \in F[X]$. Once we have 1.8, the essential uniqueness of $f = g_1 \cdots g_n$ is clear. $\square$

Theorem 1.7 implies that up to associates any nonconstant polynomial $f(X)$ can be written (essentially) uniquely in the following way:

1.9: $f(X) = g_1(X)^{e_1} \cdots g_m(X)^{e_m}$

In Equation 1.9, $g_1, \ldots, g_m$ are irreducible, $e_1, \ldots, e_m$ are positive integers, and g_i and g_j are not associates whenever $i \neq j$. Sometimes it is convenient to allow an exponent e_i in 1.9 to be zero. For example, if $f_1, \ldots, f_n$ are nonconstant polynomials, then there exists a set of irreducible polynomials $\{g_1, \ldots, g_m\}$ and nonnegative integers e_{ij} such that g_i and g_j are not associates for $i \neq j$ and

1.10: $f_i = g_1^{e_{i1}} \cdots g_m^{e_{im}}, \qquad i = 1, \ldots, n$

If $f_1, \ldots, f_n$ are factored as in equation 1.10, then a g.c.d. $(f_1, \ldots, f_n) = g_1^{z_1} \cdots g_m^{z_m}$ where $z_j = \min\{e_{1j}, \ldots, e_{nj}\}$ for all $j = 1, \ldots, m$. We say that a set of polynomials $f_1, \ldots, f_n$ are *relatively prime* if a g.c.d. $(f_1, \ldots, f_n) = 1$. Clearly, if some f_i is a nonzero constant, then $f_1, \ldots, f_n$ are relatively prime. If each f_i is a nonconstant polynomial, then 1.10 implies $f_1, \ldots, f_n$ are relatively prime if and only if the f_i have no common irreducible factor. Note that Lemma 1.4 implies the following result:

1.11: $f_1, \ldots, f_n$ are relatively prime if and only if $a_1 f_1 + \cdots + a_n f_n = 1$ for some $a_i \in F[X]$.

We shall use this remark frequently throughout the rest of this chapter.

Before closing this section, we need to say a few words about algebraically closed fields.

Definition 1.12: A field F is algebraically closed if the only irreducible polynomials in $F[X]$ are linear polynomials.

Example 1.13: Since $X^2 + 1$ is irreducible in $\mathbb{Q}[X]$ and $\mathbb{R}[X]$, we see neither $\mathbb{Q}$ nor $\mathbb{R}$ is algebraically closed. On the other hand, the field of complex numbers $\mathbb{C}$ is algebraically closed. This fact is often called the *fundamental theorem of algebra* (first proven by Gauss). It is not trivial.   □

If F is algebraically closed and $f(X)$ is a nonconstant polynomial in $F[X]$, then Theorem 1.7 implies $f(X)$ can be written uniquely in the following form:

1.14:
$$f(X) = c \prod_{i=1}^{r} (X - c_i)^{n_i}$$

In 1.14, $c \in F - \{0\}$, $c_1, \ldots, c_r$ are distinct constants in F, and $n_1 + \cdots + n_r = \partial(f)$. The constants $c_1, \ldots, c_r$ are called the *roots* of f and will henceforth be denoted $R(f)$.

We shall need the following theorem from abstract algebra:

Theorem 1.15: Let F be any field. Then there exists a field $\bar{F}$ containing F with the following properties:

(a) $\bar{F}$ is algebraically closed.
(b) If K is any algebraically closed field containing F, then there exists an F-algebra homomorphism $\psi: \bar{F} \to K$ such that $\psi(1) = 1$.

An easy exercise shows ψ must be injective in (b). It follows that any two fields containing F and satisfying (a) and (b) are isomorphic as F-algebras. For this reason, a field $\bar{F}$ satisfying (a) and (b) is called the algebraic closure of F (any other algebraic closure of F being isomorphic to $\bar{F}$). The proof of Theorem 1.15 is beyond the level of this book. The interested reader can consult [6; Thm. 32, p. 106].

If F is any field, we shall let $\bar{F}$ denote the algebraic closure of F. Our interest in $\bar{F}$ comes from equation 1.14. If $f(X)$ is a nonconstant polynomial in $F[X]$, then f may not factor into linear polynomials. Since $F \subseteq \bar{F}$, $F[X] \subseteq \bar{F}[X]$. Since $\bar{F}$ is algebraically closed, f has a unique factorization as in 1.14. If $f(X) = c \prod_{i=1}^{r}(X - c_i)^{n_i} \in \bar{F}[X]$, we shall again call $R(f) = \{c_1, \ldots, c_r\}$ the roots of f. Thus, the roots of a polynomial $f(X)$ in $F[X]$ are those elements $c \in \bar{F}$, the algebraic closure of F, such that $f(c) = 0$.

Example 1.16: Suppose $f(X) = (X^2 + 1)(X^2 + 2) \in \mathbb{R}[X]$. Since $X^2 + 1$ is irreducible in $\mathbb{R}[X]$, f has no factorization as in 1.14 over $\mathbb{R}$. It is not hard to see that $\mathbb{C}$ is the algebraic closure of $\mathbb{R}$. In $\mathbb{C}[X]$, we have $f(X) = (X + i)(X - i)(X - \sqrt{2}i)(X + \sqrt{2}i)$. The roots of f are then given by $R(f) = \{i, -i, \sqrt{2}i, -\sqrt{2}i\}$. $\square$

The fact that any field F has an algebraic closure $\bar{F}$ is often used when studying linear transformations.

Suppose $T \in \text{Hom}_F(V, V)$. Then we have the F-algebra homomorphism $\varphi \colon F[X] \to \mathscr{E}(V)$ given by $\varphi(f) = f(T)$. Suppose we wish to study the action of $f(T)$ on V for some interesting $f(X) \in F[X]$. One way to do this is to extend the scalars to $\bar{F}$ and study the natural extension of $f(T)$ to $V^{\bar{F}}$. Often information obtained about $f(T)$ on $V^{\bar{F}}$ gives us useful information about $f(T)$ on V. Recall that extending scalars to $\bar{F}$ means we form the $\bar{F}$-vector space $V^{\bar{F}} = V \otimes_F \bar{F}$. By the natural extension of T to $V^{\bar{F}}$, we mean the linear map $T \otimes_F I_{\bar{F}} \in \text{Hom}_{\bar{F}}(V^{\bar{F}}, V^{\bar{F}})$. The vector space V is imbedded in $V^{\bar{F}}$ as the F-subspace $V \otimes_F 1$. If we identify a vector $\alpha \in V$ with its image $\alpha \otimes_F 1$ in $V \otimes_F 1$, then we have $(T \otimes_F I_{\bar{F}})(\alpha) = (T \otimes_F I_{\bar{F}})(\alpha \otimes_F 1) = T(\alpha) \otimes_F 1 = T(\alpha)$. Thus, $T \otimes_F I_{\bar{F}}$ is just T on V. In the sequel, we shall identify $\alpha \otimes_F 1$ with α and set $T \otimes_F I_{\bar{F}} = T^{\bar{F}}$. When we do this, a typical vector in $V^{\bar{F}}$ is a finite sum of the form $\sum_{i=1}^{n} z_i \alpha_i$ with $z_1, \ldots, z_n \in \bar{F}$, and $\alpha_1, \ldots, \alpha_n \in V$. The action of $T^{\bar{F}}$ on such a vector is given by $T^{\bar{F}}(\sum_{i=1}^{n} z_i \alpha_i) = \sum_{i=1}^{n} z_i T(\alpha_i)$. In particular, the reader can easily check that $(f(T))^{\bar{F}} = f(T^{\bar{F}})$ for any $f(X) \in F[X]$. Now since $\bar{F}$ is algebraically closed, f can be factored in $\bar{F}[X]$ as in equation 1.14. Thus, $f(T)^{\bar{F}}$ has a particularly simple form: $f(T)^{\bar{F}} = c \prod_{i=1}^{r} (T^{\bar{F}} - c_i)^{n_i}$. We shall see how these ideas can be usefully employed in Section 5 when discussing the real Jordan canonical form of T.

EXERCISES FOR SECTION 1

(1) Prove (c) and (d) in equation 1.1.

(2) Give a proof of 1.2. [*Hint*: Proceed by induction on $\partial(f)$.]

(3) Let $f_1, \ldots, f_n \in F[X]$. Show that g.c.d. $(f_1, \ldots, f_n) = \text{g.c.d.}(\text{g.c.d.}(f_1, \ldots, f_{n-1}), f_n)$. We used this idea in Lemma 1.4.

(4) Complete the proof of Lemma 1.4 by showing that $f_r = $ g.c.d.(f, g).

(5) Give a proof of 1.8 and then complete the details in Theorem 1.7.

(6) Determine what all irreducible polynomials of degree less than or equal to three look like in $\mathbb{F}_2[X]$.

(7) Show that $f(X)g(X) = 0$ in $F[X]$ if and only if $f = 0$ or $g = 0$.

(8) Let K and F be fields and suppose $K \supseteq F$. Show that any F-algebra homomorphism $\psi: F \to K$ with $\psi(1) = 1$ must be injective. If $\psi(1) \neq 1$ is this statement true?

(9) Find $d = $ g.c.d.$(X^3 - 6X^2 + X + 4, \ X^5 - 6X + 1)$ in $\mathbb{Q}[X]$. Exhibit two polynomials A and B such that $d = A(X^3 - 6X^2 + X + 4) + B(X^5 - 6X + 1)$.

(10) Suppose $F \subseteq K$ are fields. Let $g_1, \ldots, g_n \in F[X]$. Show that if $g_1, \ldots, g_n$ are relatively prime in $F[X]$, then $g_1, \ldots, g_n$ are relatively prime in $K[X]$. Is the converse true here?

(11) In Section 2, we shall need a least common multiple, l.c.m.$(f_1, \ldots, f_n)$, of a set of polynomials $f_1, \ldots, f_n \in F[X]$. We define a l.c.m. $(f_1, \ldots, f_n)$ to be a polynomial (unique up to associates) $e(X)$ such that (a) $f_i | e$ for all $i = 1, \ldots, n$, and (b) if $f_i | g$ for $i = 1, \ldots, n$, then $e | g$. Prove that any set of polynomials $f_1, \ldots, f_n \in F[X]$ has a least common multiple.

(12) In Exercise 11, suppose each f_i is factored as in equation 1.10. For each $j = 1, \ldots, m$, set $s_j = \max\{e_{1j}, \ldots, e_{nj}\}$. Set $e = g_1^{s_1} \cdots g_m^{s_m}$. Show e is a least common multiple of $f_1, \ldots, f_n$.

(13) Prove that the product of a least common multiple and a greatest common divisor of f and g is the product fg.

(14) If $f(X) \in F[X]$ has degree n, show $|R(f)| \leqslant n$.

(15) Prove that every polynomial $f(X) \in \mathbb{R}[X]$ factors into linear and quadratic polynomials.

(16) Let $f(X) \in \mathbb{R}[X]$, and let $f'(X)$ denote the derivative of f. Show that f and f' are relatively prime in $\mathbb{R}[X]$ if and only if f has no multiple roots.

(17) Prove that $f(X) = 1 + X + X^3 + X^4$ is not irreducible over any field.

(18) Show that $p(X) = X^4 + 2X + 2 \in \mathbb{Q}[X]$ is irreducible.

(19) Let $f(X)$ be a nonconstant polynomial in $F[X]$. Set $(f) = \{gf \mid g(X) \in F[X]\}$. Show that the vector space $F[X]/(f)$ is a finite-dimensional algebra over F when multiplication is defined as follows: $(h + (f))(g + (f)) = hg + (f)$.

(20) Suppose in Exercise 19 that $f(X)$ is irreducible. Prove that $F[X]/(f)$ is a field.

2. MINIMAL AND CHARACTERISTIC POLYNOMIALS

Let us start this section with a very general result.

Theorem 2.1: Let A be an algebra over F, and assume $\dim_F(A) = m < \infty$. Then for every $x \in A$, there exists an $f(X) \in F[X]$ such that $1 \leq \partial(f) \leq m$ and $f(x) = 0$.

Proof: Consider the set $\Delta = \{1, x, \ldots, x^m\} \subseteq A$. Since $\dim_F A = m$, Δ is linearly dependent over F. Hence, there exist constants $c_0, \ldots, c_m \in F$, not all zero, with $c_0(1) + c_1 x + \cdots + c_m x^m = 0$.   Set $f(X) = c_0 + c_1 X + \cdots + c_m X^m$.   Clearly, $1 \leq \partial(f) \leq m$, and $f(x) = 0$. $\square$

   Although the proof of Theorem 2.1 is trivial, the theorem has many interesting ramifications for a $T \in \mathscr{E}(V)$. Since $\dim_F(V) = n < \infty$, $\dim_F \mathscr{E}(V) = n^2$. Thus, we can apply Theorem 2.1 to the algebra $\mathscr{E}(V)$. We conclude that there exists a nonconstant polynomial $f(X) \in F[X]$ such that $\partial(f) \leq n^2$ and $f(T) = 0$. Another way to say this is that $\ker \varphi = \{f(X) \in F[X] \mid f(T) = 0\}$, the kernel of $\varphi : F[X] \to \mathscr{E}(V)$, is not zero. Among all such nonconstant $f \in \ker \varphi$, we can select a polynomial $g(X)$ of smallest degree. Then $1 \leq \partial(g) \leq n^2$, $g(T) = 0$, and $\partial(g) \leq \partial(f)$ for any nonconstant polynomial $f \in \ker \varphi$. Suppose $\partial(g) = m$. Then $g(X) = c_m X^m + c_{m-1} X^{m-1} \cdots + c_0$ with $c_m \neq 0$.   Then $c_m^{-1} g = X^m + (c_{m-1}/c_m)X^{m-1} + \cdots + (c_0/c_m)$ is a monic polynomial of lowest degree in $\ker \varphi$. A polynomial $a_r X^r + \cdots + a_0 \in F[X]$ is said to be *monic* if $r \geq 1$ and $a_r = 1$.

   Now we claim that $c_m^{-1} g(X)$ is unique. To see this, suppose $f(X)$ is a second, nonconstant, monic polynomial in $\ker \varphi$ of smallest possible degree. Set $c_m^{-1} g(X) = h(X)$.   Applying   the   division   algorithm   1.2,   we   have $f(X) = A(X)h(X) + r(X)$ with $r = 0$ or $\partial(r) < \partial(h)$. If we apply φ to this equation, we get $0 = f(T) = A(T)h(T) + r(T) = r(T)$. If $r \neq 0$, then $r(X)$ is a nonconstant polynomial in $\ker \varphi$ of degree smaller than $\partial(g)$. This is impossible. We conclude that $r = 0$ and $f = Ah$. Now $\partial(f) = \partial(h)$ by definition. Thus, 1.1(c) implies $\partial(A) = 0$. So, A is a nonzero constant. Since f and h are both monic, we conclude that $A = 1$ and $f = h$. We have now shown that there is a unique, monic polynomial of smallest positive degree in $\ker \varphi$. This polynomial gets a special name, which we introduce below:

Definition 2.2: The unique, monic polynomial $f(X) \in F[X]$ of smallest positive degree such that $f(T) = 0$ is called the *minimal polynomial of T*. We shall henceforth denote the minimal polynomial of T by $m_T(X)$ or just m_T.

   Our discussion before Definition 2.2 implies that $m_T(X)$ exists and indeed is unique. We also know the following facts about $m_T(X)$:

2.3: (a) $1 \leq \partial(m_T) \leq n^2$.
   (b) If $f \in F[X]$ and $f(T) = 0$, then $m_T \mid f$.

(a) follows from Theorem 2.1. (b) is the same argument that was used above for the uniqueness of m_T.

We can define the minimal polynomial of an $n \times n$ matrix in a similar fashion to 2.2.

Definition 2.4: Let $A \in M_{n \times n}(F)$. The minimal polynomial $m_A(X)$ of A is the monic polynomial $f \in F[X]$ of smallest positive degree such that $f(A) = 0$.

Suppose $\underline{\alpha}$ is a basis of V, and $\Gamma(\underline{\alpha}, \underline{\alpha})(T) = A$. Since $\Gamma(\underline{\alpha}, \underline{\alpha})(\cdot): \mathscr{E}(V) \to M_{n \times n}(F)$ is an isomorphism of F-algebras, clearly $m_T(X) = m_A(X)$. Thus, we can switch back and forth freely between V [T or $m_T(X)$] and $M_{n \times n}(F)$ [A or $m_A(X)$]. The minimal polynomial is the same. Note that one consequence of the equality $m_A(X) = m_T(X)$ is that similar matrices have the same minimal polynomial. For if A and B are similar, then Theorem 3.28 of Chapter I implies that A and B are matrix representations of the same linear transformation T. Thus, $m_A(X) = m_T(X) = m_B(X)$. Let us examine a few examples.

Example 2.5: If $T = 0$, clearly $m_T(X) = X$. If $T = I_V$, then $m_T(X) = X - 1$. $\square$

Example 2.6: Let $V = \mathbb{R}^2$, and let $\underline{\delta} = \{\delta_1 = (1, 0), \delta_2 = (0, 1)\}$ be the canonical basis of $\mathbb{R}^2$. Let T be defined by $T(\delta_1) = \delta_2$ and $T(\delta_2) = -\delta_1$. Thus, T is the familiar rotation of $\mathbb{R}^2$ through a 90° angle:

$$\Gamma(\underline{\delta}, \underline{\delta})(T) = \begin{pmatrix} 0 & -1 \\ 1 & 0 \end{pmatrix} = A$$

Clearly, $A^2 + I = 0$. Thus, $X^2 + 1 \in \ker \varphi$. Since T takes no nonzero α into a multiple of itself, no linear polynomial lies in $\ker \varphi$. Thus, $m_T(X) = m_A(X) = X^2 + 1$. $\square$

Suppose we consider the same example over $\mathbb{C}$ instead of $\mathbb{R}$.

Example 2.7: Let $V = \mathbb{C}^2$, $\underline{\delta} = \{\delta_1 = (1, 0), \delta_2 = (0, 1)\}$, and suppose $T(\delta_1) = \delta_2$, $T(\delta_2) = -\delta_1$. Then again we have

$$\Gamma(\underline{\delta}, \underline{\delta})(T) = A = \begin{pmatrix} 0 & -1 \\ 1 & 0 \end{pmatrix}$$

and hence $X^2 + 1 \in \ker \varphi$. But now $T((1, -i)) = i(1, -i)$. So, the same reasoning used in Example 2.6 no longer applies. If $\partial(m_T) = 1$, then $T = zI_V$ for some $z \in \mathbb{C} - \{0\}$. But then $\delta_2 = T(\delta_1) = z\delta_1$ which is impossible. Hence $m_T(X) = X^2 + 1$ as before. $\square$

Note in Example 2.6 that the minimal polynomial $m_T(X) = X^2 + 1$ is irreducible in $\mathbb{R}[X]$. When we extend the scalars in this example to $\mathbb{C}$, we get

Example 2.7. The minimal polynomial stays the same but is no longer irreducible. $X^2 + 1 = (X - i)(X + i) \in \mathbb{C}[X]$. These examples suggest the following question: How does the minimal polynomial change under an extension of the scalars? The answer is that the minimal polynomial stays the same under all extensions of the scalars. However, as the field F gets larger, we may be able to factor $m_T(X)$ more easily, as the two examples above show.

In order to argue $m_T(X)$ remains invariant under extensions of the base field, we need to examine the kernel of φ more closely. From 2.3(b) we know that any polynomial in ker φ is a multiple of $m_T(X)$.

Lemma 2.8: Suppose g(X) is a polynomial of positive degree m in F[X]. Let $(g) = \{A(X)g(X) | A \in F[X]\}$. Then (g) is a subspace of F[X], and $\dim_F\{F[X]/(g)\} = m$.

Proof: The fact that (g) is a subspace of F[X] is obvious. We want to show $\dim_F\{F[X]/(g)\} = \partial(g)$. Let $x^i = X^i + (g)$, $i = 0, 1, \ldots, m - 1$. We shall show that $\Delta = \{1, x, \ldots, x^{m-1}\}$ is a basis of F[X]/(g). Then $\dim_F\{F[X]/(g)\} = |\Delta| = m$ and our proof will be complete.

A typical vector in F[X]/(g) is a coset of the form $f(X) + (g)$ with $f(X) \in F[X]$. Using the division algorithm 1.2, we can write $f(X) = h(X)g(X) + r(X)$ with $r = 0$ or $\partial(r) < m$. Then $f(X) + (g) = (hg + r) + (g) = r + (g)$. Suppose $r(X) = c_s X^s + \cdots + c_0$ with $s < m$. Then $r + (g) = c_s(X^s + (g)) + \cdots + c_0(1 + (g)) = c_s x^s + \cdots + c_0$. Thus, Δ spans F[X]/(g) as a vector space over F.

Suppose $\sum_{i=0}^{m-1} c_i x^i = 0$ in F[X]/(g). Then back in F[X], we have $\sum_{i=0}^{m-1} c_i X^i = A(X)g(X)$ for some polynomial A. Since $\partial(g) = m > \partial(\sum_{i=0}^{m-1} c_i X^i)$, it follows from 1.1(c) that $A = 0$. Thus $c_0 = \cdots = c_{m-1} = 0$. This proves the set Δ is linearly independent over F. Thus, Δ is a basis of F[X]/(g). □

Corollary 2.9: $0 \to (m_T) \hookrightarrow F[X] \to \text{Im } \varphi \to 0$ is an exact sequence of vector spaces over F. In particular, $\dim_F\{\text{Im } \varphi\} = \partial(m_T)$.

Proof: The curved arrow $\hookrightarrow$ between (m_T) and F[X] indicates that the map between (m_T) and F[X] is the inclusion map. We have seen that ker $\varphi = (m_T)$ by 2.3(b). The result now follows from Lemma 2.8 and Corollary 5.17 of Chapter I. □

The image of φ is just the linear span of the set $\{T^i | i = 0, 1, \ldots\}$. Corollary 2.9 says this is a subspace of $\mathscr{E}(V)$ of dimension $\partial(m_T)$.

Now suppose K is any field containing F. We can extend our scalars to K by forming $V^K = V \otimes_F K$ Let $T^K = T \otimes_F I_K$. Then we have two minimal polynomials $m_T \in F[X]$ and $m_{T^K} \in K[X]$. The statement that the minimal polynomial of T remains invariant under an extension of the base field means more precisely that $m_T(X) = m_{T^K}(X)$ in K[X].

Theorem 2.10: Let K be a field containing F, and let $T \in \text{Hom}_F(V, V)$. Then $m_T(X) = m_{T^K}(X)$ in K[X].

Proof: Since $F \subseteq K$, $F[X] \subseteq K[X]$. In particular, $m_T(X) \in K[X]$. Suppose $m_T(X) = X^r + a_{r-1}X^{r-1} + \cdots + a_0$ with $a_i \in F$. Then using Theorem 2.6 of Chapter II, we have

$$
\begin{aligned}
m_T(T^K) = m_T(T \otimes_F I_K) &= (T \otimes_F I_K)^r + a_{r-1}(T \otimes_F I_K)^{r-1} + \cdots + a_0 \\
&= (T^r \otimes_F I_K) + (a_{r-1}T^{r-1} \otimes_F I_K) + \cdots + a_0 \\
&= m_T(T) \otimes_F I_K \\
&= 0
\end{aligned}
$$

Thus, 2.3(b) implies $m_{T^K}(X) \mid m_T(X)$ in $K[X]$. Since both of these polynomials are monic, our theorem will follow if we can show that $\partial(m_T) = \partial(m_{T^K})$.

We know from Corollary 2.9 that

2.11: $\qquad\qquad 0 \to (m_T) \to F[X] \to \operatorname{Im} \varphi \to 0$

is an exact sequence of vector spaces over F. Consequently, Theorem 2.13 of Chapter II implies

2.12: $\qquad 0 \to (m_T) \otimes_F K \to F[X] \otimes_F K \to \operatorname{Im} \varphi \otimes_F K \to 0$

is an exact sequence of vector spaces over K. The reader can easily check that $F[X] \otimes_F K \cong K[X]$ (as K-algebras) under the map sending $\sum f_j(X) \otimes_F k_j$ to $\sum k_j f_j(X)$. Under this isomorphism $(m_T) \otimes_F K$ is sent to all multiples of m_T in $K[X]$. Let us call this image $(m_T)K[X]$. Then the short exact sequence in 2.12 becomes

2.13: $\qquad\qquad 0 \to (m_T)K[X] \to K[X] \to \operatorname{Im} \varphi \otimes_F K \to 0$

In particular, Lemma 2.8 implies that $\dim_K(\operatorname{Im} \varphi \otimes_F K) = \partial(m_T)$. On the other hand, using theorems from Chapter II, we have $\operatorname{Im} \varphi \otimes_F K \subseteq \mathscr{E}(V) \otimes_F K \cong \operatorname{Hom}_K(V^K, V^K)$. Under this isomorphism, $\operatorname{Im} \varphi \otimes_F K$ is clearly just the K-linear span of $\{T^i \otimes_F I_K \mid i = 0, \dots\}$. Thus, Corollary 2.9 implies $\dim_K(\operatorname{Im} \varphi \otimes_F K) = \dim_K\{L(T^i \otimes_F I_K \mid i = 0, 1, \dots)\} = \partial(m_{T^K})$. Therefore, $\partial(m_T) = \partial(m_{T^K})$ and our proof is complete. $\square$

We think of T^K as the natural extension of T to V^K. Theorem 2.10 says that the minimal polynomial of T remains the same under all extensions of the scalars. Switching to matrices, we have the following matrix version of Theorem 2.10:

Corollary 2.14: Let $A \in M_{n \times n}(F)$, and let K be any field containing F. Then the minimal polynomial, $m_{A^{(F)}}(X) \in F[X]$, of A viewed as an $n \times n$ matrix with coefficients in F is the same as the minimal polynomial, $m_{A^{(K)}}(X) \in K[X]$, of A viewed as an $n \times n$ matrix with coefficients in K.

Proof: Suppose $A = (a_{ij})$. Let $\underline{\delta} = \{\delta_1, \ldots, \delta_n\}$ be the canonical basis of F^n. Then $\Gamma(\underline{\delta}, \underline{\delta})(T) = A$ where T is the endomorphism of F^n given by $T(\delta_j) = \sum_{i=1}^{n} a_{ij}\delta_i$. We have seen in Chapter II that $\underline{\delta}$ is a K-basis of $F^n \otimes_F K \cong K^n$ and $\Gamma(\underline{\delta}, \underline{\delta})(T^K) = A$. Thus, from Theorem 2.10, we have $m_{A^{(F)}}(X) = m_T(X) = m_{T^K}(X) = m_{A^{(K)}}(X)$. $\square$

Let us summarize what we have proved about the minimal polynomial of T so far.

2.15: (a) $1 \leqslant \partial(m_T) \leqslant n^2$.

(b) $\ker \varphi = (m_T)$.

(c) The minimal polynomial remains the same under all extensions of the base field.

(d) $\partial(m_T) = \dim_F(F[T])$.

In (d), F[T] denotes the subspace of $\mathscr{E}(V)$ spanned by all powers of T.

There is one final property that $m_T(X)$ possesses that we shall discuss at this point.

Lemma 2.16: T is invertible if and only if $m_T(X)$ has a nonzero constant term.

Proof: Let $m_T(X) = X^r + a_{r-1}X^{r-1} + \cdots + a_0 \in F[X]$. Suppose $a_0 = 0$. Then $m_T(X) = Xg(X)$, where $g(X) = X^{r-1} + a_{r-1}X^{r-2} + \cdots + a_1$. Set $S = g(T)$. Then $S \in \mathscr{E}(V)$, and $0 = m_T(T) = TS$. Since $r = \partial(m_T) > \partial(g) = r - 1$, we see $S \neq 0$. Hence, there exists a vector $\alpha \in V$ such that $S(\alpha) \neq 0$. Then $T(S(\alpha)) = 0$ implies T is not invertible. In particular, if T is invertible, then $m_T(X)$ must have a nonzero constant term.

Conversely, suppose $a_0 \neq 0$. Then in $\mathscr{E}(V)$, we have

$$T\left[-\left(\frac{1}{a_0}\right)T^{r-1} - \left(\frac{a_{r-1}}{a_0}\right)T^{r-2} - \cdots - \left(\frac{a_1}{a_0}\right)\right] = I_V$$

Set

$$g(X) = -\left(\frac{1}{a_0}\right)X^{r-1} - \left(\frac{a_{r-1}}{a_0}\right)X^{r-2} - \cdots - \left(\frac{a_1}{a_0}\right)$$

Then $Tg(T) = g(T)T = I_V$, and T is invertible. $\square$

Note that the proof of Lemma 2.16 implies that if T is invertible, then $T^{-1} \in F[T]$. Thus, the inverse of T is a polynomial in T. We, of course, have a similar statement about matrices.

Corollary 2.17: If $T \in \mathscr{E}(V)$ is invertible, then $T^{-1} = f(T)$ for some $f(X) \in F[X]$. Similarly, if $A \in M_{n \times n}(F)$ is invertible, then $A^{-1} = g(A)$ for some polynomial $g(X) \in F[X]$. $\square$

We now turn our attention to the second polynomial of this section. We need to consider matrices with polynomial entries.

Definition 2.18: Let $M_{n \times n}(F[X])$ denote the set of all $n \times n$ matrices $A = (f_{ij})$ with entries $f_{ij} \in F[X]$.

Thus, an element $A \in M_{n \times n}(F[X])$ is a rectangular array of polynomials from $F[X]$, that is,

$$A = \begin{pmatrix} f_{11}(X), \ldots, f_{1n}(X) \\ \vdots \\ f_{n1}(X), \ldots, f_{nn}(X) \end{pmatrix}$$

for some choice of $f_{ij} \in F[X]$. Clearly, $M_{n \times n}(F) \subseteq M_{n \times n}(F[X])$. We can extend the algebra operations from $M_{n \times n}(F)$ to $M_{n \times n}(F[X])$ in the obvious way.

2.19: (a) $(f_{ij}) + (g_{ij}) = (f_{ij} + g_{ij})$ for $f_{ij}, g_{ij} \in F[X]$.
  (b) $c(f_{ij}) = (cf_{ij})$ for $c \in F[X]$.
  (c) $(f_{ij})(g_{ij}) = (h_{ij})$ where $h_{ij} = \sum_{k=1}^{n} f_{ik}g_{kj}$.

Thus, $M_{n \times n}(F[X])$ is an algebra over F containing $M_{n \times n}(F)$ as a subalgebra.
  We can extend the definition of the determinant to the algebra $M_{n \times n}(F[X])$ in the obvious way.

2.20:
$$\det(f_{ij}(X)) = \sum_{\sigma \in S_n} \text{sgn}(\sigma) f_{1\sigma(1)} \cdots f_{n\sigma(n)}$$

Clearly, $\det: M_{n \times n}(F[X]) \to F[X]$. Many theorems concerning the behavior of the determinant on $M_{n \times n}(F)$ pass over to $M_{n \times n}(F[X])$ with no change in proof. For example, $\det(AB) = \det(A)\det(B)$ for all $A, B \in M_{n \times n}(F[X])$. We also have the following important result:

2.21: $\text{adj}(A)A = A \, \text{adj}(A) = \det(A)I_n$ for $A \in M_{n \times n}(F[X])$

Recall that $\text{adj}(A)$ is the adjoint of A. It is defined as follows: If $A = (f_{ij}(X)) \in M_{n \times n}(F[X])$, let $M_{ij}(A)$ be the $(n-1) \times (n-1)$ matrix formed from A by deleting row i and column j from A. Thus, $M_{ij}(A) \in M_{(n-1) \times (n-1)}(F[X])$. The adjoint of A is the $n \times n$ matrix whose i, jth entry is given by $(\text{adj}(A))_{ij} = (-1)^{i+j}\det(M_{ji}(A))$. The proof of 2.21 is the same as for fields. We also have the Laplace expansion for the determinant.

2.22:
$$\det(A) = \sum_{i=1}^{n} a_{ij}(-1)^{i+j}\det(M_{ij}(A))$$

or

$$\det(A) = \sum_{j=1}^{n} a_{ij}(1)^{i+j} \det(M_{ij}(A))$$

The proof of 2.22 is the same as in the field case.

We can now introduce the characteristic polynomial of an $n \times n$ matrix with coefficients in F.

Definition 2.23: Let $A \in M_{n \times n}(F)$. Then $c_A(X) = \det(XI - A)$ is called the *characteristic polynomial of A*.

In 2.23, XI means XI_n. Thus, $XI - A \in M_{n \times n}(F[X])$, and $\det(XI - A) \in F[X]$. Expanding $\det(XI - A)$, we see that $c_A(X) = X^n + c_{n-1}X^{n-1} + \cdots + c_0$ with $c_{n-1} = -\sum_{i=1}^{n} a_{ii}$ and $c_0 = (-1)^n \det(A)$. In particular, $c_A(X)$ is a monic polynomial of degree n with coefficients in F.

Note that any matrix similar to A has the same characteristic polynomial. For suppose $B = PAP^{-1}$ in $M_{n \times n}(F)$. Then $c_B(X) = \det(XI - B) = \det(XI - PAP^{-1}) = \det(P(XI - A)P^{-1}) = \det(P)\det(XI - A)\det(P)^{-1} = \det(XI - A) = c_A(X)$. This remark allows us to extend the definition of the characteristic polynomial to any $T \in \mathscr{E}(V)$.

Definition 2.24: Let $T \in \mathscr{E}(V)$. Then $c_T(X) = \det(XI - A)$, where $A = \Gamma(\alpha, \alpha)(T)$ and α is any basis of V.

We had seen in Theorem 3.28 of Chapter I that any two matrix representations of T are similar. Hence, the definition of $c_T(X)$ does not depend on the basis α. We shall call $c_T(X)$ the characteristic polynomial of T.

Example 2.25: Let T be the linear transformation given in Example 2.6. Then

$$c_T(X) = c_A(X) = \det \begin{pmatrix} X & 1 \\ -1 & X \end{pmatrix} = X^2 + 1 = m_T(X) \quad \square$$

Note that the characteristic polynomial $c_T(X)$ is always a monic polynomial of degree $n = \dim V$. One of the most famous theorems in linear algebra is the following result, first formulated by Cayley:

Theorem 2.26 (Cayley–Hamilton): Let $A \in M_{n \times n}(F)$. Then $c_A(A) = 0$.

Proof: Suppose $c_A(X) = X^n + c_{n-1}X^{n-1} + \cdots + c_0$. Set $B = XI_n - A$. If we eliminate a row and column from B and then take the determinant, we get a polynomial in X of degree at most $n - 1$. In particular, the entries in adj(B) are all polynomials of degree at most $n - 1$. It follows that there exist unique matrices $B_0, \ldots, B_{n-1} \in M_{n \times n}(F)$ such that

2.27: $$\text{adj}(B) = B_0 + B_1 X + \cdots + B_{n-1}X^{n-1}$$

In equation 2.27, we should really write $B_i(X^i I_n)$ instead of $B_i X^i$, but the meaning of the symbol is clear.

Now from equation 2.21, we have

2.28: $\text{adj}(B)B = c_A(X)I_n = c_0 I_n + c_1 I_n X + \cdots + c_{n-1} I_n X^{n-1} + I_n X^n$

On the other hand, from equation 2.27 we have

2.29: $\text{adj}(B)B = (B_0 + \cdots + B_{n-1} X^{n-1})(X I_n - A)$

$= B_0 X + \cdots + B_{n-1} X^n - B_0 A - \cdots - B_{n-1} A X^{n-1}$

$= (-B_0 A) + (B_0 - B_1 A)X + \cdots$
$+ (B_{n-2} - B_{n-1} A)X^{n-1} + B_{n-1} X^n$

We now compare the results in 2.28 and 2.29. We have two polynomials in X with coefficients in $M_{n \times n}(F)$ that are equal. An easy argument shows that the matrices corresponding to the same powers of X in both equations must be equal. Thus, we get the following equations:

2.30: $$-B_0 A = c_0 I_n$$
$$B_0 - B_1 A = c_1 I_n$$
$$\vdots$$
$$B_{n-2} - B_{n-1} A = c_{n-1} I_n$$
$$B_{n-1} = I_n$$

If we now multiply each equation in 2.30 by successively higher powers of A, we get

2.31: $$-B_0 A = c_0 I_n$$
$$B_0 A - B_1 A^2 = c_1 A$$
$$\vdots$$
$$B_{n-2} A^{n-1} - B_{n-1} A^n = c_{n-1} A^{n-1}$$
$$B_{n-1} A^n = A^n$$

Adding the vertical columns in 2.31 gives us $0 = c_A(A)$. $\square$

Corollary 2.32: Let $T \in \mathscr{E}(V)$. Then $c_T(T) = 0$. In particular, $\partial(m_T) \leqslant \partial(c_T) = \dim V$.

Proof: Let α be a basis of V. Set $A = \Gamma(\alpha, \alpha)(T)$. Then $c_T(X) = c_A(X)$. By the Cayley–Hamilton theorem, $0 = c_A(A) = \Gamma(\alpha, \alpha)(c_T(T))$. Hence $c_T(T) = 0$. Thus, $m_T(X) | c_T(X)$ by 2.15(b). Since $\dim_F(V) = n$, $\partial(c_T(X)) = n$. Hence $\partial(m_T(X)) \leqslant n$. $\square$

We have noted in the proof of Corollary 2.32 that $m_T | c_T$. In general, these two polynomials are not equal, as the following trivial example shows:

Example 2.33: Let $A = aI_n \in M_{n \times n}(F)$. Clearly, $c_A(X) = (X - a)^n$ and $m_A(X) = X - a$. $\square$

A less trivial example is as follows:

Example 2.34: Let

$$A = \begin{bmatrix} -1 & 7 & 0 \\ 0 & 2 & 0 \\ 0 & 3 & -1 \end{bmatrix} \in M_{3 \times 3}(\mathbb{Q})$$

One can easily check that $c_A(X) = (X + 1)^2(X - 2)$ and $m_A(X) = (X + 1)(X - 2)$. $\square$

The examples above suggest that even when $m_T \neq c_T$, they always have the same irreducible factors in $F[X]$. This is indeed the case.

Theorem 2.35: Let $A \in M_{n \times n}(F)$. Then $c_A(X) | (m_A(X))^n$.

Proof: Suppose $m_A(X) = X^r + a_1 X^{r-1} + \cdots + a_r$. Note for this proof, we have changed our customary indexing on the coefficients of the minimal polynomial. Let us now form the following r matrices in $M_{n \times n}(F)$:

2.36: $$B_0 = I_n$$

and for $i = 1, \ldots, r - 1$

$$B_i = A^i + a_1 A^{i-1} + \cdots + a_{i-1}A + a_i I_n$$

Then clearly, $B_i - AB_{i-1} = a_i I$ for all $i = 1, \ldots, r - 1$. We also have $AB_{r-1} = A^r + a_1 A^{r-1} + \cdots + a_{r-1}A = m_A(A) - a_r I = -a_r I$.

Now set $C = B_0 X^{r-1} + B_1 X^{r-2} + \cdots + B_{r-2}X + B_{r-1}$. Then $C \in M_{n \times n}(F[X])$ and we have

2.37:
$$\begin{aligned}
(XI_n - A)C &= (XI_n - A)(B_0 X^{r-1} + B_1 X^{r-2} + \cdots + B_{r-2}X + B_{r-1}) \\
&= B_0 X^r + (B_1 - AB_0)X^{r-1} + \cdots \\
&\quad + (B_{r-1} - AB_{r-2})X - AB_{r-1} \\
&= X^r I_n + a_1 X^{r-1} I_n + \cdots + a_{r-1}XI_n + a_r I_n \\
&= m_A(X)I_n
\end{aligned}$$

If we now take the determinant of both sides of equation 2.37, we get $c_A(X)$ $\det C = (m_A(X))^n$. Consequently, $c_A(X) \mid (m_A(X))^n$. $\square$

Corollary 2.38: Let $T \in \mathscr{E}(V)$. Then $c_T(X)$ and $m_T(X)$ have the same set of irreducible factors in $F[X]$.

Proof: Theorem 2.26 implies $m_T(X) \mid c_T(X)$. Theorem 2.35 implies that $c_T(X) \mid (m_T(X))^n$. The result follows from 1.7. $\square$

Let us rephrase Corollary 2.38 in terms of the language used in Theorem 1.7. Suppose $m_T(X) = f_1^{d_1} \cdots f_r^{d_r}$ is the (essentially) unique factorization of the minimal polynomial of T in $F[X]$. Thus, each f_i is irreducible, f_i and f_j are not associates for $i \neq j$, and each $d_i > 0$. Then corollary 2.38 implies that the unique factorization of $c_T(X)$ (in $F[X]$) is given by $c_T(X) = f_1^{e_1} \cdots f_r^{e_r}$ with $d_i \leqslant e_i$ for all $i = 1, \ldots, r$. We must also have $\sum_{i=1}^{r} e_i \partial(f_i) = n$.

Now suppose $\bar{F}$ is the algebraic closure of F. Consider the extension $T^F = T \otimes_F I_F$ of T to V^F. We have seen in Theorem 2.10, that $m_T(X) = m_{T^{\bar{F}}}(X)$ in $\bar{F}[X]$. It is clear from the definition that $c_T(X) = c_{T^{\bar{F}}}(X)$. If we apply Corollary 2.38 to $T^F \in \mathscr{E}(V^F)$, we conclude that $c_T(X)$ and $m_T(X)$ have the same irreducible factors in $\bar{F}[X]$. Since $\bar{F}$ is algebraically closed, the only irreducible polynomials in $\bar{F}[X]$ are all linear. Hence $c_T(X)$ and $m_T(X)$ can be written in $\bar{F}[X]$ as follows:

2.39: $$m_T(X) = \prod_{i=1}^{r} (X - c_i)^{m_i} \quad \text{and} \quad c_T(X) = \prod_{i=1}^{r} (X - c_i)^{n_i}$$

In equation 2.39, we must have $0 < m_i \leqslant n_i$ for all $i = 1, \ldots, r$. Also, $n_1 + \cdots + n_r = n = \dim_F(V)$. Recall that the roots $R(f)$ of a nonconstant polynomial $f(X) \in F[X]$ are those elements $a \in \bar{F}$ such that $f(a) = 0$. Equation 2.39 implies $R(m_T) = \{c_1, \ldots, c_r\} = R(c_T)$. Thus, we have proved the following corollary:

Corollary 2.40: Let $T \in \mathscr{E}(V)$. Then the minimal polynomial and characteristic polynomial of T have the same roots in $\bar{F}$. $\square$

We shall finish this section with a brief look at invariant subspaces.

Definition 2.41: Let $T \in \mathscr{E}(V)$, and let W be a subspace of V. We say W is *invariant under T or T-invariant* if $T(\alpha) \in W$ for all $\alpha \in W$.

Thus, W is T-invariant if $T(W) \subseteq W$. Clearly, (0), V, ker T, and Im T are all T-invariant subspaces of V. In the next few sections, we shall mainly encounter T-invariant subspaces in direct sum decompositions of V. Suppose $V = V_1 \oplus \cdots \oplus V_r$ is an (internal) direct sum of subspaces $V_1, \ldots, V_r$. Let us further suppose each V_i is T-invariant. We shall denote the restriction of T to V_i by T_i. Since V_i is T-invariant, $T_i \in \mathscr{E}(V_i)$ for all $i = 1, \ldots, r$. Suppose $\alpha = \alpha_1 + \cdots + \alpha_r$ with $\alpha_i \in V_i$. Then $T(\alpha) = \sum_{i=1}^{r} T_i(\alpha_i)$.

Let $\underline{\alpha}_i$ be a basis of V_i, $i = 1, \ldots, r$. It follows from Theorem 4.16(b) of Chapter I that $\underline{\alpha} = \bigcup_{i=1}^{r} \underline{\alpha}_i$ is a basis of V. Since each V_i is T-invariant, $\Gamma(\underline{\alpha}, \underline{\alpha})(T)$ has the following form:

2.42:

$$\Gamma(\underline{\alpha}, \underline{\alpha})(T) = \begin{pmatrix} A_1 & & 0 \\ & \ddots & \\ 0 & & A_r \end{pmatrix} \quad \text{with} \quad A_i = \Gamma(\underline{\alpha}_i, \underline{\alpha}_i)(T_i)$$

Equation 2.42 gives us an immediate proof of the first half of the following theorem:

Theorem 2.43: Let $T \in \mathscr{E}(V)$ and suppose $V = V_1 \oplus \cdots \oplus V_r$ is an internal direct sum of T-invariant subspaces $V_1, \ldots, V_r$. Let T_i denote the restriction of T to V_i. Then

(a) $T_i \in \mathscr{E}(V_i)$, $i = 1, \ldots, r$.

(b) $c_T(X) = \prod_{i=1}^{r} c_{T_i}(X)$.

(c) $m_T(X)) = \mathrm{l.c.m.}(m_{T_1}(X), \ldots, m_{T_r}(X))$.

Proof: Here $c_{T_i}(X)$ is the characteristic polynomial of T_i on V_i. Similarly $m_{T_i}(X)$ is the minimal polynomial of T_i on V_i. (a) is clear. As for (b), we have $c_T(X) = c_A(X)$, where $A = \Gamma(\underline{\alpha}, \underline{\alpha})(T)$. Thus, from 2.42, we have $c_T(X) = \det(XI - A) = \prod_{i=1}^{r} \det(XI_i - A_i) = \prod_{i=1}^{r} c_{T_i}(X)$. Here I_i of course denotes the identity matrix of size the same as A_i.

For (c), let us shorten notation here and write m_i for $m_{T_i}(X)$. Recall from Exercise 11 in Section 1 that a $\mathrm{l.c.m.}(m_1, \ldots, m_r)$ is a polynomial $e \in F[X]$ with the following two properties.

2.44: (i) $m_i | e$ for $i = 1, \ldots, r$.

   (ii) If $m_i | g$ for $i = 1, \ldots, r$, then $e | g$.

We shall argue that $m_T(X)$ satisfies (i) and (ii) in 2.44.

Since $m_T(T) = 0$ on V, clearly $m_T(T_i) = 0$ on V_i. Thus, $m_i | m_T$ by 2.15(b). We have now established (i).

Suppose $g(X) \in F[X]$ such that $m_i | g$ for all $i = 1, \ldots, r$. Then $m_i(X)a_i(X) = g(X)$ for some a_i. In particular, if $\alpha_i \in V_i$, then $g(T_i)(\alpha_i) = m_i(T_i)a_i(T_i)(\alpha_i) = 0$. Now let $\alpha \in V$ and write $\alpha = \alpha_1 + \cdots + \alpha_r$ with $\alpha_i \in V_i$. Since each V_i is invariant under T, V_i is invariant under g(T). Clearly, the restriction of g(T) to V_i is nothing but $g(T_i)$. Thus, $g(T)(\alpha) = \sum_{i=1}^{r} g(T)(\alpha_i) = \sum_{i=1}^{r} g(T_i)(\alpha_i) = 0$. Therefore, $g(T) = 0$, and $m_T | g$ by 2.15(b). This proves (ii) and completes the proof of the theorem. $\square$

EXERCISES FOR SECTION 2

(1) Let $F \subseteq K$ be fields. Show that the map $\psi : F[X] \otimes_F K \to K[X]$ given by $\psi(f(X) \otimes k) = kf(X)$ is an isomorphism of K-algebras.

(2) Suppose $\sum_{i=0}^{m} A_i X^i = \sum_{i=0}^{m} B_i X^i$ in $M_{n \times n}(F[X])$ with A_i, $B_i \in M_{n \times n}(F)$ for all $i = 0, \ldots, m$. Show $A_i = B_i$ for all i.

(3) Let $A \in M_{n \times n}(F[X])$. If $\det(A) \neq 0$, does it follow that A is invertible?

(4) If $A = (a_{ij}) \in M_{n \times n}(F)$, and $c_A(X) = X^n + a_{n-1}X^{n-1} + \cdots + a_0$, show

  (i) $a_{n-1} = -\sum_{i=1}^{n} a_{ii} = -\text{Tr}(A)$.

  (ii) $a_0 = (-1)^n \det(A)$.

(5) Give an example of a vector space V and $T \in \mathscr{E}(V)$ such that $V = V_1 \oplus V_2$ with V_i a T-invariant subspace and $m_T(X) \neq m_{T_1}(X)m_{T_2}(X)$. Here T_i denotes the restriction of T to V_i.

(6) Suppose $T \in \mathscr{E}(V)$ is nilpotent, that is, $T^m = 0$ for some $m \geq 1$. Let $f(X) = a_r X^r + \cdots + a_0$ be any polynomial in $F[X]$ with $a_0 \neq 0$. Show that $f(T)$ is an invertible linear transformation on V.

(7) Find the characteristic and minimal polynomials of $T: \mathbb{R}^3 \to \mathbb{R}^3$ given by $T(\delta_1) = 6\delta_3$, $T(\delta_2) = \delta_1 - 11\delta_3$, and $T(\delta_3) = \delta_2 + 6\delta_3$. Here $\underline{\delta} = \{\delta_1, \delta_2, \delta_3\}$ is the canonical basis of $\mathbb{R}^3$.

(8) Find the characteristic and minimal polynomials of the subdiagonal matrix $A \in M_{n \times n}(F)$ given by

$$\begin{bmatrix} 0 & 0 & 0 & \cdots & 0 & 0 \\ 1 & 0 & 0 & \cdots & 0 & 0 \\ 0 & 1 & 0 & \cdots & 0 & 0 \\ \vdots & \vdots & \vdots & & \vdots & \vdots \\ 0 & 0 & 0 & \cdots & 1 & 0 \end{bmatrix}$$

(9) Let $T: \mathbb{R}^4 \to \mathbb{R}^4$ be given by $T(x_1, x_2, x_3, x_4) = (x_1 - x_4, x_1, -2x_2 - x_3 - 4x_4, 4x_2 + x_3)$.
  (a) Compute $c_T(X)$.
  (b) Compute $m_T(X)$.
  (c) Show that $\mathbb{R}^4$ is an internal direct sum of two proper T-invariant subspaces.

(10) Find the minimal polynomial of

$$\begin{bmatrix} -1 & 0 & 0 & 0 & 0 \\ 1 & -1 & 0 & 0 & 0 \\ 0 & 0 & -1 & 0 & 0 \\ 0 & 0 & 0 & 1 & 0 \\ 0 & 0 & 0 & 1 & 1 \end{bmatrix}$$

(11) Suppose W is a T-invariant subspace of V. Show that T induces a linear

transformation $\bar{T} \in \mathscr{E}(V/W)$ given by $\bar{T}(\alpha + W) = T(\alpha) + W$. What is the relationship between the minimal polynomials of T and $\bar{T}$?

(12) Let $A \in M_{n \times n}(F)$. When computing the characteristic and minimal polynomials of A, is it permissible to first row reduce A to some simpler matrix and then make the desired computations? Explain.

(13) A matrix $D = (d_{ij}) \in M_{n \times n}(F)$ is diagonal if $d_{ij} = 0$ whenever $i \neq j$. If D is a diagonal matrix, then we shall write $D = \text{diag}(a_1, \ldots, a_n)$, where $a_i = d_{ii}$ for all $i = 1, \ldots, n$. Compute $c_D(X)$ and $m_D(X)$ for any diagonal matrix D.

(14) Suppose $A \in M_{n \times n}(F)$ is a nonzero, nilpotent matrix. Thus, $A^k = 0$ for some $k \geqslant 2$. Compute $m_A(X)$ and $c_A(X)$. Show that A cannot be similar to any diagonal matrix.

(15) Let $A \in M_{n \times n}(F)$. If the degree of $m_A(X)$ is n, does it follow that A is similar to a diagonal matrix?

(16) Let $A \in M_{n \times n}(F)$. Show that A is singular if and only if zero is a root of $c_A(X)$.

(17) Let $A \in M_{n \times n}(F)$, and suppose $\bar{F}$ is an algebraic closure of F. Suppose $c_A(X) = \prod_{i=1}^{n} (X - c_i)$ in $\bar{F}[X]$. Here $c_1, \ldots, c_n$ $(\in \bar{F})$ are not necessarily distinct. Write the coefficients of $c_A(X)$ as symmetric functions of $c_1, \ldots, c_n$. The coefficients of $c_A(X)$ are functions of $c_1, \ldots, c_n$, which lie in F.

(18) Use your answer from Exercise 17 to find a matrix $A \in M_{2 \times 2}(\mathbb{R})$ such that $c_A(X) = X^2 + 2X + 5$.

(19) Suppose $A \in M_{n \times n}(F)$ is a triangular matrix. This means $a_{ij} = 0$ whenever $i > j$ (upper triangular) or $a_{ij} = 0$ whenever $i < j$ (lower triangular). Show that if $A = (a_{ij})$ is triangular, then $c_A(X) = \prod_{i=1}^{n} (X - a_{ii})$.

(20) Let $A \in M_{n \times n}(F)$ and $B \in M_{m \times m}(F)$. Consider the Kronecker product $A \otimes B$ of A and B. Prove that if $x \in R(c_A(X))$ and $y \in R(c_B(X))$, then xy is a root of $c_{A \otimes B}(X)$.

(21) Prove Corollary 2.14 directly without using tensor products. [*Hint*: Use a basis of K over F and write the coefficients of $m_{T^\kappa}(X)$ in terms of this basis.]

3. EIGENVALUES AND EIGENVECTORS

As usual, let V be a finite-dimensional vector space over F of dimension n. Let $T \in \mathscr{E}(V)$.

Definition 3.1: An element $c \in F$ is called an *eigenvalue* of T if $\ker(T - c) \neq 0$.

Thus, $c \in F$ is an eigenvalue of T if there exists a nonzero vector $\alpha \in V$ such that $T(\alpha) = c\alpha$. Eigenvalues are also called *characteristic values*, but in this text, we shall use the term "eigenvalue" exclusively. The complete set of eigenvalues for T in F will be denoted $\mathscr{S}_F(T)$ and called the *spectrum* of T (in F). Thus, $c \in \mathscr{S}_F(T)$ if and only if there exists a vector $\alpha \in V$ such that $\alpha \neq 0$ and $T(\alpha) = c\alpha$. The spectrum of T depends on the field F.

Example 3.2: Let us return to Examples 2.6 and 2.7 of the Section 2. Since $T: \mathbb{R}^2 \to \mathbb{R}^2$ represents a rotation through $90°$, no nonzero vector is taken by T into a multiple of itself. Consequently, $\mathscr{S}_{\mathbb{R}}(T) = \phi$. If we extend T to $T^{\mathbb{C}}: \mathbb{C}^2 \to \mathbb{C}^2$, then both i and $-i$ are eigenvalues of $T^{\mathbb{C}}$. We have $T^{\mathbb{C}}((1, -i)) = i(1, -i)$ and $T^{\mathbb{C}}((-1, -i)) = -i(-1, -i)$. We shall soon see that $T^{\mathbb{C}}$ can have at most two distinct eigenvalues. Therefore, $\mathscr{S}_{\mathbb{C}}(T^{\mathbb{C}}) = \{i, -i\}$. $\square$

Let us gather together some of the more obvious facts about eigenvalues.

Theorem 3.3: Let $T \in \mathscr{E}(V)$, and suppose K is a field containing F. Then

(a) $\mathscr{S}_F(T) \subseteq \mathscr{S}_K(T^K)$.

(b) $\mathscr{S}_F(T) = R(c_T) \cap F$, that is, the eigenvalues of T in F are precisely the roots of the characteristic polynomial of T that lie in F.

(c) $\mathscr{S}_F(T) = R(m_T) \cap F$.

(d) $|\mathscr{S}_F(T)| \leqslant \partial(m_T) \leqslant \dim_F(V)$.

Proof: (a) Suppose $c \in \mathscr{S}_F(T)$. Then there exists a nonzero vector $\alpha \in V$ such that $(T - c)(\alpha) = 0$. $\alpha \otimes_F 1 \neq 0$ in $V \otimes_F K$. $(T^K - c)(\alpha \otimes_F 1) = [(T - c)(\alpha)] \otimes_F 1 = 0 \otimes_F 1 = 0$. Thus, $c \in \mathscr{S}_K(T^K)$.

(b) Recall that $R(c_T)$ is the set of roots (in $\bar{F}$) of the characteristic polynomial $c_T(X)$ of T. Let $\underline{\alpha}$ be a basis of V, and set $\Gamma(\underline{\alpha}, \underline{\alpha})(T) = A$. Now $c \in \mathscr{S}_F(T)$ if and only if $\ker(T - c) \neq 0$. From Theorem 3.33(b) of Chapter I, we know $\ker(T - c) \neq 0$ if and only if $T - c$ is not an isomorphism on V. Since $\Gamma(\underline{\alpha}, \underline{\alpha})(\cdot): \mathscr{E}(V) \to M_{n \times n}(F)$ is an isomorphism of F-algebras, $T - c$ is not an isomorphism if and only if $A - c = A - cI_n$ is not invertible in $M_{n \times n}(F)$. This last statement is in turn equivalent to $\det(A - cI_n) = 0$. Now $c_T(c) = c_A(c) = \det(cI_n - A) = (-1)^n \det(A - cI_n)$. Thus, $c \in \mathscr{S}_F(T)$ if and only if c is a root of $c_T(X)$ in F. Hence, $\mathscr{S}_F(T) = R(c_T) \cap F$.

(c) We have seen in Corollary 2.40 that $R(m_T) = R(c_T)$. Hence, (c) follows from (b).

(d) From (c), we know $\mathscr{S}_F(T) \subseteq R(m_T)$. We have seen in Exercise 14 of Section 1 that $|R(m_T)| \leqslant \partial(m_T)$. The result now follows from Corollary 2.32. $\square$

Let us make a few comments about Theorem 3.3. The includion in (a) could very well be strict, as Example 3.2 shows. As we extend scalars, the extended linear transformation T^K may pick up more eigenvalues because the characteristic polynomial $c_T(X)$ may have more linear factors in $K[X]$ than it had in $F[X]$. This is precisely what is happening in Example 3.2. Over $\mathbb{R}$, $c_T(X) = X^2 + 1$. Since $X^2 + 1$ is irreducible in $\mathbb{R}[X]$, $R(c_T) \cap \mathbb{R} = \phi$. Thus $\mathscr{S}_\mathbb{R}(T) = \phi$. Over $\mathbb{C}$, $c_{T\mathbb{C}} = (X^2 + 1) = (X + i)(X - i)$. Hence, $R(c_{T\mathbb{C}}) \cap \mathbb{C} = \{i, -i\}$. Therefore, $\mathscr{S}_\mathbb{C}(T^\mathbb{C}) = \{\pm i\}$.

Theorem 3.3(b) tells us exactly how to compute the eigenvalues of T in F. Choose any matrix representation A of T, and compute the characteristic polynomial $c_A(X) = \det(XI - A)$. Then find the roots of $c_A(X)$ that lie in F. These roots are precisely the eigenvalues of T lying in F. Of course, finding the roots of $c_A(X)$ that lie in F may be very difficult. If $F = \mathbb{R}$, for instance, we can use well-known techniques from numerical analysis to at least approximate the real roots of $c_A(X)$.

Let us consider one more example before continuing.

Example 3.4: Let $T: \mathbb{R}^4 \to \mathbb{R}^4$ be given by $T(\delta_1) = \delta_1 - \delta_2$, $T(\delta_2) = 2\delta_2$, $T(\delta_3) = \delta_4$, and $T(\delta_4) = \delta_4 - \delta_3$. Here $\underline{\delta} = \{\delta_1, \delta_2, \delta_3, \delta_4\}$ is the canonical basis of $\mathbb{R}^4$. The matrix representation of T is given by

$$\Gamma(\underline{\delta}, \underline{\delta})(T) = \begin{bmatrix} 1 & 0 & 0 & 0 \\ -1 & 2 & 0 & 0 \\ 0 & 0 & 0 & -1 \\ 0 & 0 & 1 & 1 \end{bmatrix}$$

Thus, $\mathbb{R}^4 = L(\{\delta_1, \delta_2\}) \oplus L(\{\delta_3, \delta_4\})$ is a direct sum decomposition of $\mathbb{R}^4$ in terms of T-invariant subspaces. Theorem 2.43 implies $c_T(X) = (X - 1)(X - 2)(X^2 - X + 1)$. Thus, $R(c_T(X)) = \{1, 2, (1 \pm \sqrt{3}i)/2\} \subseteq \mathbb{C}$. If we now apply Theorem 3.3, we have $\mathscr{S}_\mathbb{R}(T) = \{1, 2\}$ and $\mathscr{S}_\mathbb{C}(T^\mathbb{C}) = \{1, 2, (1 \pm \sqrt{3}i)/2\}$. $\square$

Definition 3.5: Let $A \in M_{n \times n}(F)$. An element $c \in F$ is called an *eigenvalue* of A if $\det(A - cI_n) = 0$.

Clearly, c is an eigenvalue of A if and only if there exists a nonzero column vector $Y \in M_{n \times 1}(F)$ such that $AY = cY$. The set of eigenvalues of A (in F) will be denoted by $\mathscr{S}_F(A)$. If $T \in \mathscr{E}(V)$, $\underline{\alpha}$ a basis of V, and $A = \Gamma(\underline{\alpha}, \underline{\alpha})(T)$, then the proof of 3.3(b) implies that $\mathscr{S}_F(T) = \mathscr{S}_F(A)$. In particular, Theorem 3.3 remains valid with T replaced by A. We can therefore switch back and forth between T and its matrix representation A when computing eigenvalues. One corollary of this interplay is that similar matrices have the same set of eigenvalues. That is because similar matrices represent the same linear transformation on V.

Definition 3.6: A nonzero vector $\alpha \in V$ is called an *eigenvector* of T belonging to the eigenvalue c if $T(\alpha) = c\alpha$.

Thus, if $c \in \mathscr{S}_F(T)$, then any nonzero vector in $\ker(T - c)$ is an eigenvector belonging to c. We emphasize that eigenvectors are always nonzero. Let us look at some examples.

Example 3.7: Suppose $T = 0$. Then every nonzero vector in V is an eigenvector for T belonging to 0. If $T = I_V$, then every nonzero vector in V is an eigenvector for T belonging to 1. $\square$

Example 3.8: In Example 2.6, $\mathscr{S}_\mathbb{R}(T) = \phi$. Hence, T has no eigenvectors. In Example 2.7, $T^\mathbb{C}$ has $(1, -i)$ belonging to i and $(-1, -i)$ belonging to $-i$. Thus, as the reader should expect, when we enlarge the base field F by extending scalars, a given T may acquire new eigenvectors not present over F. $\square$

Example 3.9: Let $T: \mathbb{R}^4 \to \mathbb{R}^4$ be the transformation in Example 3.4. $(1, 1, 0, 0)$ is an eigenvector belonging to 1, and $(0, 1, 0, 0)$ is an eigenvector belonging to 2. $\square$

Eigenvectors are also called *characteristic vectors*. In this text, we shall not use this name. The relation between eigenvectors belonging to different eigenvalues is one of linearly independence.

Theorem 3.10: Suppose $c_1, \ldots, c_r$ are distinct eigenvalues in $\mathscr{S}_F(T)$. Let α_i be an eigenvector of T belonging to c_i for each $i = 1, \ldots, r$. Then $\Delta = \{\alpha_1, \ldots, \alpha_r\}$ is linearly independent over F.

Proof: Suppose Δ is linearly dependent over F. Then among all nontrivial relations of the form $a_1\alpha_1 + \cdots + a_r\alpha_r = 0$, we can pick a relation in which the fewest number of α_i occur. After relabeling the α_i if need be, we can assume our relation of minimal length is of the form

3.11: $$a_1\alpha_1 + \cdots + a_s\alpha_s = 0$$

In equation 3.11, $s \leqslant r$, and each a_i is a nonzero scalar in F. Multiplying 3.11 by c_1 gives

3.12: $$c_1 a_1\alpha_1 + \cdots + c_1 a_s\alpha_s = 0$$

Applying T to 3.11 gives

3.13: $$c_1 a_1\alpha_1 + \cdots + c_s a_s\alpha_s = 0$$

If we now subtract 3.12 from 3.13, we produce a nontrivial relation among the α_i that has fewer terms in it than 3.11. This is a contradiction. Thus, Δ is linearly independent over F. $\square$

There are several interesting corollaries to Theorem 3.10. In the first place, $|\mathscr{S}_F(T)| \leqslant \dim V$ since eigenvectors belonging to distinct eigenvalues must be linearly independent. We do not list this fact as a corollary since Theorem 3.3(d) is an even sharper result. Theorem 3.10 gives us sufficient conditions for representing T as a diagonal matrix.

Corollary 3.14: Suppose $T \in \mathscr{E}(V)$ such that $|\mathscr{S}_F(T)| = \dim V$. Then there exists a basis $\underline{\alpha}$ of V such that $\Gamma(\underline{\alpha}\ \underline{\alpha})(T)$ is a diagonal matrix.

Thus, if $\dim V = n$, and T has n distinct eigenvalues in F, then T can be represented by a diagonal matrix.

Proof of 3.14: Let $\mathscr{S}_F(T) = \{c_1 \ldots, c_n\}$, where $n = \dim V$. For each $i = 1, \ldots, n$, let α_i be an eigenvector belonging to c_i. Then $\underline{\alpha} = \{\alpha_1 \ldots, \alpha_n\}$ is a basis of V by Theorem 3.10. Clearly, $\Gamma(\underline{\alpha}, \underline{\alpha})(T) = \mathrm{diag}(c_1, \ldots, c_n)$. $\square$

Here and throughout the rest of the text, we shall let $\mathrm{diag}(c_1, \ldots, c_n)$ denote the $n \times n$ diagonal matrix

3.15:
$$\mathrm{diag}(c_1, \ldots, c_n) = \begin{bmatrix} c_1 & & 0 \\ & \cdot \cdot \cdot & \\ 0 & & c_n \end{bmatrix}$$

We note in passing that the converse of Corollary 3.14 is false. Namely, if some matrix representation of T is diagonal, we cannot conclude that T has n distinct eigenvalues in F. For example, $T = I_V$ is represented by the matrix I_n relative to any basis in V, but T has only one eigenvalue 1.

A slightly different version of Corollary 3.14 is worth recording here.

Corollary 3.16: Let F be an algebraically closed field and let $A \in M_{n \times n}(F)$. Suppose $c_A(X)$ has no repeated roots. Then A is similar to a diagonal matrix.

Proof: Since F is algebraically closed, $c_A(X) = \prod_{i=1}^{r}(X - c_i)^{n_i}$ in $F[X]$. Here $c_1, \ldots, c_r$ are the roots of $c_A(X)$, and we must have $n_1 + \cdots + n_r = n$. Now the statement that $c_A(X)$ has no repeated roots means each $n_i = 1$. In particular, $r = n$, and $R(c_A) = \{c_1, \ldots, c_n\}$. Thus, Theorem 3.3(b) implies $\mathscr{S}_F(A) = \{c_1, \ldots, c_n\}$.

Suppose $A = (a_{ij})$. Let $\underline{\delta} = \{\delta_1, \ldots, \delta_n\}$ be the canonical basis of F^n, and define a linear transformation $T: F^n \to F^n$ by $T(\delta_i) = \sum_{j=1}^{n} a_{ji}\delta_j = \mathrm{Col}_i(A)^t$. Then $\Gamma(\underline{\delta}, \underline{\delta})(T) = A$.

We have noted that $\mathscr{S}_F(T) = \mathscr{S}_F(A) = \{c_1, \ldots, c_n\}$. Consequently, Corollary 3.14 implies there exists a basis $\underline{\alpha}$ of F^n such that $\Gamma(\underline{\alpha}, \underline{\alpha})(T) = \mathrm{diag}(c_1, \ldots, c_n)$. Since $\Gamma(\underline{\delta}, \underline{\delta})(T)$ is similar to $\Gamma(\underline{\alpha}, \underline{\alpha})(T)$ (Theorem 3.28 of Chapter I), we conclude A is similar to $\mathrm{diag}(c_1, \ldots, c_n)$. $\square$

Clearly, a given $T \in \mathscr{E}(V)$ can be represented by a diagonal matrix if and only if V has a basis consisting of eigenvectors of T. Corollaries 3.14 and 3.16 give us sufficient conditions for such a basis of eigenvectors to exists. If T has enough eigenvalues (i.e., $\dim V$), then T can be represented as a diagonal matrix.

Example 3.17: Consider the linear transformation $T^{\mathbb{C}} : \mathbb{C}^4 \to \mathbb{C}^4$ derived from T in example 3.4. Since $\mathscr{S}_{\mathbb{C}}(T^{\mathbb{C}}) = \{1, 2, (1 \pm \sqrt{3}i)/2\}$, we conclude from Corollary 3.14 that there exists a basis $\underline{\alpha}$ of $\mathbb{C}^4$ such that

3.18: $$\Gamma(\underline{\alpha}, \underline{\alpha})(T^{\mathbb{C}}) = \begin{bmatrix} 1 & 0 & 0 & 0 \\ 0 & 2 & 0 & 0 \\ 0 & 0 & \dfrac{1 + \sqrt{3}i}{2} & 0 \\ 0 & 0 & 0 & \dfrac{1 - \sqrt{3}i}{2} \end{bmatrix}$$

We can also conclude from Corollary 3.16 that

$$A = \begin{bmatrix} 1 & 0 & 0 & 0 \\ -1 & 2 & 0 & 0 \\ 0 & 0 & 0 & -1 \\ 0 & 0 & 1 & 1 \end{bmatrix}$$

is similar in $M_{4 \times 4}(\mathbb{C})$ to the diagonal matrix given in 3.18.

Over $\mathbb{R}$, T has no diagonal representation. To see this, suppose there exists a basis $\underline{\alpha}$ of $\mathbb{R}^4$ such that $\Gamma(\underline{\alpha}, \underline{\alpha})(T) = \text{diag}(a_1, a_2, a_3, a_4)$. Then $c_T(X) = \prod_{i=1}^{4}(X - a_i) \in \mathbb{R}[X]$. Since $c_T(X) = (X - 1)(X - 2)(X^2 - X + 1)$, we can conclude $X^2 - X + 1$ factors in $\mathbb{R}[X]$. This is impossible. Thus, A is not similar in $M_{4 \times 4}(\mathbb{R})$ to any diagonal matrix. $\square$

In Example 3.17, we gave an example of a linear transformation T that cannot be diagonalized, that is, there exists no basis $\underline{\alpha}$ of $\mathbb{R}^4$ such that $\Gamma(\underline{\alpha}, \underline{\alpha})(T)$ is diagonal. The corresponding matrix statement is that A is not similar to any diagonal matrix in $M_{4 \times 4}(\mathbb{R})$. The example works because the roots of the characteristic polynomial of T (or A) do not all lie in the base field $\mathbb{R}$.

Suppose $T \in \mathscr{E}(V)$ is an arbitrary linear transformation. If there exists a basis $\underline{\alpha}$ of V such that $\Gamma(\underline{\alpha}\ \underline{\alpha})(T) = \text{diag}(a_1, \ldots, a_n)$, then clearly $R(c_T(X)) \subseteq \{a_1, \ldots, a_n\} \subseteq F$. Similarly, if $A \in M_{n \times n}(F)$ is similar to a diagonal matrix, then $R(c_A(X)) \subseteq F$. We can ask about the converse of these statements. If $R(c_T) \subseteq F$, is T diagonalizable? If $R(c_A) \subseteq F$, is A similar to a diagonal matrix in $M_{n \times n}(F)$? The answer to both of these questions is easily seen to be no. The simplest place to look for examples is the collection of nilpotent linear transformations.

Definition 3.19: A linear transformation $T \in \mathscr{E}(V)$ is said to be nilpotent of index k if $T^k = 0$ and $T^{k-1} \neq 0$. Similarly, a matrix $A \in M_{n \times n}(F)$ is nilpotent of index k if $A^k = 0$ and $A^{k-1} \neq 0$.

Suppose $\underline{\alpha}$ is a basis of V, and $A = \Gamma(\underline{\alpha}, \underline{\alpha})(T)$. Since $\Gamma(\underline{\alpha}, \underline{\alpha})(\cdot) : \mathscr{E}(V) \to M_{n \times n}(F)$ is an isomorphism of F-algebras, clearly T is nilpotent of index k if and only if A is nilpotent of index k.

Example 3.20: Let

$$N_k = \begin{bmatrix} 0 & 0 & 0 & \cdots & 0 & 0 \\ 1 & 0 & 0 & \cdots & 0 & 0 \\ 0 & 1 & 0 & \cdots & 0 & 0 \\ \vdots & \vdots & \vdots & & \vdots & \vdots \\ 0 & 0 & 0 & \cdots & 1 & 0 \end{bmatrix} \in M_{k \times k}(F)$$

An easy computation shows N_k is nilpotent of index k. □

Suppose $T \in \mathcal{E}(V)$ is nilpotent of index k. It follows readily from Exercise 6 of Section 2 that the minimal polynomial $m_T(X)$ must be a power of X alone. Since T is nilpotent of index k, $m_T(X) = X^k$. In particular, Theorem 3.3(c) implies $\mathcal{S}_F(T) = R(m_T) = \{0\}$. Thus, a nilpotent transformation or matrix has only one eigenvalue 0. Note that $m_T(X) = X^k$ implies $k \leqslant n$. The maximum index of nilpotency of any nilpotent transformation T cannot exceed $n = \dim(V)$. If T is not zero, then T cannot be diagonalized. For suppose $\Gamma(\alpha, \alpha)(T) = \mathrm{diag}(a_1, \ldots, a_n)$ for some basis α of V. Then $0 = \Gamma(\alpha, \alpha)(T^k) = [\Gamma(\alpha, \alpha)(T)]^k = [\mathrm{diag}(a_1, \ldots, a_n)]^k = \mathrm{diag}(a_1^k, \ldots, a_n^k)$. But then $a_1 = \cdots = a_n = 0$ and $T = 0$, which is impossible. Similar reasoning shows that a nonzero nilpotent matrix cannot be similar to a diagonal matrix.

Nilpotent linear transformations are the fundamental ingredients in the Jordan canonical form. We shall finish this section with a representation theorem for nilpotent transformations.

Theorem 3.21: Suppose $T \in \mathcal{E}(V)$ is a nilpotent linear transformation of index of nilpotency $k \geqslant 2$. Then there exists a basis α of V such that

$$\Gamma(\alpha, \alpha)(T) = \begin{bmatrix} N_{k_1} & & 0 \\ & \ddots & \\ 0 & & N_{k_p} \end{bmatrix}$$

where

(a) $k = k_1 \geqslant k_2 \geqslant \cdots \geqslant k_p$, and
(b) $k_1 + \cdots + k_p = n = \dim(V)$.

Let us say a few words about Theorem 3.21 before proceeding with its proof. The supposition that $k \geqslant 2$ is solely to avoid the trivial case $T = 0$. If T is nilpotent and not zero, then clearly the index of nilpotency of T is some positive integer between 2 and n. The notation for the N_{k_i} is given in Example 3.20. Thus, each N_{k_i} is a $k_i \times k_i$ matrix having ones on its subdiagonal and zeros everywhere else. Note then that $\Gamma(\alpha, \alpha)(T)$ is the next best thing to being diagonal. $\Gamma(\alpha, \alpha)(T)$ is a subdiagonal matrix with only zeros and ones appearing on its subdiagonal.

Proof of 3.21: Set $k_1 = k$. Since T is nilpotent of index k_1, we know $T^{k_1} = 0$, and $T^{k_1 - 1} \neq 0$. Hence, there exists a vector $\alpha \in V$ such that $T^{k_1 - 1}(\alpha) \neq 0$. We first claim that $\Delta = \{\alpha, T(\alpha), \ldots, T^{k_1 - 1}(\alpha)\}$ is linearly independent over F. Suppose these vectors are linearly dependent over F. Then we have

3.22:
$$c_1 \alpha + c_2 T(\alpha) + \cdots + c_{k_1} T^{k_1 - 1}(\alpha) = 0.$$

In this equation, $c_1, \ldots, c_{k_1} \in F$ and are not all zero. Suppose c_s is the first nonzero scalar in 3.22. Since $T^{k_1 - 1}(\alpha) \neq 0$, $s < k_1$. We can then rewrite 3.22 as

3.23:
$$[(c_s + c_{s+1}T + \cdots + c_{k_1}T^{k_1 - s})T^{s - 1}](\alpha) = 0$$

Since $c_s \neq 0$ and T is nilpotent, $c_s + c_{s+1}T + \cdots + c_{k_1}T^{k_1 - s}$ is invertible (see Exercise 6 of Section 2). Thus, Equation 3.23 implies $T^{s - 1}(\alpha) = 0$. But then $T^{k_1 - 1}(\alpha) = 0$, which is a contradiction.

Now let $Z_1 = L(\{\alpha, T(\alpha), \ldots, T^{k_1 - 1}(\alpha)\})$. Since Δ is linearly independent over F, $\dim_F(Z_1) = k_1$. Also note that Z_1 is a T-invariant subspace of V. If we let T_1 denote the restriction of T to Z_1, then $\Gamma(\Delta, \Delta)(T_1) = N_{k_1}$.

Suppose we can find a T-invariant complement W of Z_1. So, $V = Z_1 \oplus W$, and $T(W) \subseteq W$. Let $\hat{T}$ denote the restriction of T to W. $\hat{T}$ is clearly nilpotent since T is. The index of nilpotency k_2 of $\hat{T}$ is less than or equal to k_1 the index of nilpotency of T on V. Since $\dim_F(W) < \dim_F(V)$, we can assume (via induction on the dimension of V) that there exists a basis $\hat{\Delta}$ of W such that

3.24:
$$\Gamma(\hat{\Delta}, \hat{\Delta})(\hat{T}) = \begin{bmatrix} N_{k_2} & & 0 \\ & \ddots & \\ 0 & & N_{k_p} \end{bmatrix}$$

with $k_2 \geqslant \cdots \geqslant k_p$ and $k_2 + \cdots + k_p = \dim W$. It then follows from 2.42 that $\underline{\alpha} = \Delta \cup \hat{\Delta}$ is the required basis for V. Hence the proof of Theorem 3.21 will be complete when we argue Z_1 has a T-invariant complement.

Before constructing a complement of Z_1, we need the following technical result:

3.25: If $\beta \in Z_1$ is such that $T^{k_1 - s}(\beta) = 0$ where $0 < s \leqslant k_1$, then $\beta = T^s(\beta_0)$ for some $\beta_0 \in Z_1$.

The proof of 3.25 is as follows: Since $\beta \in Z_1$, $\beta = c_1 \alpha + c_2 T(\alpha) + \cdots + c_{k_1} T^{k_1 - 1}(\alpha)$. Thus, $0 = T^{k_1 - s}(\beta) = c_1 T^{k_1 - s}(\alpha) + \cdots + c_s T^{k_1 - 1}(\alpha)$. Since Δ is linearly independent over F, we conclude that $c_1 = \cdots = c_s = 0$. Therefore, $\beta = c_{s+1}T^s(\alpha) + \cdots + c_{k_1}T^{k_1 - 1}(\alpha) = T^s(c_{s+1}\alpha + \cdots + c_{k_1}T^{k_1 - s - 1}(\alpha))$. Since $\beta_0 = c_{s+1}\alpha + \cdots + c_{k_1}T^{k_1 - s - 1}(\alpha) \in Z_1$, the proof of 3.25 is complete.

We now claim there exists a T-invariant subspace W of V such that $Z_1 \oplus W = V$. To construct W, consider the following set $\mathcal{T} = \{W' \subseteq V \mid W' \text{ is a}$

subspace of V, $T(W') \subseteq W'$ and $Z_1 \cap W' = (0)$. Since the subspace $(0) \in \mathcal{T}$, $\mathcal{T} \neq \phi$. We can partially order the subspaces in $\mathcal{T}$ by inclusion $\subseteq$. If $\mathcal{T}_0$ is a totally ordered subset of $\mathcal{T}$, then clearly $\mathcal{T}_0$ has an upper bound $\tilde{W}$ in $\mathcal{T}$. Namely, $\tilde{W} = \bigcup_{W' \in \mathcal{T}_0} W'$. Hence, $(\mathcal{T}, \subseteq)$ is an inductive set. We can apply Zorn's lemma (2.2 of Chapter I) and choose a maximal subspace W in $\mathcal{T}$. W is clearly T-invariant, and $W \cap Z_1 = (0)$. We must argue $Z_1 + W = V$.

We shall suppose $Z_1 + W \neq V$ and derive a contradiction. If $Z_1 + W \neq V$, then there exists a $\beta \in V - (Z_1 + W)$. Since $T^{k_1} = 0$, there exists an integer u such that $0 < u \leqslant k_1$, $T^u(\beta) \in Z_1 + W$ and $T^i(\beta) \notin Z_1 + W$ for any $i < u$. Let us write $T^u(\beta) = \gamma + \delta$ with $\gamma \in Z_1$ and $\delta \in W$. Then $0 = T^{k_1}(\beta) = T^{k_1 - u}(T^u(\beta)) = T^{k_1 - u}(\gamma) + T^{k_1 - u}(\delta)$. These two vectors are in Z_1 and W, respectively, since Z_1 and W are T-invariant. Since $Z_1 \cap W = (0)$, we conclude that $T^{k_1 - u}(\gamma) = 0$ and $T^{k_1 - u}(\delta) = 0$. We can now apply 3.25 to the vector γ. We get $\gamma = T^u(\gamma_0)$ for some $\gamma_0 \in Z_1$. Therefore, $T^u(\beta) = \gamma + \delta = T^u(\gamma_0) + \delta$. Set $\beta_1 = \beta - \gamma_0$. Then $T^u(\beta_1) = T^u(\beta) - T^u(\gamma_0) = \delta \in W$. Since W is T-invariant, we can now conclude that $T^m(\beta_1) \in W$ for any $m \geqslant u$. On the other hand, if $i < u$, then $T^i(\beta_1) = T^i(\beta) - T^i(\gamma_0) \notin Z_1 + W$. $[\gamma_0 \in Z_1$ implies $T^i(\gamma_0) \in Z_1$. If $T^i(\beta) - T^i(\gamma_0) \in Z_1 + W$, then $T^i(\beta) \in Z_1 + W$. Since $i < u$ this is impossible.] We have now proved the following relations concerning β_1:

3.26:
$$T^m(\beta_1) \in W \qquad \text{for all} \quad m \geqslant u$$
$$T^i(\beta_1) \notin Z_1 + W \qquad \text{for all} \quad i < u$$

Now set $W_1 = W + L(\{\beta_1, T(\beta_1), \ldots, T^{u-1}(\beta_1)\})$. Since $u > 0$, $\beta_1 = T^0(\beta_1) \notin W$ by 3.26. Thus, W_1 properly contains W. Also it is clear using 3.26 again that W_1 is T-invariant. Since W is a maximal element of $\mathcal{T}$, $W_1 \notin \mathcal{T}$. Therefore, $W_1 \cap Z_1 \neq (0)$. Let $\beta_0 + c_1\beta_1 + c_2T(\beta_1) + \cdots + c_uT^{u-1}(\beta_1)$ be a nonzero vector in $W_1 \cap Z_1$. Here $\beta_0 \in W$, and $c_1, \ldots, c_u \in F$. Some constant c_i must be nonzero here. Otherwise, we would have $0 \neq \beta_0 \in W \cap Z_1 = (0)$. Suppose c_s is the first nonzero constant among the c_i. Thus, $1 \leqslant s \leqslant u$, and $0 \neq \beta_0 + c_sT^{s-1}(\beta_1) + \cdots + c_uT^{u-1}(\beta_1) \in W_1 \cap Z_1$.

We can rewrite this last inclusion relation as follows:

3.27:
$$0 \neq \beta_0 + (c_s + c_{s+1}T + \cdots + c_uT^{u-s})(T^{s-1}(\beta_1)) \in Z_1$$

Since $c_s \neq 0$, and T is nilpotent, $c_s + c_{s+1}T + \cdots + c_uT^{u-s}$ is invertible. It also follows from Exercise 6 of Section 2 that the inverse R of this map is a polynomial in T. In particular, W and Z_1 are invariant under R. If we apply R to Equation 3.27, we get $0 \neq R(\beta_0) + T^{s-1}(\beta_1) \in Z_1$. Since $\beta_0 \in W$, $R(\beta_0) \in W$. Thus, $T^{s-1}(\beta_1) \in W + Z_1$. But $s - 1 < u$, and we have a contradiction with 3.26. We conclude that $Z_1 + W = V$, and the proof of Theorem 3.21 is now complete. $\square$

The integers $k_1 \geqslant k_2 \geqslant \cdots \geqslant k_p$ appearing in Theorem 3.21 are called the *invariants* of T. We shall soon see that they are unique. The subspace Z_1

appearing in the proof of Theorem 3.21 is called a *T-cyclic subspace* of V. Let us formally define T-cyclic subspaces in our present context.

Definition 3.28: Let $T \in \mathscr{E}(V)$ be nilpotent. A T-invariant subspace Z of V is called a *T-cyclic subspace* if Z has a basis of the form $\Delta = \{\alpha, T(\alpha), \ldots, T^{k-1}(\alpha)\}$ for some $\alpha \in Z$ and some $k \geq 1$.

Lemma 3.29: Let $T \in \mathscr{E}(V)$ be nilpotent. Suppose Z is a T-cyclic subspace of V with basis $\Delta = \{\alpha, T(\alpha), \ldots, T^{k-1}(\alpha)\}$. Then

(a) $T^k(Z) = 0$.

(b) If T_1 denotes the restriction of T to Z, then $\Gamma(\Delta, \Delta)(T_1) = N_k$.

(c) $\dim_F(T^i(Z)) = k - i$ for all $i = 0, \ldots, k$.

Proof: (a) It is clearly enough to show $T^k(\alpha) = 0$. By definition, Z is T-invariant. Therefore $T^k(\alpha) = c_1 \alpha + c_2 T(\alpha) + \cdots + c_k T^{k-1}(\alpha)$ for some $c_1, \ldots, c_k \in F$. In particular, $(c_1 + c_2 T + \cdots + c_k T^{k-1} - T^k)(\alpha) = 0$. If $c_1 \neq 0$, then $c_1 + c_2 T + \cdots + c_k T^{k-1} - T^k$ is invertible since T is nilpotent. But then $\alpha = 0$. This is impossible since Δ is a basis of Z. Hence $c_1 = 0$. We now have $(c_2 + \cdots + c_k T^{k-2} - T^{k-1})(T(\alpha)) = 0$. The same argument shows that if $c_2 \neq 0$, then $T(\alpha) = 0$. Again this is impossible since Δ is a basis of Z. Continuing the argument, we conclude $c_1 = \cdots = c_k = 0$. Hence, $T^k(\alpha) = 0$.

(b) This assertion is obvious from (a).

(c) If $i = 0$ or k, then statement (c) is obvious. Suppose $1 \leq i \leq k - 1$. Since $\Delta = \{\alpha, T(\alpha), \ldots, T^{k-1}(\alpha)\}$ is a basis of Z, $T^i(Z)$ is spanned by $T^i(\alpha), T^{i+1}(\alpha), \ldots, T^{k+i-1}(\alpha)$. But by (a) only the first $k - i$ of these vectors are nonzero. Thus $T^i(Z) = L(\{T^i(\alpha), \ldots, T^{k-1}(\alpha)\})$. Since $\{T^i(\alpha), \ldots, T^{k-1}(\alpha)\} \subseteq \Delta$, these vectors are linearly independent over F. Hence, $\{T^i(\alpha), \ldots, T^{k-1}(\alpha)\}$ is a basis of $T^i(Z)$. This proves (c). $\square$

We can rephrase Theorem 3.21 in terms of T-cyclic subspaces as follows:

Theorem 3.30: Suppose $T \in \mathscr{E}(V)$ is a nilpotent linear transformation of index of nilpotency $k \geq 2$. Then there exist integers $k = k_1 \geq k_2 \geq \cdots \geq k_p$ and T-cyclic subspaces $Z_1, \ldots, Z_p$ of dimensions $k_1, \ldots, k_p$, respectively, such that $V = Z_1 \oplus \cdots \oplus Z_p$.

Proof: Construct Z_1 and W as in the proof of 3.21. Then Z_1 is a T-cyclic subspace of V of dimension $k = k_1$. Let T_1 be the restriction of T to (the T-invariant) subspace W. Since $T^{k_1} = 0$, $T_1^{k_1} = 0$. Thus, if k_2 is the index of nilpotency of T_1 on W, then $k_2 \leq k_1$. Proceeding as in the proof of 3.21 with T_1, we can construct a T_1-cyclic subspace Z_2 of W such that $\dim Z_2 = k_2$. Clearly, Z_2 is a T-cyclic subspace of V of dimension k_2. Again by 3.21, Z_2 has a T-invariant complement W' in W. Restrict T to W' and start again. In a finite number of steps, we construct T-cyclic subspaces $Z_1, \ldots, Z_p$ such that $V = Z_1 \oplus \cdots \oplus Z_p$, $\dim Z_i = k_i$, and $k = k_1 \geqslant k_2 \geqslant \cdots \geqslant k_p$. $\square$

Now let us turn our attention to the uniqueness of the integers $k_1 \geqslant \cdots \geqslant k_p$ appearing in Theorem 3.30.

Theorem 3.31: Let $T \in \mathscr{E}(V)$ be nilpotent of index $k \geqslant 2$. Suppose $V = Z_1 \oplus \cdots \oplus Z_p$ with each Z_i a T-cyclic subspace of dimension k_i, and $k = k_1 \geqslant k_2 \geqslant \cdots \geqslant k_p$. Suppose $V = U_1 \oplus \cdots \oplus U_s$ with each U_j a T-cyclic subspace of dimension l_j, and $l_1 \geqslant \cdots \geqslant l_s$. Then $p = s$ and $k_1 = l_1, \ldots, k_p = l_p$.

Proof: Suppose the assertion in the theorem is false. Then there is a first integer $i \geqslant 1$ where $k_i \neq l_i$. We can assume with no loss of generality that $l_i < k_i$. Then $k_1 \geqslant k_2 \geqslant \cdots \geqslant k_i$, $l_1 \geqslant l_2 \geqslant \cdots \geqslant l_i$, $k_1 = l_1, \ldots, k_{i-1} = l_{i-1}$, and $l_i < k_i$. If we apply T^{l_i} to $V = Z_1 \oplus \cdots \oplus Z_p$, we get $T^{l_i}(V) = T^{l_i}(Z_1) \oplus \cdots \oplus T^{l_i}(Z_i) \oplus \cdots \oplus T^{l_i}(Z_p)$. From Lemma 3.29(c), we have $\dim T^{l_i}(Z_1) = k_1 - l_i, \ldots$, $\dim T^{l_i}(Z_i) = k_i - l_i$. In particular,

3.32: $$\dim T^{l_i}(V) \geqslant (k_1 - l_i) + \cdots + (k_i - l_i)$$

On the other hand, $V = U_1 \oplus \cdots \oplus U_i \oplus \cdots \oplus U_s$, and Lemma 3.29(c) implies $T^{l_i}(U_j) = 0$ whenever $j \geqslant i$. Thus, $T^{l_i}(V) = T^{l_i}(U_1) \oplus \cdots \oplus T^{l_i}(U_{i-1})$. In particular,

3.33: $\dim T^{l_i}(V) = (l_1 - l_i) + \cdots + (l_{i-1} - l_i) = (k_1 - l_i) + \cdots + (k_{i-1} - l_i)$

Comparing the inequalities in 3.32 and 3.33 gives us $k_i - l_i \leqslant 0$. Thus, $k_i \leqslant l_i$, which is impossible. $\square$

We have now shown that the invariants $k_1 \geqslant \cdots \geqslant k_p$ of a nilpotent linear transformation T are unique. We can extend the definition of invariants to nilpotent matrices in the obvious way. Suppose $A \in M_{n \times n}(F)$ is nilpotent of index k. Then A defines a linear transformation $T : F^n \to F^n$, where $T(\delta_i) = \text{Col}_i(A)^t$, $i = 1, \ldots, n$. Here $\underline{\delta} = \{\delta_1, \ldots, \delta_n\}$ is as usual the canonical basis of F^n. Clearly $\Gamma(\underline{\delta}, \underline{\delta})(T) = A$. Since $\Gamma(\underline{\delta}, \underline{\delta})(\cdot) : \mathscr{E}(F^n) \to M_{n \times n}(F)$ is an algebra isomorphism, T is nilpotent of index k. We define the invariants of A to be those of T. Thus, if $k_1 \geqslant \cdots \geqslant k_p$ are the invariants of A, then by definition $k_1 \geqslant \cdots \geqslant k_p$ are the invariants of T, where $A = \Gamma(\underline{\delta}, \underline{\delta})(T)$. Since $\Gamma(\underline{\delta}, \underline{\delta})(\cdot) : \mathscr{E}(F^n) \to M_{n \times n}(F)$ is an isomorphism, this definition makes perfectly good sense. Note that if A has invariants $k_1 \geqslant \cdots \geqslant k_p$, then Theorem 3.21 implies A is similar to

$$\begin{bmatrix} N_{k_1} & & 0 \\ & \ddots & \\ 0 & & N_{k_p} \end{bmatrix}$$

We can now state the following corollary to Theorem 3.31:

Corollary 3.34: Two nonzero nilpotent matrices $A, B \in M_{n \times n}(F)$ are similar if and only if they have the same invariants.

Proof: Let $k_1 \geqslant \cdots \geqslant k_p$ and $l_1 \geqslant \cdots l_s$ be the invariants of A and B respectively. Then our comments above imply that A is similar to

$$\begin{bmatrix} N_{k_1} & & 0 \\ & \ddots & \\ 0 & & N_{k_p} \end{bmatrix}$$

and B is similar to

$$\begin{bmatrix} N_{l_1} & & 0 \\ & \ddots & \\ 0 & & N_{l_s} \end{bmatrix}$$

N')w similarity is clearly an equivalence relation $\equiv$ on $M_{n \times n}(F)$. If A and B have t e same invariants, then we have

$$A \equiv \begin{bmatrix} N_{k_1} & & 0 \\ & \ddots & \\ 0 & & N_{k_p} \end{bmatrix} = \begin{bmatrix} N_{l_1} & & 0 \\ & \ddots & \\ 0 & & N_{l_s} \end{bmatrix} \equiv B$$

T ius, A and B are similar. Conversely, suppose A and B are similar. Then

$$\begin{bmatrix} N_{k_1} & & 0 \\ & \ddots & \\ 0 & & N_{k_p} \end{bmatrix} \quad \text{and} \quad \begin{bmatrix} N_{l_1} & & 0 \\ & \ddots & \\ 0 & & N_{l_s} \end{bmatrix}$$

are similar. These two matrices then describe the same linear transformation $T \colon F^n \to F^n$ relative to two different bases, say $\underline{\alpha}$ and $\underline{\alpha}'$ of F^n (see Theorem 3.28, of Chapter I). If

$$\Gamma(\underline{\alpha}, \underline{\alpha})(T) = \begin{bmatrix} N_{k_1} & & 0 \\ & \ddots & \\ 0 & & N_{k_p} \end{bmatrix}$$

then $F^n = Z_1 \oplus \cdots \oplus Z_p$ with Z_i a T-cyclic subspace of F^n of dimension k_i. Similarly,

$$\Gamma(\underline{\alpha}', \underline{\alpha}')(T) = \begin{bmatrix} N_{l_1} & & 0 \\ & \ddots & \\ 0 & & N_{l_s} \end{bmatrix}$$

implies $F^n = U_1 \oplus \cdots \oplus U_s$ with U_j a T-cyclic subspace of F^n of dimension l_j. Theorem 3.31 now implies $p = s$ and $k_1 = l_1, \ldots, k_p = l_p$. Thus, the invariants of A and B are the same. $\square$

There is another application of Theorem 3.30 that is worth mentioning here. Suppose A is a nonzero nilpotent matrix in $M_{n \times n}(F)$. Then there exists an invertible matrix P such that

3.35:
$$PAP^{-1} = \begin{bmatrix} N_{k_1} & & 0 \\ & \ddots & \\ 0 & & N_{k_p} \end{bmatrix}$$

In 3.35, $k_1 \geqslant \cdots \geqslant k_p$ are the invariants of A. It is often important to compute P. We can find an invertible matrix P satisfying 3.35 by paying careful attention to the proof of Theorem 3.30. Define $T: F^n \to F^n$ by $T(\delta_i) = Col_i(A)^t$ as usual. Construct the T-cyclic decomposition of F^n, that is, $F^n = Z_1 \oplus \cdots \oplus Z_p$, given in Theorem 3.30. If $Z_1 = L(\{\alpha_1, T(\alpha_1), \ldots, T^{k_1-1}(\alpha_1)\}), \ldots, Z_p = L(\{\alpha_p, T(\alpha_p), \ldots, T^{k_p-1}(\alpha_p)\})$, then $\underline{\alpha} = \{\alpha_1, T(\alpha_1), \ldots, T^{k_1-1}(\alpha_1), \ldots, \alpha_p, T(\alpha_p), \ldots, T^{k_p-1}(\alpha_p)\}$ is a basis of V for which

$$\Gamma(\underline{\alpha}, \underline{\alpha})(T) = \begin{bmatrix} N_{k_1} & & 0 \\ & \ddots & \\ 0 & & N_{k_p} \end{bmatrix}$$

We now have the following equation (Theorem 3.28 of Chapter I):

3.36:
$$M(\underline{\delta}, \underline{\alpha})\Gamma(\underline{\delta}, \underline{\delta})(T)M(\underline{\delta}, \underline{\alpha})^{-1} = \Gamma(\underline{\alpha}, \underline{\alpha})(T)$$

Recall that $M(\underline{\delta}, \underline{\alpha})$ is the change of basis matrix given by $M(\underline{\delta}, \underline{\alpha}) = ([\delta_1]_{\underline{\alpha}}|\cdots|[\delta_n]_{\underline{\alpha}})$. Hence, if $P = M(\underline{\delta}, \underline{\alpha})$, then 3.36 becomes 3.35. Since we know $\underline{\alpha}$, $M(\underline{\delta}, \underline{\alpha})^{-1} = M(\underline{\alpha}, \underline{\delta})$ is the easier matrix to compute. Thus, to compute P, construct $\underline{\alpha}$ and invert $M(\underline{\alpha}, \underline{\delta})$.

Example 3.37: Suppose

$$A = \begin{bmatrix} 0 & 0 & 0 \\ 1 & 0 & 0 \\ 1 & 1 & 0 \end{bmatrix} \in M_{3 \times 3}(\mathbb{Q}).$$

Then

$$A^2 = \begin{bmatrix} 0 & 0 & 0 \\ 0 & 0 & 0 \\ 1 & 0 & 0 \end{bmatrix}$$

and $A^3 = 0$. Thus, A is nilpotent of index 3. The matrix A defines a linear transformation $T: \mathbb{Q}^3 \to \mathbb{Q}^3$ given by $T(\delta_1) = \delta_2 + \delta_3$, $T(\delta_2) = \delta_3$, and $T(\delta_3) = 0$.

In particular, $T^2(\delta_1) = \delta_3$. Thus, Z_1 in Theorem 3.30 is given by $Z_1 = L(\{\delta_1, T(\delta_1) = \delta_2 + \delta_3, T^2(\delta_1) = \delta_3\})$. $k_1 = 3$, and

$$\Gamma(\underline{\alpha}, \underline{\alpha})(T) = N_{k_1} = \begin{bmatrix} 0 & 0 & 0 \\ 1 & 0 & 0 \\ 0 & 1 & 0 \end{bmatrix}$$

$$P^{-1} = M(\underline{\alpha}, \underline{\delta}) = \begin{bmatrix} 1 & 0 & 0 \\ 0 & 1 & 0 \\ 0 & 1 & 1 \end{bmatrix} \quad \text{and} \quad P = M(\underline{\delta}, \underline{\alpha}) = \begin{bmatrix} 1 & 0 & 0 \\ 0 & 1 & 0 \\ 0 & -1 & 1 \end{bmatrix}$$

The reader can easily check that $PAP^{-1} = N_{k_1}$.   □

Finally, let us say a few words about how many similarity classes of nilpotent matrices there are in $M_{n \times n}(F)$. By Corollary 3.34, the number of similarity classes of nilpotent $n \times n$ matrices is the number of partitions of n, $n = k_1 + \cdots + k_p$, with $k_1 \geqslant \cdots \geqslant k_p$. Let us denote this number by $\mathscr{P}(n)$. The function $\mathscr{P}(n)$ has been studied intensely by combinatorialists. The value of $\mathscr{P}(n)$ for $n = 1, \ldots, 6$ is as follows: $\mathscr{P}(1) = 1$, $\mathscr{P}(2) = 2$, $\mathscr{P}(3) = 3$, $\mathscr{P}(4) = 5$, $\mathscr{P}(5) = 7$, and $\mathscr{P}(6) = 11$.

EXERCISES FOR SECTION 3

(1) Find all the eigenvectors of the map $T^{\mathbb{C}}$ given in Example 3.4.

(2) Suppose A is a lower triangular matrix of the form

$$A = \begin{bmatrix} a_{11} & 0 & \cdots & 0 \\ a_{21} & a_{22} & \cdots & 0 \\ \vdots & \vdots & \ddots & \vdots \\ a_{n1} & a_{n2} & \cdots & a_{nn} \end{bmatrix}$$

Compute the characteristic polynomial of A and $\mathscr{S}_F(A)$.

(3) Let $T \in \mathscr{E}(V)$ and $g(X) \in F[X]$. If c is an eigenvalue of T in F show that $g(c)$ is an eigenvalue of $g(T)$.

(4) Let

$$A = \begin{bmatrix} 18 & -9 & -6 \\ 17 & -9 & -5 \\ 25 & -12 & -9 \end{bmatrix} \in M_{3 \times 3}(\mathbb{Q})$$

Show that A is nilpotent. Find P such that PAP^{-1} has the subdiagonal form given in Theorem 3.21.

(5) Let $A \in M_{n \times n}(F)$ such that $R(c_A(X)) \subseteq F$. Show that A is similar to a lower triangular matrix.

(6) Let $T \in \mathscr{E}(V)$. Suppose $R(c_T) = \{0\}$. Show that T must be nilpotent.

(7) Find up to similarity all possible nilpotent matrices in $M_{6 \times 6}(\mathbb{Q})$.

(8) Find all eigenvalues and eigenvectors for

$$A = \begin{bmatrix} 2 & 1 & 1 \\ 2 & 3 & 2 \\ 1 & 1 & 2 \end{bmatrix} \in M_{3 \times 3}(\mathbb{Q})$$

(9) Let $T \in \mathscr{E}(V)$, and suppose $c_T(X) = \prod_{i=1}^{r} (X - c_i)^{n_i}$ in $F[X]$. Show that V has a basis of eigenvectors of T if and only if $\dim(\ker(T - c_i)) = n_i$ for all $i = 1, \ldots, r$.

(10) Let $A \in M_{n \times n}(F)$ be a diagonal matrix. Suppose $c_A(X) = \prod_{i=1}^{r}(X - c_i)^{n_i}$ in $F[X]$. Set $W = \{B \in M_{n \times n}(F) \,|\, BA = AB\}$. Show that $\dim(W) = n_1^2 + \cdots + n_r^2$.

(11) Let T be a projection of V onto a subspace W. Compute the spectrum $\mathscr{S}_F(T)$ of T. Do the same if T is an involution ($T^2 = I_V$).

(12) Compute the spectrum of $D: V_n \to V_n$ (notation as in Exercise 18, Section 2 of Chapter II).

(13) Let $A \in M_{n \times n}(F)$ and $B \in M_{m \times m}(F)$. Show that $\mathscr{S}_F(A \otimes B) = \{xy \,|\, x \in \mathscr{S}_F(A)$ and $y \in \mathscr{S}_F(B)\}$.

(14) Let $A, B \in M_{n \times n}(\mathbb{C})$. Show that if $AB = BA$, then A and B have a common eigenvector. Do A and B have a common eigenvalue?

(15) Suppose $A, B \in M_{n \times n}(\mathbb{C})$ and at least one of them is nonsingular. If AB is similar to a diagonal matrix, prove that BA is also similar to a diagonal matrix.

(16) Given an example that shows that if both matrices in Exercise 15 are singular, then BA need not be similar to a diagonal matrix.

(17) Use Corollary 3.34 to construct two matrices $A, B \in M_{4 \times 4}(\mathbb{R})$ such that $c_A(X) = c_B(X)$, and $m_A(X) = m_B(X)$, but A and B are not similar.

(18) If A and D are square matrices with A nonsingular, show that

$$\det\begin{pmatrix} A & B \\ C & D \end{pmatrix} = (\det A)(\det(D - CA^{-1}B))$$

(19) Let $S, T \in \mathscr{E}(V)$. Show that $\mathscr{S}_F(ST) = \mathscr{S}_F(TS)$. Do TS and ST have the same eigenvectors? (Note: Exercise 18 gives an easy proof of this problem if S and T are represented by symmetric matrices.)

4. THE JORDAN CANONICAL FORM

In this section, we shall use Theorem 3.21 to present a canonical form for those $T \in \mathscr{E}(V) = \text{Hom}_F(V, V)$ that have the property that the roots of $c_T(X)$ all lie in F. If $R(c_T(X)) \subseteq F$, then we know from Section 1 that c_T factors in $F[X]$ as follows:

4.1:
$$c_T(X) = \prod_{i=1}^{r} (X - c_i)^{n_i}$$

In equation 4.1, $\{c_1, \ldots, c_r\} = \mathscr{S}_F(T)$ and $\sum_{i=1}^{r} n_i = n = \dim_F(V)$. When we use notation as in 4.1, it is always understood that $c_1, \ldots, c_r$ are distinct elements in F. We begin with the following decomposition theorem:

Theorem 4.2: Let $T \in \mathscr{E}(V)$ and suppose $R(c_T(X)) \subseteq F$. Factor $c_T(X)$ as in equation 4.1. Set $V_i = \ker(T - c_i)^{n_i}$ for $i = 1, \ldots, r$. Then

(a) Each V_i is a nonzero, T-invariant subspace of V.
(b) $V = V_1 \oplus \cdots \oplus V_r$
(c) $\dim V_i = n_i$, $i = 1, \ldots, r$.
(d) If T_i denotes the restriction of T to V_i, then there exists a basis $\underline{\alpha}_i$ of V_i such that

4.3:
$$\Gamma(\underline{\alpha}_i, \underline{\alpha}_i)(T_i) = c_i I_{n_i} + M_i, \qquad i = 1, \ldots, r$$

In equation 4.3, M_i is a nilpotent matrix with index of nilpotency at most n_i.

Proof: (a) $\mathscr{S}_F(T) = \{c_1, \ldots, c_r\}$ by Theorem 3.3. Thus, for each c_i there exists a nonzero vector α_i such that $T(\alpha_i) = c_i \alpha_i$. In particular, $\alpha_i \in \ker(T - c_i)^{n_i} = V_i$. This shows $V_i \neq (0)$. Since T commutes with any polynomial in T, $T(T - c_i)^{n_i} = (T - c_i)^{n_i} T$. In particular, if $\beta \in V_i$, then $(T - c_i)^{n_i} T(\beta) = T(T - c_i)^{n_i}(\beta) = 0$. Hence $T(\beta) \in V_i$. This proves (a).

(b) For each $i = 1, \ldots, r$, set $h_i(X) = \prod_{j \neq i} (X - c_j)^{n_j}$. Then $h_1(X), \ldots, h_r(X)$ have no common factor in $F[X]$. In particular, g.c.d.$(h_1 \ldots, h_r) = 1$. It now follows from 1.11 that there exist polynomials $a_1(X), \ldots, a_r(X) \in F[X]$ such that $a_1 h_1 + \cdots + a_r h_r = 1$. Now consider the algebra homomorphism $\varphi: F[X] \to \mathscr{E}(V)$ given by $\varphi(X) = T$. Applying φ to $\sum_{i=1}^{r} a_i h_i = 1$ gives us $a_1(T)h_1(T) + \cdots + a_r(T)h_r(T) = I_V$. Also, $h_i(X)(X - c_i)^{n_i} = c_T(X)$. Therefore, $(T - c_i)^{n_i} h_i(T) = 0$. In particular, $\text{Im}(h_i(T)) \subseteq V_i$ for every $i = 1, \ldots, r$. Since each V_i is T-invariant, we have $\text{Im}(a_i(T)h_i(T)) \subseteq V_i$ for each $i = 1, \ldots, r$. Now let $\alpha \in V$. Then $\alpha = I_V(\alpha) = (a_1(T)h_1(T) + \cdots + a_r(T)h_r(T))(\alpha) = a_1(T)h_1(T)(\alpha) + \cdots + a_r(T)h_r(T)(\alpha) \in V_1 + \cdots + V_r$. Thus, $V = V_1 + \cdots + V_r$. To finish the proof of (b), we must show $V_i \cap (\sum_{j \neq i} V_j) = (0)$. Fix

$i = 1, \ldots, r$, and note that g.c.d.$(h_i(X), (X - c_i)^{n_i}) = 1$. Again by 1.11, there exist polynomials $p_i(X)$ and $q_i(X)$ in $F[X]$ such that $p_i h_i + q_i (X - c_i)^{n_i} = 1$. If $\alpha \in V_i \cap (\sum_{j \ne i} V_j)$, then $(T - c_i)^{n_i}(\alpha) = 0$ since $\alpha \in V_i$. $h_i(\alpha) = 0$ since $\alpha \in \sum_{j \ne i} V_j$. Thus, $\alpha = I_V(\alpha) = [p_i(T)h_i(T) + q_i(T)(T - c_i)^{n_i}](\alpha) = p_i(T)h_i(T)(\alpha) + q_i(T)(T - c_i)^{n_i}(\alpha) = 0 + 0 = 0$. This completes the proof of (b).

(c) Fix $i = 1, \ldots, r$, and let T_i denote the restriction of T to V_i. Then $V_i = \ker(T - c_i)^{n_i}$ implies $(T_i - c_i)^{n_i} = 0$. In particular, the minimal polynomial, $m_{T_i}(X)$, of T_i on V_i must divide $(X - c_i)^{n_i}$. Thus, $m_{T_i}(X) = (X - c_i)^{k_i}$ for some $k_i \leqslant n_i$. By Corollary 2.38, $c_{T_i}(X) = (X - c_i)^{l_i}$ for some $l_i \geqslant k_i$. $l_i = \dim V_i$, and by Theorem 2.43, we have

4.4:
$$\prod_{i=1}^{r} (X - c_i)^{n_i} = c_T(X) = \prod_{i=1}^{r} c_{T_i}(X) = \prod_{i=1}^{r} (X - c_i)^{l_i}$$

Theorem 1.7 now implies $l_i = n_i$ for all $i = 1, \ldots, r$.

(d) On each V_i, $T_i = c_i I_{V_i} + (T_i - c_i I_{V_i})$. We have seen in the proof of (c) that each $T_i - c_i I_{V_i}$ is nilpotent on V_i of index $k_i \leqslant n_i = \dim V_i$. By Theorem 3.21, there exists a basis $\underline{\alpha}_i$ of V_i such that

4.5:
$$\Gamma(\underline{\alpha}_i, \underline{\alpha}_i)(T_i - c_i I_{V_i}) = \begin{bmatrix} N_{k_{i1}} & & 0 \\ & \ddots & \\ 0 & & N_{k_{ip(i)}} \end{bmatrix} \doteq M_i.$$

In equation 4.5, $k_i = k_{i_1} \geqslant \cdots \geqslant k_{ip(i)}$ are the (unique) invariants of $T_i - c_i I_{V_i}$. We now have $\Gamma(\underline{\alpha}_i, \underline{\alpha}_i)(T_i) = \Gamma(\underline{\alpha}_i, \underline{\alpha}_i)(c_i I_{V_i} + (T_i - c_i I_{V_i})) = c_i \Gamma(\underline{\alpha}_i, \underline{\alpha}_i)(I_{V_i}) + \Gamma(\underline{\alpha}_i, \underline{\alpha}_i)(T_i - c_i I_{V_i}) = c_i I_{n_i} + M_i$. This gives us equation 4.3 and completes the proof of Theorem 4.2. $\square$

We have already proved our next theorem, but let us introduce a definition first.

Definition 4.6: Any matrix of the form

$$cI_k + N_k = \begin{bmatrix} c & & & 0 \\ 1 & \ddots & & \\ & \ddots & \ddots & \\ 0 & & 1 & c \end{bmatrix}$$

is called a Jordan block of size k belonging to c. A square matrix J is called a Jordan matrix if J has the form $J = \text{diag}(J_1, \ldots, J_t)$, where $J_1, \ldots, J_t$ are Jordan blocks of various sizes.

The computations after equation 4.5 show that there is a basis $\underline{\alpha}_i$ of V_i such that $\Gamma(\underline{\alpha}_i, \underline{\alpha}_i)(T_i)$ has the following form:

4.7:
$$\Gamma(\underline{\alpha}_i, \underline{\alpha}_i)(T_i) = \begin{bmatrix} B(k_{i1}) & & 0 \\ & \ddots & \\ 0 & & B(k_{ip(i)}) \end{bmatrix} \doteq J_i$$

In equation 4.7, $k_{i1} \geqslant \cdots \geqslant k_{ip(i)}$ are the invariants of $T_i - c_i I_{V_i}$ on V_i. Each $B(k_{ij})$ is a Jordan block of size k_{ij} belonging to c_i. Thus, $B(k_{ij}) = c_i I_{k_{ij}} + N_{k_{ij}}$ for $j = 1, \ldots, p(i)$.

If we now set $\underline{\alpha} = \bigcup_{i=1}^{r} \underline{\alpha}_i$, then $\underline{\alpha}$ is a basis of V. Since $V = V_1 \oplus \cdots \oplus V_r$, equation 2.42 implies

4.8:
$$\Gamma(\underline{\alpha}, \underline{\alpha})(T) = \begin{bmatrix} J_1 & & 0 \\ & \ddots & \\ 0 & & J_r \end{bmatrix} \doteq J$$

The representation of T given in equation 4.8 is called a Jordan canonical form of T. We shall see shortly that J is unique up to a permutation of its blocks $J_1, \ldots, J_r$. We have now proved the following theorem:

Theorem 4.9: Let $T \in \mathscr{E}(V)$, and suppose the roots of the characteristic polynomial of T all lie in F. Write $c_T(X)$ as in equation 4.1. Then there exists a basis $\underline{\alpha}$ of V such that

$$\Gamma(\underline{\alpha}, \underline{\alpha})(T) = \begin{bmatrix} J_1 & & 0 \\ & \ddots & \\ 0 & & J_r \end{bmatrix}$$

where each

$$J_i = \begin{bmatrix} B(k_{i1}) & & 0 \\ & \ddots & \\ 0 & & B(k_{ip(i)}) \end{bmatrix}$$

For each $i = 1, \ldots, r$, the integers $k_{i1} \geqslant \cdots \geqslant k_{ip(i)}$ are the invariants of the nilpotent transformation $T - c_i$ on $V_i = \ker(T - c_i)^{n_i}$. $B(k_{i1}), \ldots, B(k_{ip(i)})$ are Jordan blocks of sizes $k_{i1}, \ldots, k_{ip(i)}$ respectively belonging to c_i. $\quad\square$

We can restate Theorem 4.9 in terms of matrices as follows.

Corollary 4.10: Let $A \in M_{n \times n}(F)$, and suppose $c_A(X) = \prod_{i=1}^{r}(X - c_i)^{n_i}$ with $c_1, \ldots, c_r \in F$. Then A is similar to a Jordan matrix J of the following form:

$$J = \begin{bmatrix} J_1 & & 0 \\ & \ddots & \\ 0 & & J_r \end{bmatrix}$$

with

$$J_i = \begin{bmatrix} B(k_{i1}) & & 0 \\ & \ddots & \\ 0 & & B(k_{ip(i)}) \end{bmatrix}$$

For each $i = 1, \ldots, r$, $k_{i1} \geqslant \cdots \geqslant k_{ip(i)}$, and $B(k_{ij}) = c_i I_{k_{ij}} + N_{k_{ij}}$. $\square$

The constants $k_{i1} \geqslant \cdots \geqslant k_{ip(i)}$ that appear in Corollary 4.10 are computed from the natural linear transformation $T: F^n \to F^n$ associated with A. Namely, if $\underline{\delta}$ is the canonical basis of F^n, define T by $T(\delta_j) = Col_j(A)^t$. Then $\Gamma(\underline{\delta}, \underline{\delta})(T) = A$, and $k_{i1} \geqslant \cdots \geqslant k_{ip(i)}$ are the invariants of $T - c_i$ on $\ker(T - c_i)^{n_i}$.

Example 4.11: Let

$$A = \begin{bmatrix} 19 & -6 & -9 \\ 17 & -4 & -9 \\ 25 & -9 & -11 \end{bmatrix} \in M_{3 \times 3}(\mathbb{Q})$$

A simple computation shows $c_A(X) = X^3 - 4X^2 + 5X - 2 = (X - 1)^2(X - 2)$. Thus, $R(c_A) = \{1, 2\} \subseteq \mathbb{Q}$.

Let $T: \mathbb{R}^3 \to \mathbb{R}^3$ be given by $T(\delta_j) = Col_j(A)^t$. Then $\Gamma(\underline{\delta}, \underline{\delta})(T) = A$ and $c_T(X) = (X - 1)^2(X - 2)$. Theorem 4.2 implies

4.12: $\mathbb{R}^3 = \ker(T - I)^2 \oplus \ker(T - 2I)$

An easy computation shows

4.13: (a) $\Gamma(\underline{\delta}, \underline{\delta})(T - I) = A - I = \begin{bmatrix} 18 & -6 & -9 \\ 17 & -5 & -9 \\ 25 & -9 & -12 \end{bmatrix}$

(b) $\Gamma(\underline{\delta}, \underline{\delta})[(T - I)^2] = (A - I)^2 = \begin{bmatrix} -3 & 3 & 0 \\ -4 & 4 & 0 \\ -3 & 3 & 0 \end{bmatrix}$

(c) $\Gamma(\underline{\delta}, \underline{\delta})(T - 2I) = A - 2I = \begin{bmatrix} 17 & -6 & -9 \\ 17 & -6 & -9 \\ 25 & -9 & -13 \end{bmatrix}$

Now 4.13(b) implies that $V_1 = \ker(T - I)^2$ is a cyclic $(T - I)$-subspace with basis $\underline{\alpha}_1 = \{\alpha_1 = (1, 1, 1), \alpha_2 = (3, 3, 4)\}$. If T_1 denotes the restriction of T to V_1, then

$$\Gamma(\underline{\alpha}_1, \underline{\alpha}_1)(T_1) = \Gamma(\underline{\alpha}_1, \underline{\alpha}_1)(T_1 - I) + \Gamma(\underline{\alpha}_1, \underline{\alpha}_1)(I) = \begin{pmatrix} 0 & 0 \\ 1 & 0 \end{pmatrix} + \begin{pmatrix} 1 & 0 \\ 0 & 1 \end{pmatrix} = \begin{pmatrix} 1 & 0 \\ 1 & 1 \end{pmatrix}$$

Equation 4.13(c) implies $V_2 = \ker(T - 2I)$ is spanned by the single vector $\alpha_3 = (3, 4, 3)$. Thus, $\underline{\alpha} = \{\alpha_1, \alpha_2, \alpha_3\}$ is a basis of V and equation 2.42 implies

4.14:
$$\Gamma(\underline{\alpha}, \underline{\alpha})(T) = \begin{bmatrix} 1 & 0 & 0 \\ 1 & 1 & 0 \\ 0 & 0 & 2 \end{bmatrix} = J$$

J is clearly a Jordan canonical form of T (or A). If we set

$$M(\underline{\alpha}, \underline{\delta}) = \begin{bmatrix} 1 & 3 & 3 \\ 1 & 3 & 4 \\ 1 & 4 & 3 \end{bmatrix} = M(\underline{\delta}, \underline{\alpha})^{-1}$$

then equation 3.36 implies

4.15:
$$J = \begin{bmatrix} 1 & 0 & 0 \\ 1 & 1 & 0 \\ 0 & 0 & 2 \end{bmatrix} = M(\underline{\delta}, \underline{\alpha}) \begin{bmatrix} 19 & -6 & -9 \\ 17 & -4 & -9 \\ 25 & -9 & -11 \end{bmatrix} M(\underline{\delta}, \underline{\alpha})^{-1}$$

where

$$M(\underline{\delta}, \underline{\alpha})^{-1} = M(\underline{\alpha}, \underline{\delta}) = \begin{bmatrix} 1 & 3 & 3 \\ 1 & 3 & 4 \\ 1 & 4 & 3 \end{bmatrix}$$

and

$$M(\underline{\delta}, \underline{\alpha}) = \begin{bmatrix} 1 & 3 & 3 \\ 1 & 3 & 4 \\ 1 & 4 & 3 \end{bmatrix}^{-1} = \begin{bmatrix} 7 & -3 & -3 \\ -1 & 0 & 1 \\ -1 & 1 & 0 \end{bmatrix} = P$$

Thus, A is similar to J via P.   □

Let us note that if F is algebraically closed, for example, $F = \mathbb{C}$, then the hypotheses of Theorem 4.9 and Corollary 4.10 are always satisfied. Thus, any $n \times n$ matrix A is similar to a Jordan matrix of the form given in Corollary 4.10. If F is not algebraically close, for example, $F = \mathbb{R}$, then no such representation may exist.

Example 4.16: Let

$$A = \begin{pmatrix} 0 & -1 \\ 1 & 0 \end{pmatrix} \in M_{2 \times 2}(\mathbb{R})$$

Then $c_A(X) = X^2 + 1$, and consequently, $R(c_A(X)) = \{\pm i\} \not\subset \mathbb{R}$. Since $n = 2$, there are only two types of Jordan matrices in $M_{2 \times 2}(\mathbb{R})$:

$$\begin{pmatrix} a_1 & 0 \\ 0 & a_2 \end{pmatrix} \quad \text{and} \quad \begin{pmatrix} a & 0 \\ 1 & a \end{pmatrix}$$

If A is similar to either one of these forms, then $X^2 + 1 = c_A(X) = (X - a_1)(X - a_2)$ or $(X - a)^2$. This is impossible since $a, a_1, a_2 \in \mathbb{R}$. Thus, A is not similar to any Jordan form over $\mathbb{R}$. $\square$

Let us take a closer look at the Jordan canonical form of T. Suppose $T \in \mathscr{E}(V)$, and $R(c_T(X)) \subseteq F$. Write $c_T(X)$ as in equation 4.1. Then Theorem 4.2 implies $V = V_1 \oplus \cdots \oplus V_r$ with $V_i = \ker(T - c_i)^{n_i}$, and $\dim V_i = n_i$. If T_i denotes the restriction of T to the T-invariant subspace V_i, then $T_i - c_i$ is nilpotent of index $k_i \leqslant n_i$. If we let $k_i = k_{i1} \geqslant \cdots \geqslant k_{ip(i)}$ denote the unique invariants of $T_i - c_i$ on V_i, then there exists a basis $\underline{\alpha}_i$ of V_i such that

4.17:
$$\Gamma(\underline{\alpha}_i, \underline{\alpha}_i)(T_i) = J_i = \begin{bmatrix} B(k_{i1}) & & 0 \\ & \ddots & \\ 0 & & B(k_{ip(i)}) \end{bmatrix}$$

In equation 4.17, each $B(k_{ij}) = c_i I_{k_{ij}} + N_{k_{ij}}$ for $j = 1, \ldots, p(i)$. Finally, if $\underline{\alpha} = \bigcup_{i=1}^{r} \underline{\alpha}_i$, then equation 2.42 implies

4.18:
$$\Gamma(\underline{\alpha}, \underline{\alpha})(T) = J = \begin{bmatrix} J_1 & & 0 \\ & \ddots & \\ 0 & & J_r \end{bmatrix}$$

The first thing to note here is that $k_{i1} \geqslant \cdots \geqslant k_{ip(i)}$ are the unique invariants of the nilpotent transformation $T_i - c_i$ on V_i. Since $\dim V_i = n_i$, Theorem 3.21 implies $k_{i1} + \cdots + k_{ip(i)} = n_i$. But, $k_{i1} + \cdots + k_{ip(i)}$ is the size of the square matrix J_i. Thus, J_i is an $n_i \times n_i$ matrix having the eigenvalue c_i running down its diagonal. Now, n_i is the multiplicity of the root c_i in $c_T(X)$. Thus, an eigenvalue c_i of T appears as many times on the principal diagonal of J as its multiplicity in the characteristic polynomial of T. These remarks together with Theorem 3.31 readily imply that the Jordan canonical form J of T is unique up to a permutation of its blocks $J_1, \ldots, J_r$. The corresponding matrix statement is the following theorem whose proof we leave as an exercise at the end of this section:

Theorem 4.19: Let A, $B \in M_{n \times n}(F)$ be matrices whose characteristic polynomials have all of their roots in F. Then A and B are similar if and only if they have the same Jordan canonical form $J = \text{diag}\{J_1, \ldots, J_r\}$ (except possibly for a permutation of the blocks $J_1, \ldots, J_r$). $\square$

Next note that k_{i1} is the index of nilpotency of $T_i - c_i$ on V_i. Therefore, the minimal polynomial of T_i on V_i is given by $m_{T_i}(X) = (X - c_i)^{k_{i1}}$. Theorem 2.43(c)

now implies the minimal polynomial of T is given by the following equation:

4.20:
$$m_T(X) = \prod_{i=1}^{r} (X - c_i)^{k_{i1}}$$

We can now prove the following interesting result:

Theorem 4.21: Let $T \in \mathscr{E}(V)$, and assume $c_T(X) = \prod_{i=1}^{r}(X - c_i)^{n_i} \in F[X]$. Then T can be represented by a diagonal matrix if and only if $m_T(X) = \prod_{i=1}^{r}(X - c_i)$, that is, every eigenvalue of T has multiplicity one in the minimal polynomial of T.

Proof: We have seen that the Jordan canonical form of T given in equation 4.18 is unique up to a permutation of the blocks $J_1, \ldots, J_r$. If T is represented by a diagonal matrix $\text{diag}(b_1, \ldots, b_n) = B$, then B is a Jordan canonical form of T. Hence, B is J up to some permutation of the blocks $J_1, \ldots, J_r$. In particular, every block J_i is diagonal. Thus, $k_{i1} = 1$ for every $i = 1, \ldots, r$. Then equation 4.20 implies $m_T(X) = \prod_{i=1}^{r}(X - c_i)$.

Conversely, if $m_T(X) = \prod_{i=1}^{r}(X - c_i)$, then equation 4.20 implies $k_{i1} = 1$. But $k_{i1} \geqslant \cdots \geqslant k_{ip(i)}$. Therefore, $1 = k_{i1} = \cdots = k_{ip(i)}$ for all $i = 1, \ldots, r$, and J is diagonal. $\square$

Let us rephrase Theorem 4.21 slightly.

Corollary 4.22: Let $T \in \text{Hom}_F(V, V)$. T has a diagonal matrix representation over F if and only if the minimal polynomial of T is a product of distinct linear factors in $F[X]$.

Proof: Suppose $m_T(X) = \prod_{i=1}^{r}(X - c_i) \in F[X]$, where $c_1, \ldots, c_r$ are distinct. We had seen in Corollary 2.38 that $c_T(X)$ and $m_T(X)$ have the same set of irreducible factors in $F[X]$. Thus, $c_T(X) = \prod_{i=1}^{r}(X - c_i)^{n_i}$ with $n_1 + \cdots + n_r = n = \dim V$. Hence, the hypotheses of Theorem 4.21 are satisfied, and we conclude that T can be represented by a diagonal matrix relative to some basis of V.

Conversely, suppose $\Gamma(\alpha, \alpha)(T) = \text{diag}(a_1, \ldots, a_n)$ for some basis α of V. Then $c_T(X) = \prod_{i=1}^{n}(X - a_i)$. Thus, again, the hypotheses of Theorem 4.21 are satisfied. We conclude $m_T(X)$ is a product of distinct linear factors in $F[X]$. $\square$

We shall finish this section with a second version of Theorem 4.2, which will be convenient in later sections.

Theorem 4.23: Let $T \in \mathscr{E}(V)$, and suppose $R(c_T(X)) \subseteq F$. Factor $c_T(X)$ as in equation 4.1. Then there exist linear transformations $P_1, \ldots, P_r$ and $N_1, \ldots, N_r$ in $\mathscr{E}(V)$ such that

(a) $P_1, \ldots, P_r$ are pairwise orthogonal idempotents whose sum is I_V.
(b) P_i and N_i are polynomials in T for all $i = 1, \ldots, r$.

(c) $\text{Im } P_i = V_i = \ker(T - c_i)^{n_i}$ for $i = 1, \ldots, r$.

(d) $T = \sum_{i=1}^{r}(c_iP_i + N_i)$.

(e) $P_jN_i = N_iP_j = \begin{cases} N_i & \text{if } i = j \\ 0 & \text{if } i \neq j \end{cases}$

(f) N_i is nilpotent of index at most n_i for each $i = 1, \ldots, r$.

Proof: For each $i = 1, \ldots, r$, set $V_i = \ker(T - c_i)^{n_i}$ and $h_i(X) = \prod_{j \neq i}(X - c_j)^{n_j}$. Then the following facts were proven in Theorem 4.2.

4.24: (i) $V = V_1 \oplus \cdots \oplus V_r$

(ii) $\sum_{i=1}^{r} a_i(X)h_i(X) = 1$ for some $a_1, \ldots, a_r \in F[X]$

(iii) $\text{Im}(h_i(T)) \subseteq V_i$ for each $i = 1, \ldots, r$

Set $P_i = a_i(T)h_i(T)$ for $i = 1, \ldots, r$. Then each P_i is a polynomial in T, and 4.24(ii) implies $P_1 + \cdots + P_r = I_V$. Thus, $V = \text{Im } P_1 + \cdots + \text{Im } P_r$. Since V_i is T-invariant, and $\text{Im } h_i(T) \subseteq V_i$, we see $\text{Im } P_i \subseteq V_i$. In particular, $\text{Im } P_1 + \cdots + \text{Im } P_r$ is a direct sum. Thus, $V = \text{Im } P_1 \oplus \cdots \oplus \text{Im } P_r$. Equation 4.24(i) now implies $\text{Im } P_i = V_i$ for $i = 1, \ldots, r$.

Since each V_i is T-invariant and each P_j is a polynomial in T, we have $\text{Im}(P_iP_j) \subseteq V_i \cap V_j$. In particular, $P_iP_j = 0$ whenever $i \neq j$. It is now easy to see that each P_i is idempotent. Let $\alpha \in V$. Write $\alpha = \alpha_1 + \cdots + \alpha_r$ with $\alpha_i \in V_i = \text{Im } P_i$. Then $P_i(\alpha_j) = 0$ whenever $i \neq j$. Therefore, $\alpha_1 + \cdots + \alpha_r = \alpha = I_V(\alpha) = P_1(\alpha) + \cdots + P_r(\alpha) = P_1(\alpha_1 + \cdots + \alpha_r) + \cdots + P_r(\alpha_1 + \cdots + \alpha_r) = P_1(\alpha_1) + \cdots + P_r(\alpha_r)$. We conclude that $\alpha_i = P_i(\alpha_i)$ for all $i = 1, \ldots, r$. Hence, $P_i^2(\alpha) = P_i(P_i(\alpha)) = P_i(P_i(\alpha_1 + \cdots + \alpha_r)) = P_i(P_i(\alpha_i)) = P_i(\alpha_i) = P_i(\alpha)$. Thus, each P_i is idempotent. We have now established (a) and (c) and the first part of (b).

Set $N_i = (T - c_i)P_i$, $i = 1, \ldots, r$. Since each P_i is a polynomial in T so is each N_i. Polynomials in T commute. So, $P_jN_i = N_iP_j = (T - c_i)P_iP_j = 0$ if $i \neq j$ and N_i if $i = j$. We have now established (b) and (e). Notice that $N_i(V) = (T - c_i)P_i(V) = P_i(T - c_i)(V) \subseteq \text{Im } P_i = V_i$. Also, $N_i(V_j) = (T - c_i)P_i(\text{Im } P_j) = 0$ whenever $i \neq j$.

Since

$$\sum_{i=1}^{r}(c_iP_i + N_i) = \sum_{i=1}^{r}(c_iP_i + TP_i - c_iP_i) = \sum_{i=1}^{r} TP_i = T\left(\sum_{i=1}^{r} P_i\right) = TI_V = T,$$

(d) is clear. It remains to prove (f).

Fix $i = 1, \ldots, r$, and let $\alpha \in V$. Then $N_i(\alpha) = (T - c_i)P_i(\alpha)$. But $P_i(\alpha) \in V_i = \ker(T - c_i)^{n_i}$. Since P_i and $T - c_i$ commute, we have $N_i^{n_i}(\alpha) = [(T - c_i)P_i]^{n_i}(\alpha) = (T - c_i)^{n_i}P_i^{n_i}(\alpha) = 0$. Therefore, N_i is nilpotent of index at most n_i. $\square$

EXERCISES FOR SECTION 4

(1) Prove Theorem 4.19. [*Hint:* Show that $\ker(A - c_i)^{n_i} \cong \ker(B - c_i)^{n_i}$ for every eigenvalue c_i.]

(2) Show that the Jordan canonical form for a transformation T (if it exists) is unique up to a permutation of its blocks $J_1, \ldots, J_r$.

(3) Find all Jordan forms for
   (a) All 8×8 matrices having $x^2(x - 2)^3$ as minimal polynomial.
   (b) All 6×6 matrices having $(x + 2)^4(x - 5)^2$ as characteristic polynomial.

(4) Find the Jordan canonical form J of

$$A = \begin{bmatrix} 3 & 4 & 1 \\ 0 & -1 & 1 \\ -1 & -3 & 1 \end{bmatrix}$$

and find the matrix P such that $PAP^{-1} = J$ over $\mathbb{R}$.

(5) Find the Jordan canonical form J of

$$A = \begin{bmatrix} 0 & -1 & 0 & 0 \\ 1 & 0 & 0 & 0 \\ 1 & 2 & 0 & 1 \\ -1 & 3 & -1 & 0 \end{bmatrix}$$

over $\mathbb{C}$. Find P such that $PAP^{-1} = J$.

(6) Let

$$A = \begin{bmatrix} 2 & 1 & -2 & 0 \\ 0 & 0 & 1 & 0 \\ 0 & 0 & 0 & 1 \\ 1 & 0 & 0 & 0 \end{bmatrix} \in M_{4 \times 4}(\mathbb{R})$$

Find P such that PAP^{-1} is in Jordan form.

(7) Give an example of two matrices $A, B \in M_{n \times n}(F)$ such that $c_A(X) = c_B(X)$ and $m_A(X) = m_B(X)$, but A and B have different Jordan canonical forms.

(8) Let N_k be the nilpotent matrix given in Example 3.20. Show that N_k is similar to its transpose N_k^t.

(9) A slightly different version of Exercise 9 of Section 3 is the following: Let $T \in \mathscr{E}(V)$, and suppose $R(c_T(X)) \subseteq F$. Show that T can be represented by a diagonal matrix if and only if whenever $(T - c)^p(\alpha) = 0$, then $(T - c)(\alpha) = 0$.

(10) Suppose A, $B \in M_{3 \times 3}(F)$ are nilpotent. If $m_A(X) = m_B(X)$, then prove A is similar to B. Compare this result with Exercise 17 of Section 3.

(11) Use Exercise 10 to prove the following: Let A, $B \in M_{n \times n}(F)$. Suppose $c_A(X) = c_B(X) = \prod_{i=1}^{r}(X - c_i)^{n_i}$, and $m_A(X) = m_B(X)$. If $n_i \leqslant 3$ for all $i = 1, \ldots, r$, then A is similar to B.

(12) Find the Jordan canonical form of the following matrix:

$$\begin{pmatrix} 0 & 0 & 0 & 1 \\ 1 & 0 & 0 & 1 \\ -3 & 2 & 1 & 1 \\ 3 & -6 & 1 & 4 \end{pmatrix}$$

(13) Let A, $B \in M_{n \times n}(F)$. Suppose

$$\begin{pmatrix} A & 0 \\ 0 & A \end{pmatrix}$$

is similar to

$$\begin{pmatrix} B & 0 \\ 0 & B \end{pmatrix}$$

in $M_{2n \times 2n}(F)$. Show that A is similar to B.

(14) What is the Jordan form of the linear map D: $V_3 \to V_3$ (notation as in Exercise 18, Section 2 of Chapter II].

(15) Let A, $B \in M_{n \times n}(\mathbb{C})$. Suppose $AB = BA$. Show there exists a P such that PAP^{-1} and PBP^{-1} are both in Jordan canonical form.

(16) Let V be a finite-dimensional vector space over $\mathbb{C}$. Classify all $T \in \mathscr{E}(V)$ such that V has only finitely many T-invariant subspaces.

5. THE REAL JORDAN CANONICAL FORM

In this section and the next, we take up the question of what canonical form for $T \in \text{Hom}_F(V, V)$ is available when the characteristic polynomial of T does not have all of its roots in F. When $F = \mathbb{R}$, we are able to construct a form surprisingly close to the Jordan canonical form of Section 4.

For the time being, let us assume F is an arbitrary field. Let V be a finite-dimensional vector space of dimension n over F. If $T \in \text{Hom}_F(V, V)$, then we have seen that the characteristic polynomial $c_T(X)$ is a monic polynomial of degree n in $F[X]$. Using Theorem 1.7, we know $c_T(X)$ has an essentially unique factorization of the following form:

5.1: $$c_T(X) = q_1(X)^{n_1} \cdots q_r(X)^{n_r}$$

In equation 5.1, $q_1(X), \ldots, q_r(X)$ are monic, irreducible polynomials in $F[X]$. When $i \neq j$, $q_i(X)$ and $q_j(X)$ are not associates. Each $n_i \geq 1$, and $n_1 \partial(q_1) + \cdots + n_r \partial(q_r) = \partial(c_T(X)) = n$.

From Corollary 2.38, we know the minimal polynomial $m_T(X)$ can be factored in $F[X]$ in the following way:

5.2: $$m_T(X) = q_1(X)^{m_1} \cdots q_r(X)^{m_r}$$

In equation 5.2, $1 \leq m_i \leq n_i$ for every $i = 1, \ldots, r$.

We shall need the following generalization of Theorem 4.2.

Theorem 5.3 (Primary Decomposition Theorem): Let $T \in \mathscr{E}(V)$, and write the characteristic and minimal polynomials of T as in equations 5.1 and 5.2. For each $i = 1, \ldots, r$, set $V_i = \ker(q_i(T)^{n_i})$. Then

(a) Each V_i is a nonzero, T-invariant subspace of V.

(b) $V = V_1 \oplus \cdots \oplus V_r$.

(c) $V_i = \ker(q_i(T)^{m_i})$, $i = 1, \ldots, r$.

(d) $\dim_F(V_i) = n_i \partial(q_i)$, $i = 1, \ldots, r$.

Proof: (a) T commutes with $q_i(T)^{n_i}$, and, thus, V_i is T-invariant. For each $i = 1, \ldots, r$, set $h_i(X) = \prod_{j \neq i} q_j(X)^{n_j}$. Since $h_i(X)$ is missing the factor $q_i(X)$, $m_T(X) \nmid h_i(X)$. In particular, $h_i(T) \neq 0$. Hence there exists a vector $\alpha \in V$ such that $h_i(T)(\alpha) \neq 0$. But $q_i(T)^{n_i}(h_i(T)(\alpha)) = (q_i(T)^{n_i}h_i(T))(\alpha) = c_T(T)(\alpha) = 0$. Thus, $h_i(T)(\alpha)$ is a nonzero vector in V_i. We have now proven (a) as well as the fact that $\mathrm{Im}[h_i(T)] \subseteq V_i$ for all $i = 1, \ldots, r$.

(b) The polynomials $h_1(X), \ldots, h_r(X)$ are clearly relatively prime in $F[X]$. It follows from 1.11 that there exist polynomials $a_1(X), \ldots, a_r(X) \in F[X]$ such that $a_1 h_1 + \cdots + a_r h_r = 1$. In particular, we have for any vector $\alpha \in V$, $\alpha = I_V(\alpha) = a_1(T)h_1(T)(\alpha) + \cdots + a_r(T)h_r(T)(\alpha) \in V_1 + \cdots + V_r$.

Suppose $\alpha \in V_i \cap (\sum_{j \neq i} V_j)$. $q_i(X)^{n_i}$ and $h_i(X)$ are obviously relatively prime. Hence, $Aq_i^{n_i} + Bh_i = 1$ for some $A, B \in F[X]$. Then $\alpha = I_V(\alpha) = A(T)q_i(T)^{n_i}(\alpha) + B(T)h_i(T)(\alpha) = 0 + 0 = 0$. This proves (b).

(c) Since $m_i \leq n_i$, $\ker(q_i(T)^{m_i}) \subseteq \ker(q_i(T)^{n_i}) = V_i$. Now the same arguments used in (a) and (b) when applied to $q_1(X)^{m_1}, \ldots, q_r(X)^{m_r}$ show that each subspace $\ker(q_i(T)^{m_i})$ is nonzero, and $V = \ker(q_1(T)^{m_1}) \oplus \cdots \oplus \ker(q_r(T)^{m_r})$. Comparing dimensions gives us $\ker(q_i(T)^{m_i}) = \ker(q_i(T)^{n_i})$ for all $i = 1, \ldots, r$.

(d) Let T_i denote the restriction of T to V_i. Then $q_i(T_i)^{n_i} = 0$. In particular, the minimal polynomial m_{T_i} of T_i must divide and hence be a power of $q_i(X)$. Corollary 2.38 then implies $c_{T_i}(X) = q_i(X)^{p_i}$ for some $p_i \geqslant 1$. Since $\dim V_i = \partial(c_{T_i})$, we have $\dim V_i = p_i \partial(q_i)$. On the other hand, Theorem 2.43(b) implies

$$q_1(X)^{n_1} \cdots q_r(X)^{n_r} = c_T(X) = \prod_{i=1}^{r} c_{T_i}(X) = \prod_{i=1}^{r} q_i(X)^{p_i}$$

Hence, $p_i = n_i$ for every $i = 1, \ldots, r$. $\square$

In this section, we shall use Theorem 5.3 in the following form:

Corollary 5.4: Let $T \in \mathscr{E}(V)$, and suppose c is a root of $c_T(X)$ in F. Write $c_T(X) = (X - c)^m q(X)$ in $F[X]$ with $X - c$ and $q(X)$ relatively prime. Then $V_1 = \ker(T - c)^m$ is a nonzero, T-invariant subspace of dimension m. Furthermore, $V = V_1 \oplus W$ for some T-invariant subspace W of V.

Proof: The complete factorization of $c_T(X)$ given in 5.1 has $(X - c)^m$ as one of its terms $q_i(X)^{n_i}$. We may assume $q_1(X)^{n_1} = (X - c)^m$. The result now follows from Theorem 5.3 with $W = \ker(q_2(T)^{n_2}) \oplus \cdots \oplus \ker(q_r(T)^{n_r})$. $\square$

We can now take up the question of what canonical form is available for T when $R(c_T(X)) \nsubseteq F$. The first thing one might try is to pass to the algebraic closure $\bar{F}$ of F. Thus, consider the extended map $T^{\bar{F}}$ on $V^{\bar{F}} = V \otimes_F \bar{F}$. As we have seen in Section 2, the characteristic polynomial $c_T(X)$ is the same as $c_{T^{\bar{F}}}(X)$. Since $\bar{F}$ is algebraically closed, $c_T(X)$ decomposes into linear factors in $\bar{F}[X]$ and Theorem 4.2 implies $T^{\bar{F}}$ has a Jordan canonical form $J \in M_{n \times n}(\bar{F})$. Of course, the entries in J may not all lie in F, but we can hope that if the relationship between F and $\bar{F}$ is special enough, we may be able to use J to produce some reasonable canonical form for T in $M_{n \times n}(F)$. This is precisely what happens when $F = \mathbb{R}$. Then $\bar{F} = \mathbb{C}$. By using complex conjugation $\sigma \in \text{Hom}_{\mathbb{R}}(\mathbb{C}, \mathbb{C})$ (See Section 2, Chapter II), we can convert a Jordan canonical form for $T^{\mathbb{C}}$ to a reasonable form for T itself over $\mathbb{R}$.

Let us set up the notation we shall use for the rest of this section. V will denote a vector space over $\mathbb{R}$ of dimension n. $V^{\mathbb{C}} = V \otimes_{\mathbb{R}} \mathbb{C}$ is the complexification of V. We shall identify a vector $\alpha \in V$ with its image $\alpha \otimes_{\mathbb{R}} 1$ in $V^{\mathbb{C}}$. Then, $V^{\mathbb{C}} = \{\sum_{k=1}^{p} z_k \alpha_k | z_k \in \mathbb{C}, \alpha_k \in V\}$. Note then that any vector $\alpha \in V^{\mathbb{C}}$ can be written uniquely in the form $\alpha = \mu + i\lambda$, where $\mu, \lambda \in V$ and $i = \sqrt{-1}$. We had also seen in Chapter II that σ extends to an $\mathbb{R}$-isomorphism $I_V \otimes_{\mathbb{R}} \sigma : V^{\mathbb{C}} \cong V^{\mathbb{C}}$. Recall that the value of $I_V \otimes_{\mathbb{R}} \sigma$ on a typical vector $\sum_{k=1}^{p} z_k \alpha_k \in V^{\mathbb{C}}$ is given by $(I_V \otimes_{\mathbb{R}} \sigma)(\sum_{k=1}^{p} z_k \alpha_k) = \sum_{k=1}^{p} \bar{z}_k \alpha_k$. We shall shorten our notation here and write $(I_V \otimes_{\mathbb{R}} \sigma)(\beta) = \bar{\beta}$ for any $\beta \in V^{\mathbb{C}}$. Thus, if $\beta = \sum_{k=1}^{p} z_k \alpha_k$, then $\bar{\beta} = \sum_{k=1}^{p} \bar{z}_k \alpha_k$. Equivalently, if $\beta = \mu + i\lambda$ with $\mu, \lambda \in V$, then $\bar{\beta} = \mu - i\lambda$. It is important to keep in mind here that $\beta \to \bar{\beta}$ is an $\mathbb{R}$-isomorphism of $V^{\mathbb{C}}$, but not a $\mathbb{C}$-isomorphism because it is not a $\mathbb{C}$-linear map.

Now suppose $T \in \text{Hom}_{\mathbb{R}}(V, V)$. As usual, we set $T^{\mathbb{C}} = T \otimes_{\mathbb{R}} I_{\mathbb{C}}$. Thus $T^{\mathbb{C}}$ is the $\mathbb{C}$-linear transformation on $V^{\mathbb{C}}$ given by $T^{\mathbb{C}}(\sum_{k=1}^{p} z_k \alpha_k) = \sum_{k=1}^{p} z_k T(\alpha_k)$ or equivalently $T^{\mathbb{C}}(\mu + i\lambda) = T(\mu) + iT(\lambda)$. We had observed in Theorem 2.26 of Chapter II that $T^{\mathbb{C}}$ commutes with conjugation $\beta \to \bar{\beta}$ on $V^{\mathbb{C}}$. More generally, we have the following fact:

5.5: Let $f(X) = \sum_{k=0}^{m} z_k X^k \in \mathbb{C}[X]$. Then for every $\alpha \in V^{\mathbb{C}}$, $\overline{f(T^{\mathbb{C}})(\alpha)} = \bar{f}(T^{\mathbb{C}})(\bar{\alpha})$.

In 5.5, $\bar{f} = \sum_{k=0}^{m} \bar{z}_k X^k$ is the conjugate of f. To prove 5.5, it clearly suffices to assume α is a vector of the form $\alpha = \beta \otimes_{\mathbb{R}} z$ for some $\beta \in V$, $z \in \mathbb{C}$. We then have

$$\overline{f(T^{\mathbb{C}})(\alpha)} = \overline{f(T^{\mathbb{C}})(\beta \otimes_{\mathbb{R}} z)} = \overline{\left\{ \sum_{k=0}^{m} T^k(\beta) \otimes_{\mathbb{R}} z_k z \right\}} = \sum_{k=0}^{m} \{ T^k(\beta) \otimes_{\mathbb{R}} \bar{z}_k \bar{z} \}$$

$$= \bar{f}(T^{\mathbb{C}})(\beta \otimes_{\mathbb{R}} \bar{z}) = \bar{f}(T^{\mathbb{C}})(\bar{\alpha})$$

One interesting application of equation 5.5 is the following lemma.

Lemma 5.6: If $z \in \mathscr{S}_{\mathbb{C}}(T^{\mathbb{C}})$, then $\bar{z} \in \mathscr{S}_{\mathbb{C}}(T^{\mathbb{C}})$.

Proof: If z is an eigenvalue of $T^{\mathbb{C}}$, then $T^{\mathbb{C}}(\alpha) = z\alpha$ for some nonzero eigenvector $\alpha \in V^{\mathbb{C}}$. $\bar{\alpha}$ is also nonzero and equation 5.5 implies $T^{\mathbb{C}}(\bar{\alpha}) = \bar{z}\bar{\alpha}$. Thus, if α is an eigenvector of $T^{\mathbb{C}}$ with corresponding eigenvalue z, then $\bar{\alpha}$ is an eigenvector of $T^{\mathbb{C}}$ with corresponding eigenvalue $\bar{z}$. $\square$

Now since $\mathbb{C}$ is algebraically closed, the characteristic polynomial of $T^{\mathbb{C}}$ (which is the same as the characteristic polynomial of T) factors into linear factors. Lemma 5.6 implies that if z is a root of $c_T(X)$, then $\bar{z}$ is also a root of $c_T(X)$. Thus, the spectrum of $T^{\mathbb{C}}$ can be written as follows:

5.7: $$\mathscr{S}_{\mathbb{C}}(T^{\mathbb{C}}) = \{ c_1, \ldots, c_r, z_1, \bar{z}_1, \ldots, z_t, \bar{z}_t \}$$

In equation 5.7, $c_1 \ldots, c_r$ denote the (distinct) real roots of $c_T(X)$, and z_1, $\bar{z}_1, \ldots, z_t, \bar{z}_t$ denote the (distinct) complex roots which are not real. We have $r \geqslant 0$, $t \geqslant 0$, and $r + 2t \leqslant n = \partial(c_T)$.

If $t = 0$, then $c_T(X)$ has only real roots. In this case, T has a Jordan canonical representation $J \in M_{n \times n}(\mathbb{R})$ and there is nothing left to say. Hence, throughout the rest of this discussion, we shall assume $t \geqslant 1$. Equation 5.7 implies the characteristic polynomial of $T^{\mathbb{C}}$ factors in $\mathbb{C}[X]$ as follows:

5.8: $$c_T(X) = \prod_{j=1}^{r} (X - c_j)^{n_j} \prod_{l=1}^{t} (X - z_l)^{p_l}(X - \bar{z}_l)^{q_l}.$$

Since the coefficients of $c_T(X)(= c_{T\mathbb{C}}(X))$ are all real, conjugating $c_T(X)$ merely interchanges z_l and $\bar{z}_l$ in equation 5.8. In particular, $p_l = q_l$ for all $l = 1, \ldots, t$. Thus, the multiplicity of z_l and $\bar{z}_l$ in $c_T(X)$ is the same. We can now rewrite equation 5.8 as follows:

5.9:
$$c_T(X) = \prod_{j=1}^{r} (X - c_j)^{n_j} \prod_{l=1}^{t} (X - z_l)^{p_l}(X - \bar{z}_l)^{p_l}$$

Since $\partial(c_T) = \dim_{\mathbb{R}} V = n$, we must have $\sum_{j=1}^{r} n_j + 2\sum_{l=1}^{t} p_l = n$.

Let us fix $l = 1, \ldots, t$ and consider the two $T^{\mathbb{C}}$-invariant subspaces $\ker(T^{\mathbb{C}} - z_l)^{p_l}$ and $\ker(T^{\mathbb{C}} - \bar{z}_l)^{p_l}$ in $V^{\mathbb{C}}$. These two subspaces have dimension p_l over $\mathbb{C}$ by Theorem 4.2. Furthermore, conjugation $\alpha \to \bar{\alpha}$ is an $\mathbb{R}$-isomorphism from $\ker(T^{\mathbb{C}} - z_l)^{p_l} \to \ker(T^{\mathbb{C}} - \bar{z}_l)^{p_l}$. This leads to an important observation.

Lemma 5.10: The nilpotent transformation $T^{\mathbb{C}} - z_l$ on $\ker(T^{\mathbb{C}} - z_l)^{p_l}$ has the same invariants as $T^{\mathbb{C}} - \bar{z}_l$ on $\ker(T^{\mathbb{C}} - \bar{z}_l)^{p_l}$.

Proof: If W is a $\mathbb{C}$-subspace of $V^{\mathbb{C}}$, let us denote the conjugate of W by $\bar{W}$. Thus, $\bar{W} = \{\bar{\alpha} \mid \alpha \in W\}$. The proof of 5.10 consists of the following observations, which are all easy to prove. If W is a $\mathbb{C}$-subspace of $\ker(T^{\mathbb{C}} - z_l)^{p_l}$, then $\bar{W}$ is a $\mathbb{C}$-subspace of $\ker(T^{\mathbb{C}} - \bar{z}_l)^{p_l}$. If W is $T^{\mathbb{C}}$-invariant, so is $\bar{W}$. If $W_1 \oplus \cdots \oplus W_s = \ker(T^{\mathbb{C}} - z_l)^{p_l}$, then $\bar{W}_1 \oplus \cdots \oplus \bar{W}_s = \ker(T^{\mathbb{C}} - \bar{z}_l)^{p_l}$. If W is a $(T^{\mathbb{C}} - z_l)$-cyclic subspace of $\ker(T^{\mathbb{C}} - z_l)^{p_l}$ with basis $\{\alpha, (T^{\mathbb{C}} - z_l)(\alpha), \ldots, (T^{\mathbb{C}} - z_l)^{k-1}(\alpha)\}$, then $\bar{W}$ is a $(T^{\mathbb{C}} - \bar{z}_l)$-cyclic subspace of $\ker(T^{\mathbb{C}} - \bar{z}_l)^{p_l}$ with basis $\{\bar{\alpha}, (T^{\mathbb{C}} - \bar{z}_l)(\bar{\alpha}), \ldots, (T^{\mathbb{C}} - \bar{z}_l)^{k-1}(\bar{\alpha})\}$. The proof of the lemma now follows from Theorem 3.21 and the definition of invariants. $\square$

If we now combine Theorem 4.9 with Lemma 5.10, we get the following important corollary:

Corollary 5.11: Let $\beta_l = \{\underline{\beta}_{l_1}, \ldots, \beta_{l_{p_l}}\}$ be a basis of $\ker(T^{\mathbb{C}} - z_l)^{p_l}$ such that if $T_l^{\mathbb{C}}$ denotes the restriction of $T^{\mathbb{C}}$ to $\ker(T^{\mathbb{C}} - z_l)^{p_l}$, then

$$\Gamma(\beta_l, \beta_l)(T_l^{\mathbb{C}}) = \begin{bmatrix} B(k_{l1}) & & 0 \\ & \ddots & \\ 0 & & B(k_{lq(l)}) \end{bmatrix} = J_l$$

Here $k_{l1} \geqslant \cdots \geqslant k_{lq(l)}$, and $B(k_{lj}) = z_l I_{k_{lj}} + N_{k_{lj}}$ for all $j = 1, \ldots, q(l)$. Then $\bar{\beta}_l = \{\bar{\beta}_{l1}, \ldots, \bar{\beta}_{lp_l}\}$ is a basis of $\ker(T^{\mathbb{C}} - \bar{z}_l)^{p_l}$. Furthermore, if $\hat{T}_l^{\mathbb{C}}$ denotes the restriction of $T^{\mathbb{C}}$ to $\ker(T^{\mathbb{C}} - \bar{z}_l)^{p_l}$, then

$$\Gamma(\bar{\beta}_l, \bar{\beta}_l)(\hat{T}_l^{\mathbb{C}}) = \begin{bmatrix} \overline{B(k_{l1})} & & 0 \\ & \ddots & \\ 0 & & \overline{B(k_{lq(l)})} \end{bmatrix} = \bar{J}_l$$

Here $\overline{B(k_{lj})} = \bar{z}_l I_{k_{lj}} + N_{k_{lj}}$ for all $j = 1, \ldots, q(l)$. $\square$

Thus, the complex blocks in the Jordan canonical form of T^C occur in conjugate pairs. Let us now introduce the following subspaces of V^C:

5.12:
$$U_j = \ker(T^C - c_j)^{n_j}, \qquad j = 1, \ldots, r$$
$$U_{r+l} = \ker(T^C - z_l)^{p_l} \oplus \ker(T^C - \bar{z}_l)^{p_l}, \qquad l = 1, \ldots, t$$

From Theorem 4.2, we know each U_j, $j = 1, \ldots, r$, is a T^C-invariant subspace of V^C of dimension n_j over $\mathbb{C}$. Each U_{r+l} is a T^C-invariant subspace of V^C of dimension $2p_l$ over $\mathbb{C}$, and $V^C = U_1 \oplus \cdots \oplus U_r \oplus U_{r+1} \oplus \cdots \oplus U_{r+t}$. We claim that each of the two types of subspaces in 5.12 have bases contained in $V(= V \otimes_\mathbb{R} 1)$.

For $j = 1, \ldots, r$, c_j is a real root of $c_T(X)$. It follows from Corollary 5.4 that $V_j = \ker(T - c_j)^{n_j}$ is a T-invariant subspace of V of dimension n_j over $\mathbb{R}$. Clearly, $V_j \otimes_\mathbb{R} \mathbb{C} = U_j$, and, thus, any $\mathbb{R}$-basis of V_j is a $\mathbb{C}$-basis of U_j. By Theorem 3.21, there exists an $\mathbb{R}$-basis $\underline{\alpha}_j$ of V_j such that

5.13:
$$\Gamma(\underline{\alpha}_j, \underline{\alpha}_j)(T_j) = \begin{bmatrix} B(y_{j1}) & & 0 \\ & \ddots & \\ 0 & & B(y_{js(j)}) \end{bmatrix} = J_j$$

As usual, in equation 5.13, T_j denotes the restriction of T to V_j. $y_{j1} \geqslant \cdots \geqslant y_{js(j)}$ are the invariants of $T_j - c_j$ on V_j, and $B(y_{jq}) = c_j I_{y_{jq}} + N_{y_{jq}}$ for all $q = 1, \ldots, s(j)$. By Theorem 2.17 of Chapter 2, $\underline{\alpha}_j$ is a $\mathbb{C}$-basis of U_j. The representation of T_j^C (the restriction of T^C to U_j) with respect to $\underline{\alpha}_j$ is identical to 5.13.

Now for $l = 1, \ldots, t$, let $z_l = a_l + ib_l$ with $a_l, b_l \in \mathbb{R}$. Let $\beta_l = \{\beta_{l1}, \ldots, \beta_{lp_l}\}$ be a basis of $\ker(T^C - z_l)^{p_l}$ such that

5.14:
$$\Gamma(\beta_l, \beta_l)(T_l^C) = \begin{bmatrix} B(k_{l1}) & & 0 \\ & \ddots & \\ 0 & & B(k_{lq(l)}) \end{bmatrix}$$

In equation 5.14, T_l^C denotes the restriction of T^C to $\ker(T^C - z_l)^{p_l}$. $k_{l1} \geqslant \cdots \geqslant k_{lq(l)}$ are the invariants of $T_l^C - z_l$ on $\ker(T^C - z_l)^{p_l}$. The rest of the notation is the same as in Corollary 5.11.

It now follows from Corollary 5.11 that $\beta_l \cup \bar{\beta}_l$ is a basis of U_{r+l}. Write each $\beta_{lj} \in \beta_l$ as follows: $\beta_{lj} = \mu_{lj} + i\lambda_{lj}$ with $\mu_{lj}, \lambda_{lj} \in V$. Set $\Delta_l = \{\mu_{l1}, \lambda_{l1}, \ldots, \mu_{lp_l}, \lambda_{lp_l}\}$. Then $\Delta_l \subseteq V$. Clearly Δ_l spans the $\mathbb{C}$-vector space U_{r+l}. Since $\dim_\mathbb{C}(U_{r+l}) = 2p_l$, we conclude Δ_l is a basis for U_{r+l} over $\mathbb{C}$.

For each $l = 1, \ldots, t$, let $V_{r+l} = L_\mathbb{R}(\Delta_l) \subseteq V$. Then $V_{r+l} \otimes_\mathbb{R} \mathbb{C} \cong U_{r+l}$. In particular, $\dim_\mathbb{R}(V_{r+l}) = 2p_l$. Since $V^C = U_1 \oplus \cdots \oplus U_{r+t}$, $\Delta = \{\bigcup_{j=1}^r \underline{\alpha}_j\} \cup \{\bigcup_{l=1}^t \Delta_l\}$ is a $\mathbb{C}$-basis of V^C. It now follows that Δ is a basis of V over $\mathbb{R}$, and, in particular, $V = V_1 \oplus \cdots \oplus V_r \oplus V_{r+1} \oplus \cdots \oplus V_{r+t}$.

We have already noted that $V_1, \ldots, V_r$ are T-invariant subspaces of $\mathbb{R}$ with the restriction of T to V_j being represented by equation 5.13 relative to α_j. Equation 5.14 readily implies each V_{r+l} is a T-invariant subspace. To see this, consider the first block $B(k_{l1})$ in $\Gamma(\beta_l, \beta_l)(T_l^C)$. Let us simplify notation here and write $k_{l1} = k$. Then $B(k)$ corresponds to the T^C-invariant subspace of $\ker(T^C - z_l)^{p_l}$ spanned by the first k vectors $\beta_{l1}, \ldots, \beta_{lk}$ of β_l. We then have the following equations:

5.15:

$$T(\mu_{l1}) + iT(\lambda_{l1}) = T^C(\beta_{l1}) = z_l\beta_{l1} + \beta_{l2} = (a_l + ib_l)(\mu_{l1} + i\lambda_{l1}) + (\mu_{l2} + i\lambda_{l2})$$

$$T(\mu_{l2}) + iT(\lambda_{l2}) = T^C(\beta_{l2}) = z_l\beta_{l2} + \beta_{l3} = (a_l + ib_l)(\mu_{l2} + i\lambda_{l2}) + (\mu_{l3} + i\lambda_{l3})$$

$$\vdots$$

$$T(\mu_{lk-1}) + iT(\lambda_{lk-1}) = T^C(\beta_{lk-1}) = z_l\beta_{lk-1} + \beta_{lk}$$
$$= (a_l + ib_l)(\mu_{lk-1} + i\lambda_{lk-1}) + (\mu_{lk} + i\lambda_{lk})$$

and

$$T(\mu_{lk}) + iT(\lambda_{lk}) = T^C(\beta_{lk}) = z_l\beta_{lk} = (a_l + ib_l)(\mu_{lk} + i\lambda_{lk})$$

If we now equate the real and imaginary parts in equation 5.15, we get the following equations:

5.16:

$$T(\mu_{l1}) = a_l\mu_{l1} - b_l\lambda_{l1} + \mu_{l2}$$

$$T(\lambda_{l1}) = b_l\mu_{l1} + a_l\lambda_{l1} + \lambda_{l2}$$

$$T(\mu_{l2}) = a_l\mu_{l2} - b_l\lambda_{l2} + \mu_{l3}$$

$$T(\lambda_{l2}) = b_l\mu_{l2} + a_l\lambda_{l2} + \lambda_{l3}$$

$$\vdots$$

$$T(\mu_{lk}) = a_l\mu_{lk} - b_l\lambda_{lk}$$

$$T(\lambda_{lk}) = b_l\mu_{lk} + a_l\lambda_{lk}$$

Thus, the subspace spanned by the first 2k vectors $\mu_{l1}, \lambda_{l1} \ldots, \mu_{lk}, \lambda_{lk}$ of Δ_l form a T-invariant subspace of V_{r+l}. The representation of T on this subspace is given by the 2k × 2k matrix

5.17:

$$\begin{bmatrix} D & & & & \\ I_2 & D & & & 0 \\ & I_2 & & \ddots & \\ 0 & & \ddots & \ddots & \\ & & & I_2 & D \end{bmatrix}$$

where

$$D = \begin{pmatrix} a_l & b_l \\ -b_l & a_l \end{pmatrix}$$

In equation 5.17, there are $k = k_{l1}$ 2×2 matrices D running down the diagonal and the 2×2 identity matrix I_2 on the subdiagonal. In the case that $k_{l1} = 1$, then 5.17 simplifies to just D.

Clearly, each block $B(k_{lj})$ in equation 5.14 gives us the corresponding $2k_{lj} \times 2k_{lj}$ matrix as in 5.17. [Each diagonal element z_l in $B(k_{lj})$ is replaced by D and every 1 on the subdiagonal of $B(k_{lj})$ (if any) is replaced with I_2.] We have now proved that V_{r+l} is T-invariant and if $\hat{T}_l$ denotes the restriction of T to V_{r+l}, then

5.18:
$$\Gamma(\Delta_l, \Delta_l)(\hat{T}_l) = \begin{bmatrix} H(k_{l1}) & & 0 \\ & \ddots & \\ 0 & & H(k_{lq(l)}) \end{bmatrix}$$

In 5.18, $H(k_{lj})$ is the $2k_{lj} \times 2k_{lj}$ matrix constructed from $B(k_{lj})$ as in equation 5.17.

We can now put all of this material together. We have proved the following theorem.

Theorem 5.19: Let $T \in \mathrm{Hom}_{\mathbb{R}}(V, V)$ and suppose

$$c_T(X) = \prod_{j=1}^{r} (X - c_j)^{n_j} \prod_{l=1}^{t} (X - z_l)^{p_l}(X - \bar{z}_l)^{p_l} \in \mathbb{C}[X]$$

Here $c_1, \ldots, c_r$ are the distinct real roots of c_T and $z_1, \bar{z}_1, \ldots, z_t, \bar{z}_t$ are the nonreal roots. Let $T^{\mathbb{C}}$ denote the complexification of T. Then

(a) $T^{\mathbb{C}}$ has a Jordan canonical form of the following type:

$$\begin{bmatrix} J_1 & & & & & & & \\ & \ddots & & & & & & \\ & & J_r & & & & 0 & \\ & & & J_{r+1} & & & & \\ & & & & \bar{J}_{r+1} & & & \\ & & & & & \ddots & & \\ & 0 & & & & & J_{r+t} & \\ & & & & & & & \bar{J}_{r+t} \end{bmatrix}$$

For $j = 1, \ldots, r$,

$$J_j = \begin{bmatrix} B(y_{j1}) & & 0 \\ & \ddots & \\ 0 & & B(y_{js(j)}) \end{bmatrix}$$

Here $y_{j1} \geqslant \cdots \geqslant y_{js(j)}$, $y_{j1} + \cdots + y_{js(j)} = n_j$, and $B(y_{jm}) = c_j I_{y_{jm}} + N_{y_{jm}}$ for $m = 1, \ldots, s(j)$. For $l = 1, \ldots, t$,

$$J_{r+l} = \begin{bmatrix} B(k_{l1}) & & 0 \\ & \ddots & \\ 0 & & B(k_{lq(l)}) \end{bmatrix}$$

Here $k_{l1} \geqslant \cdots \geqslant k_{lq(l)}$, $k_{l1} + \cdots + k_{lq(l)} = p_l$, and $B(k_{lm}) = z_l I_{k_{lm}} + N_{k_{lm}}$ for $m = 1, \ldots, q(l)$. For $l = 1, \ldots, t$,

$$\bar{J}_{r+l} = \begin{bmatrix} \overline{B(k_{l1})} & & 0 \\ & \ddots & \\ 0 & & \overline{B(k_{lq(l)})} \end{bmatrix}$$

Here $\overline{B(k_{lm})} = \bar{z}_l I_{k_{lm}} + N_{k_{lm}}$ for $m = 1, \ldots, q(l)$.

For each $l = 1, \ldots, t$ and $m = 1, \ldots, q(l)$, let $H(k_{lm})$ be the $2k_{lm} \times 2k_{lm}$ matrix given in equation 5.17. Thus, $H(k_{lm})$ is formed from $B(k_{lm})$ by replacing each z_l by

$$\begin{pmatrix} a_l & b_l \\ -b_l & a_l \end{pmatrix}$$

where $z_l = a_l + ib_l$ and each 1 by I_2.

(b) There exists a basis Δ of V such that

5.20: $$\Gamma(\Delta, \Delta)(T) = \begin{bmatrix} J_1 & & & & & \\ & \ddots & & & 0 & \\ & & J_r & & & \\ & 0 & & K_1 & & \\ & & & & \ddots & \\ & & & & & K_t \end{bmatrix}$$

For $j = 1, \ldots, r$, J_j is the same as in (a). For $l = 1, \ldots, t$,

$$K_l = \begin{bmatrix} H(k_{l1}) & & 0 \\ & \ddots & \\ 0 & & H(k_{lq(l)}) \end{bmatrix} \qquad \square$$

The matrix representation given in equation 5.20 is called a real Jordan canonical form of T. Our disscussion before Theorem 5.19 tells us how to construct a basis Δ of V that gives a real Jordan canonical form. Namely, find a real basis (satisfying 5.13) of each $\ker(T^C - c_j)^{n_j}$ and add to these vectors the real and imaginary parts of a basis (satisfying 5.14) of each $\ker(T^C - z_l)^{p_l}$.

Example 5.21: Let $T: \mathbb{R}^2 \to \mathbb{R}^2$ be given by $T(\delta_1) = \delta_2$ and $T(\delta_2) = -\delta_1$. Then

$$\Gamma(\underline{\delta}, \underline{\delta})(T) = \begin{pmatrix} 0 & -1 \\ 1 & 0 \end{pmatrix}$$

and $c_T(X) = X^2 + 1 = (X - i)(X + i)$. Thus, in the notation of Theorem 5.19, $r = 0, t = 1, z_1 = i, \bar{z}_1 = -i$. We know from Example 3.2 that T^C is represented by the diagonal matrix

$$\Gamma(\underline{\beta}, \underline{\beta})(T^C) = \begin{pmatrix} i & 0 \\ 0 & -i \end{pmatrix}$$

where $\underline{\beta} = \{\beta_1 = (1, -i), \beta_2 = \bar{\beta}_1 = (1, i)\}$. In the notation of 5.19,

$$J = \begin{pmatrix} i & 0 \\ 0 & -i \end{pmatrix}$$

is the Jordan canonical form of T^C with $J_1 = (i)$ and $\bar{J}_1 = (-i)$. The real and imaginary parts of i are 0 and 1. Therefore,

$$D = \begin{pmatrix} 0 & 1 \\ -1 & 0 \end{pmatrix} = H(k_{11})$$

$\beta_1 = \mu_1 + i\lambda_1$ for $\mu_1 = (1, 0)$, and $\lambda_1 = (0, -1)$. Thus, $\Delta = \{\mu_1, \lambda_1\}$ is basis for $\mathbb{R}^2$, and $\Gamma(\Delta, \Delta)(T) = D$ is a real Jordan form of T. $\square$

If $A \in M_{n \times n}(\mathbb{R})$, then, as usual, we associate with A a linear transformation $T: \mathbb{R}^n \to \mathbb{R}^n$ given by $T(\delta_j) = \text{Col}_j(A)^t, j = 1, \ldots, n$. Then $\Gamma(\underline{\delta}, \underline{\delta})(T) = A$. By a *real Jordan canonical form* of A, we shall mean a real Jordan canonical form of T. Thus, if we fix an ordering of the eigenvalues of A as in equation 5.7, then A is similar to a real Jordan canonical matrix of the form given in equation 5.20.

Example 5.22: Let

$$A = \begin{bmatrix} 14 & -3 & -9 \\ 15 & -3 & -10 \\ 13 & -2 & -9 \end{bmatrix} \in M_{3 \times 3}(\mathbb{R})$$

A simple calculation shows $c_A(X) = X^3 - 2X^2 + X - 2 = (X - 2)(X^2 + 1)$. Thus, $c_A(X) = (X - 2)(X - i)(X + i) \in \mathbb{C}[X]$. Corollary 4.22 then implies

$$J = \begin{bmatrix} 2 & 0 & 0 \\ 0 & i & 0 \\ 0 & 0 & -i \end{bmatrix}$$

is the Jordan canonical form of A in $M_{3 \times 3}(\mathbb{C})$. Then the computations in Example 5.21 imply

$$\begin{bmatrix} 2 & 0 & 0 \\ 0 & 0 & 1 \\ 0 & -1 & 0 \end{bmatrix}$$

is a real Jordan canonical form of A.   □

We note in passing that the analog of Theorem 4.19 is true for real Jordan canonical forms. Let A, B $\in M_{n \times n}(\mathbb{R})$. Then A and B are similar if and only if A and B have the same real Jordan canonical form, that is, there is an ordering of the eigenvalues of A (and B) such that the resulting real Jordan canonical forms are the same. We leave this remark as an exercise at the end of this section. Since a real Jordan canonical form of T (or A) is unique up to similarity, authors often refer to "the" real Jordan canonical form of T (or A).

In the remainder of this section, we discuss one of the more important applications of the real Jordan canonical form, that is, solving systems of linear differential equations. Suppose I is some open interval containing 0 in $\mathbb{R}$. Let $x_1(t), \ldots, x_n(t) \in C^1(I)$. We are interesting in solving a system of linear differential equations of the following type:

5.23:
$$\frac{dx_1}{dt} = a_{11}x_1 + \cdots + a_{1n}x_n$$

$$\vdots$$

$$\frac{dx_n}{dt} = a_{n1}x_1 + \cdots + a_{nn}x_n$$

In equation 5.23, $a_{ij} \in \mathbb{R}$ for $i, j = 1, \ldots, n$. We are interested in finding a solution to 5.23 subject to some initial condition $x_1(0) = c_1, \ldots, x_n(0) = c_n$.

Let us introduce the obvious vector notation here. Set $A = (a_{ij}) \in M_{n \times n}(\mathbb{R})$, $x = (x_1 \ldots, x_n)^t$, and $C = (c_1, \ldots, c_n)^t$. Then 5.23 can be rewritten as follows:

5.24:
$$x' = Ax \quad \text{with} \quad x(0) = C$$

Here x' of course means the $n \times 1$ matrix $(x_1', \ldots, x_n')^t$ where x_i' denotes the derivative of $x_i(t)$.

Now the solution procedure in 5.24 is to replace A with the simplest matrix similar to A that we can find. Suppose $J = PAP^{-1}$ for some invertible matrix $P \in M_{n \times n}(\mathbb{R})$. Set $y = Px$. Then $y' = Px'$, and $y(0) = PC$. Also, $Jy = PAP^{-1}(Px) = PAx = Px' = y'$. Thus, to find a solution to 5.24 we need only solve the following equation:

5.25: $$y' = Jy \quad \text{with} \quad y(0) = PC$$

If y is a solution to 5.25, then $x = P^{-1}y$ is a solution to 5.24. For, $x' = P^{-1}y' = P^{-1}Jy = P^{-1}(PAP^{-1})Px = Ax$. Also, $x(0) = P^{-1}y(0) = P^{-1}PC = C$.

If we let J be the real Jordan canonical form of A, then the equations we get in 5.25 are easy to solve. In the first place, we have seen in Theorem 5.19 that J is a series of diagonal blocks.

5.26:
$$J = \begin{bmatrix} B_1 & & 0 \\ & \ddots & \\ 0 & & B_K \end{bmatrix}$$

Each block in 5.26 has one of two possible forms:

5.27:
$$B_j = \begin{bmatrix} c & & & & 0 \\ 1 & \ddots & & & \\ & \ddots & \ddots & & \\ 0 & & & 1 & c \end{bmatrix} \quad \text{or} \quad B_j = \begin{bmatrix} D & & & 0 \\ I_2 & \ddots & & \\ & \ddots & \ddots & \\ 0 & & I_2 & D \end{bmatrix}$$

In equation 5.27, D is a 2×2 matrix of the form

$$\begin{pmatrix} a & b \\ -b & a \end{pmatrix}$$

The notation is meant to include the two trivial cases $B_j = (c)$ or $B_j = D$. Suppose the size of each B_j is $n_j \times n_j$. Then clearly, $y' = Jy$ decomposes into K sets of equations $y'_j = B_j y_j$, $j = 1, \ldots, K$. Here $y_1 = (y_1, \ldots, y_{n_1})^t$, $y_2 = (y_{n_1+1}, \ldots, y_{n_1+n_2})^t$, etc. Thus, to find a solution to equation 5.25, it suffices to know how to solve equations of the following two types:

5.28:
$$\begin{bmatrix} x'_1 \\ \cdot \\ \cdot \\ \cdot \\ x'_n \end{bmatrix} = \begin{bmatrix} c & \cdots & 0 & 0 \\ 1 & & 0 & 0 \\ 0 & & 0 & 0 \\ & & \vdots & \vdots \\ 0 & \cdots & 1 & c \end{bmatrix} \begin{bmatrix} x_1 \\ \cdot \\ \cdot \\ \cdot \\ x_n \end{bmatrix} \quad \text{with} \quad x_i(0) = c_i, \quad i = 1, \ldots, n$$

and

5.29:
$$\begin{bmatrix} x_1' \\ \cdot \\ \cdot \\ \cdot \\ x_{2n}' \end{bmatrix} = \begin{bmatrix} D & \cdots & 0 & 0 \\ I_2 & & 0 & 0 \\ 0 & & \vdots & \vdots \\ \vdots & & & \\ 0 & \cdots & I_2 & D \end{bmatrix} \begin{bmatrix} x_1 \\ \cdot \\ \cdot \\ \cdot \\ x_{2n} \end{bmatrix} \qquad \text{with } x_i(0) = c_i, \quad i = 1, \dots, 2n,$$

and $\quad D = \begin{pmatrix} a & b \\ -b & a \end{pmatrix}$

Before proceeding with a solution to these two types of equations, we need to recall a few facts about exponentials of matrices. If $A \in M_{n \times n}(\mathbb{R})$, then e^A is the $n \times n$ matrix defined by the following equation:

5.30:
$$e^A = \sum_{k=0}^{\infty} \frac{A^k}{k!}$$

The reader can easily argue that the partial sums $S_n = \sum_{k=0}^{n} A_k/k!$ of the series in 5.30 converge to a well-defined $n \times n$ matrix we denote by e^A. We shall need the following facts:

Theorem 5.31: (a) If $Q = PAP^{-1}$, then $e^Q = Pe^A P^{-1}$.
  (b) If $AB = BA$, then $e^{A+B} = e^A e^B$.
  (c) $e^{-A} = (e^A)^{-1}$.
  (d) If

$$A = \begin{pmatrix} a & b \\ -b & a \end{pmatrix}$$

  then

$$e^A = e^a \begin{pmatrix} \cos b & \sin b \\ -\sin b & \cos b \end{pmatrix}$$

  (e) If $c \in \mathscr{S}_{\mathbb{R}}(A)$, then $e^c \in \mathscr{S}_{\mathbb{R}}(e^A)$.
  (f) $d(e^{tA})/dt = Ae^{tA}$.

Proof: All six of these assertions are easy computations, which we leave to the exercises. $\square$

Theorem 5.31(f) provides us with a unique solution to equation 5.24, namely $x = e^{tA}C$. For, $x' = d(e^{tA}C)/dt = Ae^{tA}C = Ax$, and $x(0) = e^{0A}C = C$. The fact that $e^{tA}C$ is the only solution to 5.24 is a simple computation. We want to see what form this solution takes in our two special cases 5.28 and 5.29.

Let us consider equation 5.28 first. Set

$$B = \begin{bmatrix} c & & & 0 \\ 1 & & & \\ & \ddots & \ddots & \\ 0 & & 1 & c \end{bmatrix} \quad \text{and} \quad N = \begin{bmatrix} 0 & & & 0 \\ 1 & & & \\ & \ddots & \ddots & \\ 0 & & 1 & 0 \end{bmatrix}$$

So, $B, N \in M_{n \times n}(\mathbb{R})$. We had seen in Section 3 that N is nilpotent of index n. Since cI_n commutes with N, Theorem 5.31(b) implies $e^{tB} = e^{tcI_n}e^{tN} = e^{ct}e^{tN}$. Thus, the solution to 5.28 is $x = e^{tB}C = e^{ct}e^{tN}C$. Using the definition of e^{tN}, we have

5.32:
$$e^{tN} = \sum_{k=0}^{n-1} \frac{(tN)^k}{k!} = \begin{bmatrix} 1 & 0 & 0 & \cdots & 0 & 0 \\ t & 1 & 0 & \cdots & 0 & 0 \\ \dfrac{t^2}{2!} & t & 1 & \cdots & 0 & 0 \\ \vdots & \vdots & \vdots & \ddots & \vdots & \vdots \\ \dfrac{t^{n-2}}{(n-2)!} & \dfrac{t^{n-3}}{(n-3)!} & \dfrac{t^{n-4}}{(n-4)!} & \cdots & 1 & 0 \\ \dfrac{t^{n-1}}{(n-1)!} & \dfrac{t^{n-2}}{(n-2)!} & \dfrac{t^{n-3}}{(n-3)!} & \cdots & t & 1 \end{bmatrix}$$

If we now substitute 5.32 into $x = e^{ct}e^{tN}C$, we get

5.33:
$$x_j = e^{ct} \sum_{k=0}^{j-1} \frac{t^k}{k!} c_{j-k} \quad \text{for} \quad j = 1, \ldots, n$$

Now let us consider equation 5.29. If the matrix in equation 5.29 is just D, then Theorem 5.31(d) implies the solution is

5.34:
$$x = e^{tD}C = e^{ta} \begin{pmatrix} \cos bt & \sin bt \\ -\sin bt & \cos bt \end{pmatrix} \begin{pmatrix} c_1 \\ c_2 \end{pmatrix}$$

So, we can assume $n > 1$ and proceed with the general case.

Set $\mu = a - bi$. For $j = 1, \ldots, n$, let $z_j = x_{2j-1} + ix_{2j}$ and $w_j = c_{2j-1} + ic_{2j}$. Then equation 5.29 becomes the following system:

5.25:
$$\begin{bmatrix} \mu & \cdots & 0 & 0 \\ 1 & & 0 & 0 \\ 0 & & 0 & 0 \\ \vdots & & \vdots & \vdots \\ 0 & \cdots & 1 & \mu \end{bmatrix} \begin{bmatrix} z_1 \\ \cdot \\ \cdot \\ \cdot \\ z_n \end{bmatrix} = \begin{bmatrix} z_1' \\ \cdot \\ \cdot \\ \cdot \\ z_n' \end{bmatrix} \quad \text{with} \quad z_i(0) = w_i \quad \text{for} \quad i = 1, \ldots, n$$

The equations in 5.35 are solved in the same manner as equation 5.28. Thus,

5.36:
$$z_j(t) = e^{\mu t} \sum_{k=0}^{j-1} \frac{t^k}{k!} w_{j-k}, \quad j = 1, \ldots, n$$

Now recall $e^{\mu t} = e^{at-bti} = e^{at}(\cos bt - i \sin bt)$. Substituting this expression into equation 5.36 and letting $w_j = c_{2j-1} + ic_{2j}$ gives us our final solution:

5.37:

$$x_{2j-1}(t) = e^{at} \sum_{k=0}^{j-1} \frac{t^k}{k!} [c_{2(j-k)-1} \cos bt + c_{2(j-k)} \sin bt]$$

$$x_{2j}(t) = e^{at} \sum_{k=0}^{j-1} \frac{t^k}{k!} [c_{2(j-k)} \cos bt - c_{2(j-k)-1} \sin bt], \qquad j = 1, \ldots, n$$

Note in either case 5.28 or 5.29 the solution is linear combinations (coefficients in $\mathbb{R}[t]$) of exponentials, sines, and cosines. Thus, we have the following theorem:

Theorem 5.38: Let $A \in M_{n \times n}(\mathbb{R})$, and let $x(t)$ be a solution to the differential equation $Ax = x'$. Then each coordinate $x_j(t)$ of x is a linear combination of functions of the form $t^k e^{at} \cos bt$, $t^l e^{at} \sin bt$, where $a + bi$ runs through the eigenvalues of A. $\square$

Example 5.39: Consider the following system of differential equations:

5.40:

$$x_1' = 2x_1 + x_2 + x_3$$
$$x_2' = x_2 + x_3$$
$$x_3' = -x_2 + x_3$$
$$x_4' = x_1 + x_2 + 2x_4 \qquad \text{with} \quad x(0) = (c_1, c_2, c_3, c_4)^t$$

The matrix of this system is

$$A = \begin{bmatrix} 2 & 1 & 1 & 0 \\ 0 & 1 & 1 & 0 \\ 0 & -1 & 1 & 0 \\ 1 & 1 & 0 & 2 \end{bmatrix}$$

The first order of business is to find the real Jordan canonical form of A and the matrix P for which $PAP^{-1} = J$.

The characteristic polynomial of A is given by

5.41: $\qquad c_A(X) = (X - 2)^2(X^2 - 2X + 2) = (X - 2)^2(X - z_1)(X - \bar{z}_1)$

In equation 5.41, $z_1 = 1 + i$ and $\bar{z}_1 = 1 - i$. We conclude that

5.42
$$\begin{bmatrix} 2 & 0 & 0 & 0 \\ 1 & 2 & 0 & 0 \\ 0 & 0 & z_1 & 0 \\ 0 & 0 & 0 & \bar{z}_1 \end{bmatrix}$$

is the Jordan canonical form of A in $M_{4 \times 4}(\mathbb{C})$ and

5.43:
$$J = \begin{bmatrix} 2 & 0 & 0 & 0 \\ 1 & 2 & 0 & 0 \\ 0 & 0 & 1 & 1 \\ 0 & 0 & -1 & 1 \end{bmatrix}$$

is the real Jordan canonical form of A.

5.44:

$$(A - 2)^2 = \begin{bmatrix} 0 & -2 & 0 & 0 \\ 0 & 0 & -2 & 0 \\ 0 & 2 & 0 & 0 \\ 0 & 0 & 2 & 0 \end{bmatrix} \quad \text{and} \quad A - z_1 = \begin{bmatrix} 1-i & 1 & 1 & 0 \\ 0 & -i & 1 & 0 \\ 0 & -1 & -i & 0 \\ 1 & 1 & 0 & 1-i \end{bmatrix}$$

Equation 5.44 readily implies $\{(1, 0, 0, 0)^t, (0, 0, 0, 1)^t\}$ is a basis of $\ker(A - 2)^2$, and $\{(1, i, -1, -i)\}$ is a basis of $\ker(A - z_1))$ Since $(1, i, -1, -i) = (1, 0, -1, 0) + i(0, 1, 0, -1)$, we conclude that

$$\Delta = \left\{ \begin{bmatrix} 1 \\ 0 \\ 0 \\ 0 \end{bmatrix}, \begin{bmatrix} 0 \\ 0 \\ 0 \\ 1 \end{bmatrix}, \begin{bmatrix} 1 \\ 0 \\ -1 \\ 0 \end{bmatrix}, \begin{bmatrix} 0 \\ 1 \\ 0 \\ -1 \end{bmatrix} \right\}$$

is a basis for $M_{4 \times 4}(\mathbb{R})$ giving the real canonical form J of A. It now follows from equation 3.36 that

5.45: $P = \begin{bmatrix} 1 & 0 & 1 & 0 \\ 0 & 1 & 0 & 1 \\ 0 & 0 & -1 & 0 \\ 0 & 1 & 0 & 0 \end{bmatrix} \quad \text{and} \quad P^{-1} = \begin{bmatrix} 1 & 0 & 1 & 0 \\ 0 & 0 & 0 & 1 \\ 0 & 0 & -1 & 0 \\ 0 & 1 & 0 & -1 \end{bmatrix}$

Our solutions to equations 5.28 and 5.29 imply that the system $Jy = y'$ with

$y(0) = PC$ has solutions given by

5.46:
$$y_1 = e^{2t}(c_1 + c_3)$$
$$y_2 = e^{2t}(t(c_1 + c_3) + (c_2 + c_4))$$
$$y_3 = e^{t}[c_2 \sin t - c_3 \cos t]$$
$$y_4 = e^{t}[c_3 \sin t + c_2 \cos t]$$

Thus, $x(t) = P^{-1}y$ is given by

$$x_1(t) = e^{2t}(c_1 + c_3) + e^{t}[c_2 \sin t - c_3 \cos t]$$
$$x_2(t) = e^{t}[c_3 \sin t + c_2 \cos t]$$
$$x_3(t) = -e^{t}[c_2 \sin t - c_3 \cos t]$$
$$x_4(t) = e^{2t}[t(c_1 + c_3) + (c_2 + c_4)] - e^{t}(c_3 \sin t + c_2 \cos t) \quad \square$$

EXERCISES FOR SECTION 5

(1) Find an invertible matrix P such that

$$PAP^{-1} = \begin{bmatrix} 2 & 0 & 0 \\ 0 & 0 & 1 \\ 0 & -1 & 0 \end{bmatrix}$$

for the matrix A given in Example 5.22.

(2) Give a detailed proof of the four assertions in Lemma 5.10.

(3) Let $A, B \in M_{n \times n}(\mathbb{R})$. Show that A and B are similar if and only if there is an ordering of the eigenvalues of A and B so that the resulting real Jordan canonical forms of A and B are the same.

(4) Find the real Jordan canonical form J of

$$A = \begin{bmatrix} 0 & 0 & 0 & -8 \\ 1 & 0 & 0 & 16 \\ 0 & 1 & 0 & -14 \\ 0 & 0 & 1 & 6 \end{bmatrix}$$

Also compute a matrix P such that $PAP^{-1} = J$.

(5) Find the real Jordan canonical form J of

$$A = \begin{bmatrix} 1 & -2 & -1 & 1 \\ 0 & 2 & -3 & 0 \\ 0 & 3 & 2 & 0 \\ -1 & -1 & 2 & 1 \end{bmatrix}$$

and compute P such that $PAP^{-1} = J$.

(6) Let $A \in M_{n \times n}(\mathbb{R})$. Set $S_j = \sum_{k=0}^{j} A^k/k!$. Show $\{S_j\}$ is a Cauchy sequence in $M_{n \times n}(\mathbb{R})$ and hence converges to some n × n matrix.

(7) Prove Theorem 5.31.

(8) Show that $e^{tA}C$ is the unique solution to equation 5.24.

(9) Solve the following system of differential equations:

$$x_1' = 14x_1 - 3x_2 - 9x_3$$
$$x_2' = 15x_1 - 3x_2 - 10x_3$$
$$x_3' = 13x_1 - 2x_2 - 9x_3$$

(10) Solve the following system of differential equations:

$$x_1' = -8x_4$$
$$x_2' = x_1 + 16x_4$$
$$x_3' = x_2 - 14x_4$$
$$x_4' = x_3 + 6x_4$$

(11) Solve the following system of differential equations:

$$x_1' = x_1 - 2x_2 - x_3 + x_4$$
$$x_2' = 2x_2 - 3x_3$$
$$x_3' = 3x_2 + 2x_3$$
$$x_4' = -x_1 - x_2 + 2x_3 + x_4$$

(12) Solve the following system of differential equations:

$$x_1' = 3x_2 - 2x_3$$
$$x_2' = x_1 - 2x_2 + 2x_4$$
$$x_3' = 2x_1$$
$$x_4' = x_1 - 4x_2 + x_3 + 2x_4$$

(13) Let $A \in M_{n \times n}(\mathbb{C})$. Prove that $\det(e^A) = e^{\text{Tr}(A)}$.

(14) Let $T: \mathbb{R}^3 \to \mathbb{R}^3$ be a linear transformation represented by the matrix

$$\begin{pmatrix} 6 & -3 & -2 \\ 4 & -1 & -2 \\ 10 & -5 & -3 \end{pmatrix}$$

Compute $c_T(X)$, $m_T(X)$ and the subspaces V_i in Theorem 5.3.

(15) Given the setting in Theorem 5.3, suppose W is a T-invariant subspace of V. Prove that $W = \bigoplus_{i=1}^r (V_i \cap W)$.

6. THE RATIONAL CANONICAL FORM

In this section, we continue our theme from the last section. $T \in \text{Hom}_F(V, V)$, and we want to discuss what canonical form may be available for T. We have seen in Section 5 that if $F = \mathbb{R}$, then the Jordan canonical form for $T^{\mathbb{C}}$ can be used to construct a real Jordan canonical form for T. For a general field F and its algebraic closure $\bar{F}$, no such special relations as those used in Section 5 exist. Thus, the Jordan canonical form of T^F in $M_{n \times n}(\bar{F})$ does not give us any particular form for T in $M_{n \times n}(F)$.

In this section, F is an arbitrary field, and we stay in $M_{n \times n}(F)$. We work with the minimal polynomial of T and construct a canonical form for T based on m_T. We shall assume as always that V is a finite-dimensional vector space over F with $\dim_F V = n$. Let $T \in \text{Hom}_F(V, V)$. We shall factor the minimal polynomial $m_T(X)$ of T as in equation 5.2. Then the primary decomposition theorem implies

6.1: $$V = V_1 \oplus \cdots \oplus V_r$$

Each $V_i = \ker(q_i(T)^{m_i})$ is a nonzero T-invariant subspace of V. If T_i denotes the restriction of T to V_i, then Theorem 2.43 implies $m_{T_i}(X) = q_i(X)^{m_i}$ for $i = 1, \ldots, r$.

Thus, in constructing a canonical form for T, it suffices to consider T_i on V_i. Hence, we can assume the minimal polynomial of T is just a power of a single irreducible polynomial $q(X) \in F[X]$. We shall need the following definition:

Definition 6.2: A T-invariant subspace Z of V is said to be *T-cyclic* if there exists a vector $\alpha \in Z$ such that $Z = \{f(T)(\alpha) \mid f \in F[X]\}$.

We have seen T-cyclic subspaces before in the context of nilpotent transformations. If T is nilpotent and Z is a T-cyclic subspace of V in the sense of Definition 3.28, then clearly Z is T-cyclic. In our present context, we do not assume T is nilpotent.

Lemma 6.3: Let Z be a T-cyclic subspace of V. Suppose $\dim Z = m > 0$. Then

(a) Z has a basis of the form $\Delta = \{\alpha, T(\alpha), \ldots, T^{m-1}(\alpha)\}$ for some $\alpha \neq 0$ in Z.

(b) If $\hat{T}$ denotes the restriction of T to Z, then

$$
\textbf{6.4:} \qquad \Gamma(\Delta, \Delta)(\hat{T}) =
\begin{bmatrix}
0 & 0 & 0 & \cdots & 0 & c_0 \\
1 & 0 & 0 & \cdots & 0 & c_1 \\
0 & 1 & 0 & \cdots & 0 & c_2 \\
\vdots & \vdots & \vdots & \ddots & \vdots & \vdots \\
0 & 0 & 0 & \cdots & 0 & c_{m-2} \\
0 & 0 & 0 & \cdots & 1 & c_{m-1}
\end{bmatrix}
$$

where $T^m(\alpha) = c_0\alpha + c_1 T(\alpha) + \cdots + c_{m-1}T^{m-1}(\alpha)$.

(c) The minimal polynomial of $\hat{T}$ on Z is given by

$$
m_{\hat{T}}(X) = X^m - c_{m-1}X^{m-1} - \cdots - c_0
$$

Proof: (a) Since Z is cyclic, there exists a nonzero $\alpha \in Z$ such that $Z = \{f(T)(\alpha) \mid f \in F[X]\}$. In particular, $Z = L(\{\alpha, T(\alpha), T^2(\alpha), \ldots\})$. If $T^k(\alpha) = 0$, then clearly $T^l(\alpha) = 0$ for all $l \geqslant k$, and $Z = L(\{\alpha, \ldots, T^{k-1}(\alpha)\})$. Since $\dim Z = m$, we conclude that none of the first m vectors $\alpha, T(\alpha), \ldots, T^{m-1}(\alpha)$ is zero. We claim that $\Delta = \{\alpha, T(\alpha), \ldots, T^{m-1}(\alpha)\}$ is linearly independent over F. Suppose not. Then there is a linear dependence relation among the vectors of Δ of the following type:

$$
\textbf{6.5:} \qquad c_0\alpha + c_1 T(\alpha) + \cdots + c_k T^k(\alpha) = 0
$$

In equation 6.5, $1 \leqslant k \leqslant m - 1$ and $c_k \neq 0$. Dividing by c_k, we can rewrite equation 6.5 as follows:

$$
\textbf{6.6:} \qquad T^k(\alpha) = b_0\alpha + b_1 T(\alpha) + \cdots + b_{k-1}T^{k-1}(\alpha)
$$

Thus, $T^k(\alpha) \in L(\{\alpha, (T\alpha), \ldots, T^{k-1}(\alpha)\})$. But then $T^{k+1}(\alpha) = T(T^k(\alpha)) = b_0 T(\alpha) + \cdots + b_{k-2}T^{k-1}(\alpha) + b_{k-1}T^k(\alpha) \in L(\{\alpha, T(\alpha), \ldots, T^{k-1}(\alpha)\})$. Continuing with this argument, we get $T^l(\alpha) \in L(\{\alpha, T(\alpha), \ldots, T^{k-1}(\alpha)\})$ for every $l \geqslant k$. Therefore, $Z \subseteq L(\{\alpha, T(\alpha), \ldots, T^{k-1}(\alpha)\})$. This is impossible since $k < m = \dim Z$. Thus, Δ is linearly independent over F and the proof of (a) is complete.

(b) $T^m(\alpha) \in L(\Delta)$ implies $T^m(\alpha) = c_0\alpha + c_1 T(\alpha) + \cdots + c_{m-1}T^{m-1}(\alpha)$. The constants $c_0 \ldots, c_{m-1}$ are unique, and equation 6.4 is now obvious.

(c) Let $g = m_{\hat{T}}(X)$, the minimal polynomial of $\hat{T}$ on Z. For any $f \in F[X]$, $f(T)(Z) = 0$ if and only if $f(T)(\alpha) = 0$. Thus, g is the monic polynomial of smallest positive degree for which $g(T)(\alpha) = 0$. Since Δ is a basis of

Z, $\partial(g) \geqslant m$. But $(T^m - c_{m-1}T^{m-1} - \cdots - c_0)(\alpha) = 0$ from (b). Therefore, $g \mid (X^m - c_{m-1}X^{m-1} - \cdots - c_0)$. Since g is monic, we conclude $g = X^m - c_{m-1}X^{m-1} - \cdots - c_0$. $\square$

Note that Lemma 6.3 implies the dimension of a T-cyclic subspace Z of V is exactly the degree of the minimal polynomial of T restricted to Z. We shall use this fact many times in what follows. We need to give a formal name to the matrix appearing in equation 6.4.

Definition 6.7: Let $g(X) = X^m - c_{m-1}X^{m-1} - \cdots - c_0$ be a monic polynomial of degree m in $F[X]$. The $m \times m$ matrix

$$\begin{bmatrix} 0 & 0 & 0 & \cdots & 0 & c_0 \\ 1 & 0 & 0 & \cdots & 0 & c_1 \\ 0 & 1 & 0 & \cdots & 0 & c_2 \\ \vdots & \vdots & \vdots & \ddots & \vdots & \vdots \\ 0 & 0 & 0 & \cdots & 0 & c_{m-2} \\ 0 & 0 & 0 & \cdots & 1 & c_{m-1} \end{bmatrix}$$

is called the companion matrix of $g(X)$. We shall henceforth denote the companion matrix of g by $C(g(X))$.

Thus, the matrix of $\hat{T}$ appearing in equation 6.4 is just the companion matrix of the minimal polynomial of $\hat{T}$. We can restate Lemma 6.3 as follows:

Corollary 6.8: Let Z be a T-cyclic subspace of V, and suppose $\dim Z = m > 0$. Then Z has a basis $\Delta = \{\alpha, T(\alpha), \ldots, T^{m-1}(\alpha)\}$ such that $\Gamma(\Delta, \Delta)(\hat{T}) = C(m_{\hat{T}})$.
$\square$

In our next theorem, we shall argue that each $V_i = \ker(q_i(T)^{m_i})$ in equation 6.1 is a direct sum of T-cyclic subspaces. If $V_i = Z_{i1} \oplus \cdots \oplus Z_{ip(i)}$ is such a decomposition, then Corollary 6.8 implies each Z_{ij} has a basis Δ_{ij} such that $\Gamma(\Delta_{ij}, \Delta_{ij})(T_{ij}) = C(m_{T_{ij}})$. Here T_{ij} is the restriction of T to Z_{ij}. Now $q_i(T_{ij})^{m_i} = 0$. Hence $m_{T_{ij}} \mid q_i(X)^{m_i}$. Since $q_i(X)$ is irreducible, $m_{T_{ij}}(X) = q_i(X)^{e_{ij}}$ for some $e_{ij} \leqslant m_i$. Thus, there is a matrix representation of T on V_i consisting of blocks of companion matrices of various powers of $q_i(X)$. The main theorem we need to prove is the following:

Theorem 6.9: Let $T \in Hom_F(V, V)$, and suppose $m_T(X) = q(X)^e$, where q is a (monic) irreducible polynomial over F. Then $V = Z_1 \oplus \cdots \oplus Z_p$, where each Z_i is a T-cyclic subspace of V.

Proof: We proceed by induction on $n = \dim_F(V)$. If $n = 1$, then $V = L(\alpha)$ for some $\alpha \neq 0$. $T(\alpha) = c\alpha$ for some $c \in F$, and, thus, V itself is T-cyclic. We therefore may assume $\dim V = n > 1$.

Since $m_T(X) = q(X)^e$, $q(T)^{e-1} \neq 0$. Hence, there exists a nonzero vector $\alpha_1 \in V$ such that $q(T)^{e-1}(\alpha_1) \neq 0$. Let $Z_1 = L(\{\alpha_1, T(\alpha_1), T^2(\alpha_1), \dots\})$. Thus, Z_1 is the T-cyclic subspace of V generated by α_1. Let $d = \partial(q)$, and let T_1 denote the restriction of T to Z_1. Our previous discussion shows $m_{T_1} = q(X)^l$ for some $l \leq e$. But $q(T)^{e-1}(\alpha_1) \neq 0$. Therefore, $m_{T_1}(X) = q(X)^e$.

Lemma 6.3 implies $\dim_F(Z_1) = \partial(m_{T_1}) = de$. If $de = n$, then $Z_1 = V$ and our proof is complete.

Let us assume $de < n$. Since Z_1 is T-invariant, T induces a linear transformation $\bar{T}: V/Z_1 \to V/Z_1$ given by

6.10:
$$\bar{T}(\beta + Z_1) = T(\beta) + Z_1$$

The fact that $\bar{T}$ is a well-defined linear transformation is exercise 11 in Section 2. If $f(X) \in F[X]$, then clearly we have

6.11:
$$f(\bar{T})(\beta + Z_1) = f(T)(\beta) + Z_1$$

We get two important facts from equation 6.11. First, $m_{\bar{T}}(X) \mid m_T(X)$. Second, if W is a T-cyclic subspace of V generated by a vector β and $q(X)^l$ is the minimal polynomial of T on W, then $q(X)^l$ must be a multiple of the minimal polynomial of $\bar{T}$ on the $\bar{T}$-cyclic subspace of V/Z_1 generated by $\beta + Z_1$.

Since $m_{\bar{T}}(X) \mid m_T(X)$, $m_{\bar{T}}(X) = q(X)^{e_1}$ with $e_1 \leq e$. Also, $\dim Z_1 \geq 1$ implies $\dim\{V/Z_1\} < n$. Hence, we may apply our induction hypothesis to $\bar{T}$ on V/Z_1 and conclude that $V/Z_1 = \bar{Z}_2 \oplus \cdots \oplus \bar{Z}_p$. Each $\bar{Z}_i$ is a $\bar{T}$-cyclic subspace of V/Z_1. The minimal polynomial of $\bar{T}$ on $\bar{Z}_i$ has the form $q(X)^{e_i}$, and we may assume $e \geq e_1 \geq e_2 \geq \cdots \geq e_p$.

We shall complete the proof of the theorem by constructing T-cyclic subspaces $Z_2, \dots, Z_r$ in V such that

6.12: (a) Each Z_i is isomorphic to $\bar{Z}_i$ for $i = 2, \dots, p$.
   (b) The minimal polynomial of T on Z_i is the same as the minimal polynomial $q(X)^{e_i}$ of $\bar{T}$ on $\bar{Z}_i$.
   (c) $V = Z_1 \oplus \cdots \oplus Z_r$.

Each subspace $\bar{Z}_i$ of V/Z_1 has the form $\bar{Z}_i = \{\alpha + Z_1 \mid \alpha \in W_i\}$. Here W_i is some subspace of V containing Z_1. Let us suppose $\bar{Z}_i$ is generated as a $\bar{T}$-cyclic subspace of V/Z_1 by the coset $\alpha_i + Z_1$ for $i = 2, \dots, p$. Fix $i = 2, \dots, p$. Then Lemma 6.3 implies $\bar{Z}_i = L(\{\alpha_i + Z_1, T(\alpha_i) + Z_1, \dots, T^{e_id-1}(\alpha_i) + Z_1\})$. Since $q(\bar{T})^{e_i}(\alpha_i + Z_1) = 0$, $q(T)^{e_i}(\alpha_i) \in Z_1$. Since Z_1 is cyclic, we have $q(T)^{e_i}(\alpha_i) = f(T)(\alpha_1)$ for some $f \in F[X]$.

Now we claim there exists a vector $\beta_i \in Z_1$ such that $q(T)^{e_i}(\alpha_i + \beta_i) = 0$. To see this, first note $0 = q(T)^e(\alpha_i) = [q(T)^{e-e_i}q(T)^{e_i}](\alpha_i) = q(T)^{e-e_i}f(T)(\alpha_1)$. We have seen from the first part of this proof that the minimal polynomial of T on Z_1 is $q(X)^e$. Thus, $0 = [q(T)^{e-e_i}f(T)](\alpha_1)$ implies $q(X)^e \mid q(X)^{e-e_i}f(X)$. Hence, there exists a polynomial $h(X) \in F[X]$ such that $q(X)^e h(X) = q(X)^{e - e_i} f(X)$. Clearly, $f(X) = h(X)q(X)^{e_i}$. Set $\beta_i = -h(T)(\alpha_1) \in Z_1$. Then $q(T)^{e_i}(\alpha_i + \beta_i) = q(T)^{e_i}(\alpha_i) + q(T)^{e_i}(\beta_i) = f(T)(\alpha_1) - q(T)^{e_i} h(T)(\alpha_1) = f(T)(\alpha_1) - f(T)\alpha_1 = 0$.

Now let Z_i be the T-cyclic subspace of V generated by $\alpha_i + \beta_i$. Since $(\alpha_i + \beta_i) + Z_1 = \alpha_i + Z_1$ generates the $\bar{T}$-cyclic subspace $\bar{Z}_i$ of V/Z_1, and the minimal polynomial of $\bar{T}$ on $\bar{Z}_i$ is $q(X)^{e_i}$, our remarks after equation 6.11 imply the minimal polynomial of T on Z_i is a multiple of $q(X)^{e_i}$. But $q(T)^{e_i}(\alpha_i + \beta_i) = 0$. Thus, $q(X)^{e_i}$ is the minimal polynomial of T on Z_i. It now follows from Lemma 6.3 that Z_i and $\bar{Z}_i$ have the same dimension $e_i d$. Since the natural map $\gamma \to \gamma + Z_1$ induces a surjective map from Z_i to $\bar{Z}_i$, we conclude from Theorem 3.33 of Chapter I that the natural map $\gamma \to \gamma + Z_1$ is an isomorphism of Z_i onto $\bar{Z}_i$. We have now proven (a) and (b) of 6.12.

Let us denote the natural map from V to V/Z_1 by π. Thus, $\pi(\gamma) = \gamma + Z_1$. We have seen in the previous paragraph that π restricted to each Z_i, $i = 2$, ..., p, is an isomorphism of Z_i onto $\bar{Z}_i$. This fact easily implies $Z_1 + \cdots + Z_p = Z_1 \oplus \cdots \oplus Z_p$. To see this, we need only show that if $\gamma_1 + \cdots + \gamma_p = 0$ with $\gamma_i \in Z_i$, then $\gamma_1 = \cdots = \gamma_p = 0$ (Theorem 4.16 of Chapter I). If $\gamma_1 + \cdots + \gamma_p = 0$, then $\pi(\gamma_2) + \cdots + \pi(\gamma_p) = 0$ in $V/Z_1 = \bar{Z}_2 \oplus \cdots \oplus \bar{Z}_p$. Thus, $\pi(\gamma_2) = \cdots = \pi(\gamma_p) = 0$. Then $\gamma_2 = \cdots = \gamma_p = 0$ since $\pi: Z_i \cong \bar{Z}_i$ for $i \geq 2$.

We claim $V = Z_1 + \cdots + Z_p$. Let $\gamma \in V$. Then $\pi(\gamma) \in V/Z_1 = \bar{Z}_2 + \cdots + \bar{Z}_p$. Hence there exist vectors $\gamma_i \in Z_i$, $i = 2, \ldots, p$, such that $\pi(\gamma) = \pi(\gamma_2) + \cdots + \pi(\gamma_p)$. This last equation implies $\gamma - (\gamma_2 + \cdots + \gamma_p) \in Z_1 = \ker \pi$. Thus, $\gamma \in Z_1 + \cdots + Z_p$. This completes the proof of (c) in 6.12.

Of course, the completion of the proof of 6.12(c) also completes the proof of the theorem since each Z_i is T-cyclic. $\square$

We should point out here that Theorem 6.9 gives us a different proof of Theorem 3.21. If T is nilpotent, then $m_T(X) = X^k$. The companion matrix of X^k is just the matrix N_k defined in 3.20.

We can now prove our main result in this section.

Theorem 6.13: Let $T \in \mathscr{E}(V)$ have minimal polynomial given by $m_T(X) = q_1(X)^{m_1} \cdots q_r(X)^{m_r}$ as in equation 5.2. Then V is a finite direct sum of T-cyclic subspaces, $V = Z_{11} \oplus \cdots \oplus Z_{1p(1)} \oplus \cdots \oplus Z_{r1} \oplus \cdots \oplus Z_{rp(r)}$. The subspaces Z_{ij} satisfy the following properties:

(a) The minimum polynomial of T restricted to Z_{ij} is $q_i(X)^{e_{ij}}$, where $m_i = e_{i1} \geq \cdots \geq e_{ip(i)}$ for all $i = 1, \ldots, r$.

(b) $\dim_F(Z_{ij}) = e_{ij} \partial(q_i)$.

(c) There exists a basis Δ_{ij} of Z_{ij} such that if $\Delta = \bigcup_{i,j} \Delta_{ij}$, then

$$\Gamma(\Delta, \Delta)(T) = \begin{bmatrix} R_1 & & 0 \\ & \ddots & \\ 0 & & R_r \end{bmatrix}$$

where

$$R_i = \begin{bmatrix} C(q_i^{e_{i1}}) & & 0 \\ & \ddots & \\ 0 & & C(q_i^{e_{ip(i)}}) \end{bmatrix}$$

for all $i = 1, \ldots, r$.

Proof: We have virtually proved everything here already. The primary decomposition theorem implies $V = V_1 \oplus \cdots \oplus V_r$ with $V_i = \ker(q_i(T)^{m_i})$. Theorem 2.43 implies the minimal polynomial of T restricted to V_i is given by $q_i(X)^{m_i}$. Hence by Theorem 6.9, each $V_i = Z_{i1} \oplus \cdots \oplus Z_{ip(i)}$, where Z_{ij} is a T-cyclic subspace of V. We had seen in the proof of 6.9 that the Z_{ij} can be chosen such that the restriction of T to Z_{i1} has minimal polynomial $q_i(X)^{m_i}$, and the restriction of T to the remaining Z_{ij} has minimal polynomial $q_i(X)^{e_{ij}}$ where $m_i = e_{i1} \geqslant e_{i2} \geqslant \cdots \geqslant e_{ip(i)}$. Thus, we have established (a). (b) follows from Lemma 6.3. Lemma 6.3 also implies there exists a basis Δ_{ij} of Z_{ij} such that $\Gamma(\Delta_{ij}, \Delta_{ij})(T_{ij}) = C(q_i^{e_{ij}})$. Here T_{ij} denotes the restriction of T to Z_{ij}. (c) now follows from equation 2.42. $\square$

The matrix constructed in 6.13(c) is called a rational canonical form of T. As we shall soon see, it is unique up to a permutation of the R_i. Let us consider some examples.

Example 6.14: Suppose $T: \mathbb{Q}^2 \to \mathbb{Q}^2$ is given by $T(\delta_1) = \delta_2$, and $T(\delta_2) = -\delta_1$. Then

$$\Gamma(\underline{\delta}, \underline{\delta})(T) = \begin{pmatrix} 0 & -1 \\ 1 & 0 \end{pmatrix}$$

$c_T(X) = X^2 + 1$, which is irreducible in $\mathbb{Q}[X]$. Therefore, $m_T(X) = X^2 + 1$. The rational canonical form of T is just the companion matrix

$$C(X^2 + 1) = \begin{pmatrix} 0 & -1 \\ 1 & 0 \end{pmatrix} \quad \square$$

As usual, if $A \in M_{n \times n}(F)$, then the rational canonical form of A is the rational canonical form of $T: F^n \to F^n$, where $\Gamma(\underline{\delta}, \underline{\delta})(T) = A$.

Example 6.15: Let

$$A = \begin{bmatrix} -1 & 7 & 0 \\ 0 & 2 & 0 \\ 0 & 3 & -1 \end{bmatrix} \in M_{3 \times 3}(\mathbb{Q})$$

We had seen in Example 2.34 that $c_A(X) = (X + 1)^2(X - 2)$ and $m_A(X) = (X + 1)(X - 2)$. The companion matrix of any linear polynomial $X - a$ is the 1×1 matrix (a). We conclude that the rational canonical form of A must be the diagonal matrix $D = \text{diag}(-1, -1, 2)$. $\square$

Example 6.16: Let

$$A = \begin{bmatrix} 1 & 1 & 2 & 0 \\ -1 & 4 & 5 & -4 \\ 0 & -1 & -1 & 1 \\ 0 & 4 & 2 & -4 \end{bmatrix} \in M_{4 \times 4}(\mathbb{Q})$$

A simple calculation shows $c_A(X) = (X^2 + 2)(X^2 + 1)$. Thus, the rational canonical form R of A is given by

$$R = \begin{bmatrix} 0 & -2 & 0 & 0 \\ 1 & 0 & 0 & 0 \\ 0 & 0 & 0 & -1 \\ 0 & 0 & 1 & 0 \end{bmatrix} \quad \square$$

Let us turn our attention to the uniqueness of the polynomials $q_1^{e_{11}}, \ldots, q_1^{e_{1p(1)}}, \ldots, q_r^{e_{r1}}, \ldots, q_r^{e_{rp(r)}}$, which appear in Theorem 6.13.

Theorem 6.17: Let $T \in \mathscr{E}(V)$ have minimal polynomial given by $m_T(X) = q_1(X)^{m_1} \cdots q_r(X)^{m_r}$ as in equation 5.2.

Let $V = Z_{11} \oplus \cdots \oplus Z_{rp(r)}$ be the T-cyclic decomposition given in Theorem 6.13. Suppose we have a second T-cyclic decomposition $V = W_{11} \oplus \cdots \oplus W_{1u(1)} \oplus \cdots \oplus W_{r1} \oplus \cdots \oplus W_{ru(r)}$, where the minimal polynomial of T restricted to each W_{ij} is given by $q_i(X)^{f_{ij}}$, and $f_{i1} \geqslant \cdots \geqslant f_{iu(i)}$. Then $u(i) = p(i)$ for every $i = 1, \ldots, r$, and $e_{ij} = f_{ij}$ for every $j = 1, \ldots, p(i)$.

Before proving Theorem 6.17, we note that Theorem 2.43 implies that any decomposition of $V = U_1 \oplus \cdots \oplus U_N$ into T-cyclic subspaces such that the restriction of T to U_i has minimal polynomial a multiple of some q_j must involve all factors $q_1, \ldots, q_r$ of m_T. Thus, for every $i = 1, \ldots, r$ there must exist some $j = 1, \ldots, N$ such that the minimal polynomial of T on U_j is a power of q_i. Hence, Theorem 6.17 is the natural sort of uniqueness statement one would expect.

Proof of 6.17: If $V = W_{11} \oplus \cdots \oplus W_{ru(r)}$, then Theorem 2.43 implies that $m_T(X) = q_1^{f_{11}} \cdots q_r^{f_{r1}}$. It follows that $f_{11} = m_1, \ldots, f_{r1} = m_r$. Also, for each $i = 1, \ldots, r$, $W_{i1} \oplus \cdots \oplus W_{iu(i)} \subseteq V_i = \ker q_i(T)^{m_i}$. The primary decomposition theorem now implies $W_{i1} \oplus \cdots \oplus W_{iu(i)} = V_i = Z_{i1} \oplus \cdots \oplus Z_{ip(i)}$ for all $i = 1, \ldots, r$. Thus, without loss of generality, we may assume $r = 1$, and argue

$u(1) = p(1)$, and $e_{1j} = f_{1j}$ for all $j = 1, \ldots, p(1)$. We simplify notation by dropping the 1 and write $V = Z_1 \oplus \cdots \oplus Z_p$ with $e_1 \geqslant \cdots \geqslant e_p$, and $V = W_1 \oplus \cdots \oplus W_u$ with $f_1 \geqslant \cdots \geqslant f_u$.

Suppose the integers $\{e_1, \ldots, e_p\}$ and $\{f_1, \ldots, f_u\}$ are not the same. Then there is a first integer m, where $e_m \neq f_m$. From our comments above, we know $e_1 = f_1$. Therefore, $1 < m \leqslant \min\{u, p\}$. We can suppose $e_m > f_m$ with no loss in generality. Then $e_1 = f_1, \ldots, e_{m-1} = f_{m-1}$, and $e_m > f_m$. $m_T(X) = q(X)^{e_1}$. Since $f_m \geqslant f_i$ for $i \geqslant m$, we know $q(T)^{f_m}(W_i) = 0$ whenever $i \geqslant m$. Thus, we have

6.18:
$$q(T)^{f_m}(V) = q(T)^{f_m}(W_1) \oplus \cdots \oplus q(T)^{f_m}(W_{m-1})$$

Now each W_i is a T-cyclic subspace of dimension $f_i \partial(q)$. A simple computation shows $\dim(q(T)^{f_m}(W_i)) = \partial(q)(f_i - f_m)$ for $i = 1, \ldots, m - 1$. Thus, equation 6.18 implies

6.19:
$$\dim\{q(T)^{f_m}(V)\} = \sum_{i=1}^{m-1} \partial(q)(f_i - f_m)$$

On the other hand, $q(T)^{f_m}(V) \supseteq \bigoplus_{i=1}^{m} q(T)^{f_m}(Z_i)$, and $\dim\{q(T)^{f_m}(Z_i)\} = \partial(q)(e_i - f_m)$ for $i \leqslant m$. Therefore, $\dim\{q(T)^{f_m}(V)\} \geqslant \sum_{i=1}^{m} \partial(q)(e_i - f_m)$. If we now substitute this inequality into equation 6.19 and use the fact that $e_1 = f_1, \ldots, e_{m-1} = f_{m-1}$, we get $e_m \leqslant f_m$. This is impossible and the proof of Theorem 6.17 is complete. $\square$

The polynomials $q_1(X)^{e_{11}}, \ldots, q_1(X)^{e_{1p(1)}}, \ldots, q_r(X)^{e_{r1}}, \ldots, q_r(X)^{e_{rp(r)}}$ are called the *elementary divisors* of T. Since each $q_i^{e_{ij}}$ is the minimal polynomial of T on Z_{ij}, Lemmas 2.8 and 6.3 imply $Z_{ij} \cong F[X]/(q_i^{e_{ij}})$. Thus, we have the following corollary to Theorem 6.13.

Corollary 6.20: Let $T \in \mathscr{E}(V)$ have elementary divisors $q_1^{e_{11}}, \ldots, q_1^{e_{1p(1)}}, \ldots, q_r^{e_{r1}}, \ldots, q_r^{e_{rp(r)}} \in F[X]$. Then

$$V \cong \bigoplus_{i=1}^{r} \bigoplus_{j=1}^{p(i)} \{F[X]/(q_i(X)^{e_{ij}})\} \quad \square$$

Note that Corollary 6.20 implies $\prod_{i=1}^{r} \prod_{j=1}^{p(i)} q_i(X)^{e_{ij}} = c_T(X)$. This is why the $q_i^{e_{ij}}$ are called elementary divisors. The matrix version of Theorem 6.17 is easily stated. We leave the proof as an exercise at the end of this section.

Theorem 6.21: Two matrices $A, B \in M_{n \times n}(F)$ are similar if and only if A and B have the same elementary divisors. $\square$

We shall finish this section with another similarity result. Before stating it, we need the following definition:

Definition 6.22: Let $A \in M_{n \times n}(F)$. For each $i = 1, \ldots, n$, let g_i be a greatest common divisor of all i-rowed minors of $XI_n - A$. The polynomials g_1, $g_2 g_1^{-1}, \ldots, g_n g_{n-1}^{-1}$ are called the invariant factors of $XI_n - A$ (or the invariant factors of A).

We note that the usual theory of determinants implies $g_{i-1} | g_i$ in $F[X]$. Thus, each invariant factor is a well-defined polynomial in $F[X]$. The invariant factors of A are important because they determine the similarity class of A.

Theorem 6.23: Let $A, B \in M_{n \times n}(F)$. Then A and B are similar if and only if the invariant factors of $XI - A$ and $XI - B$ are the same.

Proof: The statement that the invariant factors are the same is of course up to units, i.e. nonzero constants in F. If A and B are similar, then so are $XI - A$ and $XI - B$. It then easily follows that the invariant factors of $XI - A$ and $XI - B$ are the same. Let us suppose A and B have the same set of invariant factors.

In $M_{n \times n}(F[X])$, we can perform the same elementary row (and column) operations the reader is familiar with in $M_{n \times n}(F)$. We can interchange two rows (or columns) of a matrix $C \in M_{n \times n}(F[X])$. We can multiply one row (or column) of C by a polynomial $h(X)$ and add the result to another row (or column). Both of these operations are performed by multiplying C on the left (or right) by suitable invertible matrices in $M_{n \times n}(F[X])$. We can also multiply a row (or column) of C by a nonzero constant from F.

Now by applying suitable row and column operations to $XI - A$, it is not difficult to see that $XI - A$ is equivalent to a diagonal matrix of the form $\mathrm{diag}(g_n g_{n-1}^{-1}, \ g_{n-1} g_{n-2}^{-1}, \ldots, g_1)$. Hence, there exist invertible matrices $R, S \in M_{n \times n}(F[X])$ such that $R(XI - A)S = \mathrm{diag}(g_n g_{n-1}^{-1}, \ldots, g_1)$. We ask the reader to provide a proof of this fact in the exercises at the end of this section. Being equivalent is clearly an equivalence relation $\equiv$ on $M_{n \times n}(F[X])$. Thus, if $XI - A$ and $XI - B$ have the same invariant factors, then $XI - A$ and $XI - B$ are equivalent. Hence, there exist invertible matrices $P, Q \in M_{n \times n}(F[X])$ such that $P(XI - A)Q = XI - B$.

We claim there exist invertible matrices $p, q \in M_{n \times n}(F)$ such that $p(XI - A)q = XI - B$. To see this, we first note that P and Q can be viewed as polynomials in X with coefficients in the algebra $M_{n \times n}(F)$. We can then divide $XI - B$ into P and Q (in a process entirely analogous to 1.2) and write $P = (XI - B)p_1 + p$ and $Q = q_1(XI - B) + q$. Here $p_1, q_1 \in M_{n \times n}(F[X])$ and $p, q \in M_{n \times n}(F)$. Then a tedious computation shows $XI - B = P(XI - A)Q = (XI - B)R(XI - B) + p(XI - A)q$. Here $R = p_1 P^{-1} + Q^{-1} q_1 - p_1(XI - A)q_1$. If we carefully analyze the powers of X in this relation, we see $R = 0$. Therefore, $p(XI - A)q = XI - B$. Again comparing powers of X, we see $pq = I$.

So, if $XI - A$ and $XI - B$ have the same invariant factors, then $p(XI - A)q = XI - B$ for invertible matrices $p, q \in M_{n \times n}(F)$. But we have seen in the previous paragraph that $q = p^{-1}$. Again comparing powers of X, we see $pAp^{-1} = B$. Thus, A and B are similar. $\square$

We can now state the following interesting corollary to Theorem 6.23.

Corollary 6.24: Let $F \subseteq K$ be fields, and suppose A, $B \in M_{n \times n}(F)$. Then A and B are similar in $M_{n \times n}(K)$ if and only if A and B are similar in $M_{n \times n}(F)$.

Proof: $M_{n \times n}(F) \subseteq M_{n \times n}(K)$. Thus, if A and B are similar in $M_{n \times n}(F)$, they are similar in $M_{n \times n}(K)$.

Suppose A and B are similar in $M_{n \times n}(K)$. By Theorem 6.23, $XI - A$ and $XI - B$ have the same invariant factors in $K[X]$. But the invariant factors of $XI - A$ and $XI - B$ when computed in $F[X]$ are the same as those in $K[X]$. (See Exercise 20 at the end of this section.) Hence, $XI - A$ and $XI - B$ have the same invariant factors in $F[X]$. In particular, A and B are similar in $M_{n \times n}(F)$. $\square$

A couple of comments are in order here. We have shown in Corollary 6.24 that if A, $B \in M_{n \times n}(F)$, and $PAP^{-1} = B$ for some $P \in M_{n \times n}(K)$, then A and B are similar in $M_{n \times n}(F)$. Hence, there exists a $P_1 \in M_{n \times n}(F)$ such that $P_1 A P_1^{-1} = B$. It is not in general true that $P_1 = P$.

We have now discussed two different sets of polynomials in $F[X]$ which can be associated with a matrix $A \in M_{n \times n}(F)$. The first set is the set of elementary divisors $\{q_1^{e_{11}}, \ldots, q_r^{e_{rp(r)}}\}$ of the matrix A. These polynomials obviously depend on the field F. The second set is the set of invariant factors $\{g_1, \ldots, g_n g_{n-1}^{-1}\}$ of A. This set depends only on A and not on the particular F for which $A \in M_{n \times n}(F)$. Both sets determine the similarity class of A. There are formulas connecting the elementary divisors with the invariant factors of A since $XI - A$ is similar to $XI - R$. Here R is a rational canonical form of A. We invite the reader to determine the relationships between these two sets.

EXERCISES FOR SECTION 6

(1) Find an invertible matrix $P \in M_{3 \times 3}(\mathbb{Q})$ such that PAP^{-1} is the rational canonical form of A when A is the matrix given in Example 6.15.

(2) Find a $P \in M_{4 \times 4}(\mathbb{Q})$ such that $PAP^{-1} = R$ for the matrix A in Example 6.16.

(3) In the proof of Theorem 6.17, we claimed that $\dim\{q(T)^{f_m}(W_i)\} = \partial(q)(f_i - f_m)$ for $i \leqslant m$. Give a proof of this statement.

(4) Prove Theorem 6.21.

(5) Find the rational canonical form R of

$$A = \begin{bmatrix} 0 & -1 & 0 & 0 \\ 1 & 0 & 0 & 0 \\ 1 & 2 & 0 & 1 \\ -1 & 3 & -1 & 0 \end{bmatrix} \in M_{4 \times 4}(\mathbb{Q})$$

Also, find P such that $PAP^{-1} = R$.

(6) Suppose $A \in M_{6 \times 6}(\mathbb{Q})$ with minimal polynomial $m_A(X) = (X - 2)$ $\times (X^2 + 1)^2$. Find all possible rational canonical forms and elementary divisors for A.

(7) Show that A is similar to A^t for any $A \in M_{n \times n}(F)$.

(8) Find the rational canonical form of

$$\begin{pmatrix} \sin \theta & -\cos \theta \\ \cos \theta & \sin \theta \end{pmatrix}.$$

(9) Show that the rational canonical form of a diagonal matrix is itself.

(10) A matrix $A \in M_{n \times n}(F)$ is said to be *indecomposable* if A is not similar to any matrix of the form $\text{diag}(A_1, A_2)$ for some smaller square matrices A_1 and A_2. A is *nonderogatory* if $m_A(X) = c_A(X)$. Show that A is indecomposable if and only if A is nonderogatory and $m_A(X) = q(X)^e$ with $q(X)$ irreducible in $F[X]$.

(11) Let $T \in \mathscr{E}(V)$, and let $T^* \in \mathscr{E}(V^*)$ denote the dual of T. Show that $c_T(X) = c_{T^*}(X)$ and $m_T(X) = m_{T^*}(X)$.

(12) Let $T \in \mathscr{E}(V)$, and suppose $\dim(V) = 2$. Show that V is T-cyclic or $T = xI_V$ for some $x \in F$.

(13) Let A be the companion matrix of $g(X) = X^m - c_{m-1}X^{m-1} - \cdots - c_0$. Show directly that $c_A(X) = g(X)$.

(14) Let $p_1(X), \ldots, p_n(X)$ be a set of monic, primary polynomials, all of degree at least one in $F[X]$. Show there exists a matrix A whose nontrivial elementary divisors are $p_1, \ldots, p_n$. [A polynomial p is primary if p is a power of an irreducible polynomial].

(15) Let

$$A = \begin{bmatrix} 1 & 3 & 3 \\ 3 & 1 & 3 \\ -3 & -3 & -5 \end{bmatrix}$$

Find a P such that $P^{-1}AP$ is in rational canonical form.

(16) Let $A \in M_{n \times n}(\mathbb{C})$. Suppose $\mathscr{S}_{\mathbb{C}}(A) \subseteq \mathbb{R}$. Prove that A is similar to a matrix in $M_{n \times n}(\mathbb{R})$.

(17) Suppose $A \in M_{n \times n}(\mathbb{R})$. If $A^2 = -I$, show $n = 2k$ for some integer k. Prove that A is similar over $\mathbb{R}$ to

$$\begin{pmatrix} 0 & -I_k \\ I_k & 0 \end{pmatrix}$$

(18) Let $T \in \mathscr{E}(V)$. Show that V is T-cyclic if and only if every $S \in \mathscr{E}(V)$ that commutes with T is a polynomial in T.

(19) An endomorphism $T \in \mathscr{E}(V)$ is said to be semisimple if every T-invariant subspace of V has a T-invariant complement. If $m_T(X)$ is irreducible in $F[X]$, prove that T is semisimple.

(20) In the proof of Corollary 6.24, we use the fact that if $g = \text{g.c.d.}(f_1, \ldots, f_r)$ in $F[X]$, then $g = \text{g.c.d.}(f_1, \ldots, f_r)$ in $K[X]$. Give a proof of this fact.

(21) In theorem 6.23, argue that $XI - A$ is equivalent to $\text{diag}(g_n g_{n-1}^{-1}, \ldots, g_1)$.

Chapter IV

Normed Linear Vector Spaces

1. BASIC DEFINITIONS AND EXAMPLES

In this chapter and the next, we shall take a brief look at some of the more important functions which are often present on a vector space V. A great deal of what we have to say in the present chapter can be done over any suitably ordered field F. However, we shall simplify our discussion and assume throughout that $F = \mathbb{R}$, the field of real numbers. Hence, V will denote a real vector space in this chapter. We do not assume V is finite dimensional over $\mathbb{R}$. Let us begin with the definition of a norm on V.

Definition 1.1: A norm on V is a function $f: V \to \mathbb{R}$ such that

(a) $f(\alpha) > 0$ if $\alpha \in V - \{0\}$.
(b) $f(x\alpha) = |x| f(\alpha)$ for all $\alpha \in V$ and $x \in \mathbb{R}$.
(c) $f(\alpha + \beta) \leqslant f(\alpha) + f(\beta)$ for all $\alpha, \beta \in V$.

In Definition 1.1, the notation $|x|$ means the absolute value of the real number x. In previous portions of this text, we have used the notation $|A|$ to denote the cardinality of the set A. The use of the symbol $|\ |$ will always be clear from the context and will cause no confusion in the sequel. Note that 1.1(b) implies $f(0) = 0$.

The most familiar example of a norm on a real vector space V is $V = \mathbb{R}$, and $f(\alpha) = |\alpha|$. We shall give more interesting examples in a moment. If f is a norm on V, then we shall adopt the standard notation of the subject matter and write $f(\alpha) = \|\alpha\|$ for all $\alpha \in V$. Thus, the symbol $\|\ \|$ indicates a real valued function on V that satisfies (a), (b), and (c) of 1.1.

Definition 1.2: A normed linear vector space is a real vector space V together with some fixed norm $\| \ \|: V \to \mathbb{R}$.

Obviously a given vector space V can be viewed in different ways as a normed linear vector space by specifying different norms on V. Thus, a normed linear vector space is actually an ordered pair $(V, \| \ \|)$ consisting of a real vector space V and a real valued function $\| \ \|: V \to \mathbb{R}$ satisfying the axioms of 1.1. When it is not important to specify the exact nature of $\| \ \|$, we shall drop this part of the notation and simply refer to V itself as a normed linear space. Let us consider some nontrivial examples.

Example 1.3: Let n be a positive integer, and set $V = \mathbb{R}^n$. V has at least three important norms whose definitions are given by the following equations:

1.4: (a) $\|(x_1, \ldots, x_n)\|_1 = \sum_{i=1}^{n} |x_i|$

(b) $\|(x_1, \ldots, x_n)\| = \left(\sum_{i=1}^{n} x_i^2 \right)^{1/2}$

(c) $\|(x_1, \ldots, x_n)\|_\infty = \sup\{|x_i| \mid i = 1, \ldots, n\}$

The fact that $\| \ \|_1$, $\| \ \|$, and $\| \ \|_\infty$ satisfy the axioms in definition 1.1 is a straightforward exercise. [To argue $\| \ \|$ satisfies 1.1(c), we need to use Schwarz's inequality in Chapter V]. The norm $\| \ \|$ given in 1.4(b) is just the usual Euclidean norm on $\mathbb{R}^n$. The norm $\| \ \|_\infty$ given in 1.4(c) is called the *uniform norm* on $\mathbb{R}^n$. It is the easiest of the three norms to use when making computations. Note that when $n = 1$, all three of these norms reduce to the absolute value $|\ |$ on $\mathbb{R}$. □

Example 1.5: Let $V = C([a, b])$ denote the set of continuous, real valued functions on a closed interval $[a, b]$. V also has at least three important norms that are used in analysis:

1.6: (a) $\|f\|_1 = \int_a^b |f(t)| \, dt$
(b) $\|f\| = (\int_a^b f(t)^2 \, dt)^{1/2}$
(c) $\|f\|_\infty = \sup\{|f(t)| \mid t \in [a, b]\}$

Once again the reader can check that 1.6 defines norms on $C([a, b])$. The norm $\| \ \|_\infty$ is usually called the *uniform norm* on $C([a, b])$. □

Example 1.7: Let $V = \bigoplus_{i=1}^{\infty} \mathbb{R}$. A typical vector in V is an infinite sequence $\alpha = (x_1, x_2, \ldots)$ having at most a finite number of nonzero components x_i. Thus, the norms $\| \ \|_1$, $\| \ \|$, and $\| \ \|_\infty$ given in equation 1.4 can be extended to norms on V in the obvious way:

1.8: (a) $\|(x_1, x_2, \ldots)\|_1 = \sum_{i=1}^{\infty} |x_i|$
(b) $\|(x_1, x_2, \ldots)\| = (\sum_{i=1}^{\infty} x_i^2)^{1/2}$
(c) $\|(x_1, x_2, \ldots)\|_\infty = \sup\{|x_i| \mid i = 1, 2, \ldots\}$

Since any two vectors in V sit in $\mathbb{R}^n$ for some n sufficiently large, it is clear that the equations in 1.8 define norms on V. ☐

Example 1.9: Let $V = \{(x_1, x_2, \ldots) \in \mathbb{R}^{\mathbb{N}} | \sum_{i=1}^{\infty} x_i^2 < \infty\}$. We had seen in Exercise 6, Section 1 of Chapter I that V is a subspace of $\mathbb{R}^{\mathbb{N}}$. We can define a norm on V as follows:

1.10:
$$\|(x_1, x_2, \ldots)\| = \left(\sum_{i=1}^{\infty} x_i^2 \right)^{1/2}$$

V with the norm defined in equation 1.10 is a well-known Hilbert space, usually denoted ι^2. ☐

We can construct other normed linear vector spaces by considering injective maps into a given normed space $(V, \| \ \|)$. If W is a subspace of V, then clearly $\| \ \|$ restricted to W is a norm on W. More generally, if $T: X \to V$ is an injective linear transformation, then $\|\alpha\|' = \|T(\alpha)\|$ defines a norm on X.

Now suppose $(V, \| \ \|)$ is an arbitrary normed linear vector space. We can use the norm $\| \ \|$ on V to measure the distance $d(\alpha, \beta)$ between two vectors α and β in V. Let us introduce the following topological notions:

Definition 1.11: Let A be a subset of V, and let α and β be vectors in V. Then

(a) $d(\alpha, \beta) = \|\alpha - \beta\|$.

(b) $B_r(\alpha) = \{\gamma | \|\alpha - \gamma\| < r\}$.

(c) We say A is bounded if $A \subseteq B_r(0)$ for some positive number r.

(d) The distance $d(\alpha, A)$ between α and the set A is the number $d(\alpha, A) = \inf\{\|\alpha - \gamma\| | \gamma \in A\}$.

(e) α is said to be an interior point of A if there exists an $r > 0$ such that $B_r(\alpha) \subseteq A$.

(f) The set of interior points of A will be denoted A^0.

(g) A is said to be *open* if $A^0 = A$.

(h) A is said to be *closed* if the complement A^c of A (i.e., $A^c = V - A$) is open.

All these definitions depend on the particular norm being used in V. The function $d: V \times V \to \mathbb{R}$ defined in 1.11(a) is called the *distance function* (relative to the norm $\| \ \|$). The reader can readily verify that d satisfies the following properties:

1.12: (a) $d(\alpha, \beta) > 0$ if $\alpha \neq \beta$ and $d(\alpha, \alpha) = 0$.

(b) $d(\alpha, \beta) = d(\beta, \alpha)$ for all $\alpha, \beta \in V$.

(c) $d(\alpha, \gamma) \leqslant d(\alpha, \beta) + d(\beta, \gamma)$ for all $\alpha, \beta, \gamma \in V$.

Any set V together with a function $d: V \times V \to \mathbb{R}$ that satisfies the conditions in 1.12 is called a *metric space*. Thus, any normed linear vector space is a metric space with distance function given by the norm.

The set $B_r(\alpha)$ introduced in 1.11(b) is called the *ball of radius r around* α. Its exact shape, of course, depends on the particular norm being used.

Example 1.13: Let $V = \mathbb{R}^2$, and consider the three norms given in equation 1.4. If $\alpha = 0$ and $r = 1$, then the ball $B_1(0)$ is the following set:

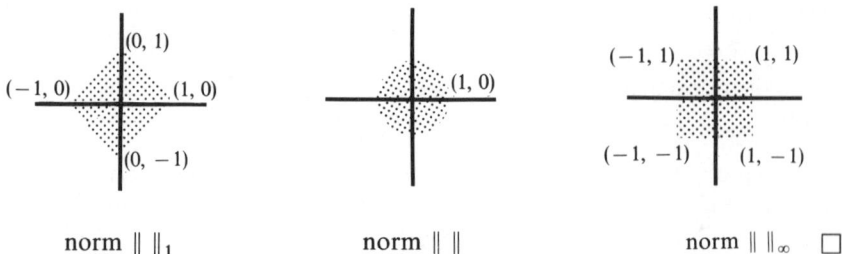

norm $\|\ \|_1$ norm $\|\ \|$ norm $\|\ \|_\infty$

The reader can easily check that $B_r(\alpha)$ is an open set in V. In fact, it is clear from the definitions that a set A in V is open if and only if for every $\alpha \in A$ there exists an $r > 0$ such that $B_r(\alpha) \subseteq A$. The collection of subsets $\mathcal{U} = \{A \,|\, A \text{ is open}$ in $V\}$ forms a topology on V. This means that ϕ and $V \in \mathcal{U}$, and finite intersections and arbitrary unions of sets from $\mathcal{U}$ are again sets in $\mathcal{U}$.

We can also introduce the familiar concepts of limits, continuity, and so on by using the distance function d. In what follows, it is assumed that V and W are normed linear vector spaces. We shall use the same symbol $\|\ \|$ to denote the norm in both spaces V and W. It will always be clear from the context when the symbol $\|\ \|$ is being used to represent the norm on V and when it is being used to represent the norm on W.

Definition 1.14: Let V and W be normed linear vector spaces and suppose A is a subset of V. Let f: A $\to$ W be a function and suppose $\alpha \in V$.

(a) $\lim_{\xi \to \alpha} f(\xi) = \beta$ if for every $r > 0$ there exists an $s > 0$ such that for every $\xi \in A$, $0 < \|\xi - \alpha\| < s \Rightarrow \|f(\xi) - \beta\| < r$.

(b) If $\alpha \in A$, and $\lim_{\xi \to \alpha} f(\xi) = f(\alpha)$, then we say f is *continuous* at α.

(c) f is continuous on A if f is continuous at every point of A.

(d) f is *Lipschitz on A* if there exists a positive constant c such that $\|f(\xi) - f(\eta)\| \leqslant c\|\xi - \eta\|$ for all $\xi, \eta \in A$.

We shall mainly be considering functions whose domain is all of V. If f: V $\to$ W is such a function, then 1.14(a) and 1.14(c) can be rewritten in terms of open sets as follows:

1.15: (a) $\lim_{\xi \to \alpha} f(\xi) = \beta$ if for every open set U in W containing β, there exists an open set U$'$ in V containing α such that $f(U') \subseteq U$.

(b) f is *continuous on V* (or just continuous) if for every open set U in W, $f^{-1}(U)$ is an open set in V.

Definitions 1.14(a)–1.14(c) describe ideas that are familiar from the calculus. The notion of a function being Lipschitz is more peculiar to analysis coming from normed linear spaces. Note that any f: A → W that is Lipschitz on A is certainly continuous on A. Our most important example of a Lipschitz function is the norm itself.

Lemma 1.16: Let $(V, \| \ \|)$ be any normed linear vector space. Then the norm $\| \ \|$ is a Lipschitz function from $(V, \| \ \|)$ to $(\mathbb{R}, | \ |)$.

Proof: If $\alpha, \beta \in V$, then $\|\alpha\| = \|\alpha - \beta + \beta\| \leqslant \|\alpha - \beta\| + \|\beta\|$. Therefore, $\|\alpha\| - \|\beta\| \leqslant \|\alpha - \beta\|$. Reversing the roles of α and β gives us $\|\beta\| - \|\alpha\| \leqslant \|\beta - \alpha\| = \|\alpha - \beta\|$. Thus, $\big| \|\alpha\| - \|\beta\| \big| \leqslant \|\alpha - \beta\|$. This is precisely the statement that $\| \ \|$ is Lipschitz on V. $\square$

Since Lipschitz functions are continuous, we conclude from Lemma 1.16 that the norm $\| \ \|$ is a continuous, real valued function on V. We shall use this fact often in the sequel.

Now suppose T: V → W is a linear transformation. It follows easily from the definition that T is Lipschitz on V if and only if there exists a c > 0 such that $\|T(\xi)\| \leqslant c\|\xi\|$ for all $\xi \in V$. Linear transformations that are Lipschitz on V are very important in analysis and algebra alike. These types of transformations are called *bounded linear operators*. Let us formally introduce the notation we shall use for the set of bounded linear operators.

Definition 1.17: A bounded linear operator T: V → W is a linear transformation that is Lipschitz on V. The set of all bounded linear operators from V to W will be denoted $\mathscr{B}(V, W)$.

Thus, $T \in \mathscr{B}(V, W)$ if and only if $T \in \mathrm{Hom}_{\mathbb{R}}(V, W)$, and there exists a positive constant c such that $\|T(\xi)\| \leqslant c\|\xi\|$ for all $\xi \in V$. Let us consider a few examples before continuing.

Example 1.18: Let $V = \mathbb{R}^n$ with $n \geqslant 2$, and consider the norm $\| \ \|_1$ given in Example 1.3. Let θ_i and π_i denote the canonical injections and projections of $\mathbb{R}^n$. Thus, $\theta_i(x) = (0, \ldots, x, \ldots, 0)$ and $\pi_i(x_1, \ldots, x_n) = x_i$ for all $i = 1, \ldots, n$.

Let $\mathbb{R}$ have the standard norm $\|x\| = |x|$. Then $\theta_i \in \mathrm{Hom}_{\mathbb{R}}(\mathbb{R}, V)$. Since $\|\theta_i(x)\|_1 = \|(0, \ldots, x, \ldots, 0)\|_1 = |x| = \|x\|$, each θ_i is a bounded linear operator. Therefore, $\theta_i \in \mathscr{B}(\mathbb{R}, \mathbb{R}^n)$.

Each projection π_i is a linear transformation from $\mathbb{R}^n$ to $\mathbb{R}$. If $(x_1, \ldots, x_n) \in \mathbb{R}^n$, then $\|\pi_i(x_1, \ldots, x_n)\| = |x_i| \leqslant \sum_{j=1}^n |x_j| = \|(x_1, \ldots, x_n)\|_1$. Thus, $\pi_i \in \mathscr{B}(\mathbb{R}^n, \mathbb{R})$. $\square$

The reader can easily verify that θ_i and π_i are also bounded linear operators with respect to the Euclidean norm $\| \ \|$ and the uniform norm $\| \ \|_\infty$ on $\mathbb{R}^n$.

Example 1.19: Let $V = C([a, b])$ with norm $\| \, \|_\infty$ given in equation 1.6. For every $f \in V$, set $T(f) = \int_a^b f(t) \, dt$. Then $T \in \text{Hom}_\mathbb{R}(V, \mathbb{R})$. As usual, let the norm on $\mathbb{R}$ be the absolute value $| \, |$. Set $m = \|f\|_\infty = \sup\{|f(t)| \, | \, t \in [a, b]\}$. Then $|T(f)| = |\int_a^b f(t) \, dt| \leqslant \int_a^b |f(t)| \, dt \leqslant m(b - a) = (b - a)\|f\|_\infty$. Thus, $T \in \mathcal{B}(C([a, b]), \mathbb{R})$. $\square$

Let us give an example of a linear transformation that is not a bounded linear operator.

Example 1.20: Let $V = C([0, 1]) = W$. Then $T = I_V$, the identity map on V, is a linear transformation from V to W. Norm V with $\| \, \|_1$ and W with $\| \, \|_\infty$ (notation as in 1.6). Then we claim $T : (C([0, 1]), \| \, \|_1) \to (C([0, 1]), \| \, \|_\infty)$ is not a bounded linear operator. To see this, suppose T were bounded. Then there would exist a $c > 0$ such that $\|T(f)\|_\infty \leqslant c\|f\|_1$ for all $f \in C([0, 1])$. If $f = t^n$, then this inequality would become $1 \leqslant c/(n + 1)$. That relation is clearly impossible for all n. Thus, T is not bounded. $\square$

Example 1.20 shows that $\mathcal{B}(V, W)$ is in general a proper subset of $\text{Hom}_\mathbb{R}(V, W)$. Another example, a bit more algebraic, is the following:

Example 1.21: Set $V = W = \bigoplus_{i=1}^\infty \mathbb{R}$, and again let T be the identity map from V to W. Norm V with $\| \, \|_\infty$ and W with $\| \, \|_1$ (notation as in 1.8). If T were a bounded linear operator, then there would exist a $c > 0$ such that $\|\alpha\|_1 \leqslant c\|\alpha\|_\infty$ for all $\alpha \in V$. For $\alpha = (1, \ldots, 1, 0 \ldots)$, this inequality implies $n \leqslant c$. This is clearly impossible for all n. $\square$

The reader will note that the last two examples were both infinite dimensional. If both V and W are finite dimensional over $\mathbb{R}$, then every linear transformation from V to W is a bounded linear operator. Thus, $\mathcal{B}(V, W) = \text{Hom}(V, W)$ whenever $\dim(V), \dim(W) < \infty$. This fact is not obvious. We shall prove it in Section 3 of this chapter.

Now suppose V and W are arbitrary normed linear vector spaces, and let $T \in \text{Hom}(V, W)$. The relation between being bounded and being continuous is easily stated.

Theorem 1.22: Let V and W be normed linear vector spaces and suppose $T \in \text{Hom}(V, W)$. Then the following are equivalent:

(a) T is continuous at one point of V.

(b) T is continuous on all of V.

(c) T is a bounded linear operator.

Proof: Lipschitz maps are continuous. Thus, the implications (c) $\Rightarrow$ (b) $\Rightarrow$ (a) are all obvious. Hence, it suffices to show (a) $\Rightarrow$ (c). Suppose T is continuous at $\alpha \in V$. Then there exists an $s > 0$ such that $\|\xi - \alpha\| < s \Rightarrow \|T(\xi) - T(\alpha)\| < 1$.

Now any vector $\beta \in V$ can be written in the form $\beta = \xi - \alpha$ (set $\xi = \beta + \alpha$). Thus, $\|\beta\| < s \Rightarrow \|T(\beta)\| < 1$. If β is any nonzero vector in V, then $\|\beta\| \neq 0$, and $\|s\beta/2\|\beta\| \| = s/2 < s$. Therefore, $\|T(s\beta/2\|\beta\|)\| < 1$. This last inequality is equivalent to $\|T(\beta)\| < (2/s)\|\beta\|$. We conclude that T is bounded. $\square$

Let $T \in \mathscr{B}(V, W)$. Then there exists a positive constant c such that $\|T(\alpha)\| \leqslant c\|\alpha\|$ for all $\alpha \in V$. Such a number c is called a *bound* for the linear operator T. In Example 1.19, for instance, we showed that $b - a$ was a bound for $\int_a^b(\ )\,dt$.

If c is a bound for T, and if $\xi \in V$ such that $\|\xi\| = 1$, then $\|T(\xi)\| \leqslant c$. In particular, the set of bounds for T is a subset of $\mathbb{R}$ that is bounded below by the nonnegative number $\|T(\xi)\|$. We conclude that the number $\inf\{c \,|\, c$ is a bound for T$\}$ exists. This number is commonly called the *norm* of T. We shall henceforth denote it $\|T\|$. Thus, we have the following definition:

Definition 1.23: Let $T \in \mathscr{B}(V, W)$. Then $\|T\| = \inf\{c \,|\, c$ is a bound for T$\}$.

There are a couple of alternative definitions for the norm of T that are worth recording here.

Lemma 1.24: Let $T \in \mathscr{B}(V, W)$. Then

(a) $\|T\| = \sup\{\|T(\alpha)\|/\|\alpha\| \,|\, \alpha \in V - \{0\}\}$.
(b) $\|T\| = \sup\{\|T(\alpha)\| \,|\, \|\alpha\| = 1\}$. $\square$

The proofs of (a) and (b) in 1.24 are straightforward. We leave them for exercises at the end of this section. Let us consider an example.

Example 1.25: We return to Example 1.19. We have already noted that $b - a$ is a bound for $T = \int_a^b(\ )\,dt$. Thus, $\|T\| \leqslant b - a$. Consider the constant function $1 \in C([a, b])$. $\|1\|_\infty = 1$, and $T(1) = \int_a^b 1\,dt = b - a$. Thus, $\|T(1)\| = |b - a| = b - a$. In particular, Lemma 1.24(b) implies $\|T\| \geqslant b - a$. Therefore, $\|T\| = b - a$. $\square$

Now suppose S and T are bounded linear operators from V to W. If c_1 and c_2 are bounds for S and T, respectively, then for all x, $y \in \mathbb{R}$, we have $\|(xS + yT)(\alpha)\| = \|xS(\alpha) + yT(\alpha)\| \leqslant |x|\|S(\alpha)\| + |y|\|T(\alpha)\| \leqslant (c_1|x| + c_2|y|)\|\alpha\|$. Thus, $xS + yT$ is a bounded linear operator with bound $c_1|x| + c_2|y|$. In particular, $\mathscr{B}(V, W)$ is a subspace of Hom(V, W). We can use the definition of $\|T\|$ given in 1.23 to put a norm on $\mathscr{B}(V, W)$.

Theorem 1.26: $\mathscr{B}(V, W)$ is a normed linear vector space if $\|T\|$ is defined as in 1.23.

Proof: We have already noted that $\mathscr{B}(V, W)$ is a vector space under the usual operations $xS + yT$ in $\text{Hom}(V, W)$. It remains to verify that axioms (a), (b), and (c) of Definition 1.1 are satisfied.

(a) Let $T \in \mathscr{B}(V, W)$. Lemma 1.24(b) implies $\|T\| \geqslant 0$. Suppose $\|T\| = 0$. Then 1.24(a) implies $\|T(\alpha)\| = 0$ for all $\alpha \in V$. But then $T(\alpha) = 0$ for all $\alpha \in V$ and, consequently, $T = 0$.

(b) Let $T \in \mathscr{B}(V, W)$ and $x \in \mathbb{R}$. Using 1.24(b), we have

$$\|xT\| = \sup\{\|xT(\alpha)\| \mid \|\alpha\| = 1\} = \sup\{|x| \, \|T(\alpha)\| \mid \|\alpha\| = 1\}$$
$$= |x|\sup\{\|T(\alpha)\| \mid \|\alpha\| = 1\} = |x| \, \|T\|.$$

(c) Let $S, T \in \mathscr{B}(V, W)$. Again using 1.24(b), we have

$$\begin{aligned}
\|S + T\| &= \sup\{\|(S + T)(\alpha)\| \mid \|\alpha\| = 1\}\\
&= \sup\{\|S(\alpha) + T(\alpha)\| \mid \|\alpha\| = 1\}\\
&\leqslant \sup\{\|S(\alpha)\| + \|T(\alpha)\| \mid \|\alpha\| = 1\}\\
&\leqslant \sup\{\|S(\alpha)\| \mid \|\alpha\| = 1\} + \sup\{\|T(\alpha)\| \mid \|\alpha\| = 1\}\\
&= \|S\| + \|T\| \quad \square
\end{aligned}$$

The norm introduced in Definition 1.23 is usually referred to as the *uniform norm* on $\mathscr{B}(V, W)$. Note that if $T \in \mathscr{B}(V, W)$ has uniform norm $\|T\|$, then $\|T(\alpha)\| \leqslant \|T\| \, \|\alpha\|$ for all $\alpha \in V$. We conclude this section with another useful fact about the uniform norm.

Theorem 1.27: If $T \in \mathscr{B}(V, W)$, and $S \in \mathscr{B}(W, Z)$, then $ST \in \mathscr{B}(V, Z)$. Furthermore, $\|ST\| \leqslant \|S\| \, \|T\|$.

Proof: For any $\alpha \in V$, we have $\|ST(\alpha)\| = \|S(T(\alpha))\| \leqslant \|S\| \, \|T(\alpha)\| \leqslant \|S\| \, \|T\| \, \|\alpha\|$. $\square$

EXERCISES FOR SECTION 1

(1) Show that equations 1.4(a) and 1.4(c) and 1.6(a) and 1.6(c) define norms on $\mathbb{R}^n$ and $C([a, b])$, respectively.

(2) Let W denote a real vector space. A function $f: W \to \mathbb{R}$ is called a *seminorm* on W if f satisfies the following properties:
   (a) $f(\alpha) \geqslant 0$ for all $\alpha \in W$.
   (b) $f(x\alpha) = |x|f(\alpha)$ for all $x \in \mathbb{R}$ and $\alpha \in W$.
   (c) $f(\alpha + \beta) \leqslant f(\alpha) + f(\beta)$ for all $\alpha, \beta \in W$.

Show that if f is a seminorm on W and $T \in \text{Hom}(V, W)$, then fT is a seminorm on V. Use this fact to construct a seminorm on W that is not a norm.

In the rest of these exercises, V and W will denote normed linear vector spaces.

(3) Lipschitz functions on V need not be linear transformations. Give at least two examples.

(4) Show that $B_r(\alpha)$ is an open set in V.

(5) Show that $\mathcal{U} = \{A \mid A \text{ is open in } V\}$ is closed under arbitrary unions and finite intersections. Give an example in $\mathbb{R}$ that shows that arbitrary intersections of open sets need not be open.

(6) Give an example of a set A in V such that A is neither open nor closed.

(7) Let $A \subseteq V$. A vector α is said to be in the closure of A if $B_r(\alpha) \cap A \neq \phi$ for all $r > 0$. Let $\bar{A} = \{\alpha \in V \mid \alpha \text{ is in the closure of } A\}$. $\bar{A}$ is called the closure of the set A.

(a) Show that $\bar{A}$ is the smallest closed set in V containing A.

(b) Give an example where $A \neq \bar{A}$.

(c) Show that $\alpha \in \bar{A}$ if and only if $d(\alpha, A) = 0$.

(8) The boundary A^∂ of a set A in V is the difference between the closure of A and its interior. Thus, $A^\partial = \bar{A} - A^0$.

(a) Show $\alpha \in A^\partial$ if and only if for all $r > 0$, $B_r(\alpha) \cap A \neq \phi$ and $B_r(\alpha) \cap A^c \neq \phi$.

(b) Compute the boundary of $B_1(0)$ in the examples in 1.13.

(9) Show that $B_r(\alpha) + B_s(\beta) = B_{r+s}(\alpha + \beta)$.

(10) Give a detailed proof of the assertion that 1.15(a) and 1.15(b) are equivalent to 1.14(a) and 1.14(c), respectively.

(11) Show that θ_i and π_i are bounded linear operators when $\| \ \|_1$ is replaced by either $\| \ \|$ or $\| \ \|_\infty$ (notation as in 1.4) in $\mathbb{R}^n$.

(12) Let $T \in \mathcal{B}(V, W)$. Show $\|T(\xi)\| \leqslant \|T\| \|\xi\|$ for all $\xi \in V$.

(13) Let $\xi \in V$, and consider the map $E_\xi : \mathcal{B}(V, W) \to W$ given by $E_\xi(T) = T(\xi)$. Show that E_ξ is a bounded linear operator.

(14) Let $\xi \in V$. Show that there exists an $f \in \mathcal{B}(V, \mathbb{R})$ such that $\|f\| = 1$ and $|f(\xi)| = \|\xi\|$.

(15) Use Exercise 14 to show that $\|E_\xi\| = \|\xi\|$ in Exercise 13.

(16) If $A = (a_{ij}) \in M_{n \times n}(\mathbb{R})$, define $\|A\| = \sup\{\sum_{j=1}^n |a_{ij}| \mid i = 1, \ldots, n\}$. Show that this equation defines a norm on $M_{n \times n}(\mathbb{R})$ for which $\|AB\| \leqslant \|A\| \|B\|$ for all matrices A and B.

(17) Consider $(\mathbb{R}^n, \| \ \|_\infty)$. Let $T \in \mathscr{B}(\mathbb{R}^n, \mathbb{R}^n)$. If $A = \Gamma(\underline{\delta}, \underline{\delta})(T)$, then show $\|T\| = \|A\|$. Here $\|A\|$ is the norm given in Exercise 16.

(18) Prove Lemma 1.24. (You will need Exercise 12).

(19) Prove that every ball $B_r(\alpha)$ in V is a convex set. (A set S in V is convex if β, $\delta \in S \Rightarrow x\beta + (1 - x)\delta \in S$ for all $x \in [0, 1]$.)

(20) Consider $(\mathbb{R}^n, \| \ \|_1)$, and let $\alpha = (x_1, \ldots, x_n) \in \mathbb{R}^n$. Define a map $T: \mathbb{R}^n \to \mathbb{R}$ by $T(y_1, \ldots, y_n) = \sum_{k=1}^{n} x_k y_k$. Show that $T \in \mathscr{B}(\mathbb{R}^n, \mathbb{R})$, and compute $\|T\|$.

(21) Use Exercise 9 to show that the sum of any two bounded sets is bounded.

(22) Formulate the appropriate notion of a function $f: V \to W$ being Lipschitz at a point $\alpha \in V$. Consider the function $f: \mathbb{R} \to \mathbb{R}$ given by $f(x) = |x|^{1/2}$. Show that f is Lipschitz at 1, but is not Lipschitz at 0.

2. PRODUCT NORMS AND EQUIVALENCE

Two normed linear vector spaces V and W are said to be *norm isomorphic* if there exists an isomorphism $T: V \cong W$ such that T and T^{-1} are bounded linear operators, that is, $T \in \mathscr{B}(V, W)$ and $T^{-1} \in \mathscr{B}(W, V)$. For example, we have seen in Exercise 17 of Section 1 that $\mathscr{B}(\mathbb{R}^n, \mathbb{R}^n)$ and $M_{n \times n}(\mathbb{R})$ are norm isomorphic when $\mathbb{R}^n$ is given the uniform norm $\| \ \|_\infty$. If two normed linear vector spaces are norm isomorphic, they are for all practical purposes identical. Thus, in the sequel, we shall identify spaces that are norm isomorphic whenever it is convenient to do so.

For two different norms on the same space V, we have the following important definition:

Definition 2.1: Two norms $\| \ \|$ and $\| \ \|'$ on the same real vector space V are said to be *equivalent* if there exist positive constants a and b such that $\|\alpha\| \leqslant a\|\alpha\|'$, and $\|\alpha\|' \leqslant b\|\alpha\|$ for all $\alpha \in V$.

Thus, two norms $\| \ \|$ and $\| \ \|'$ on V are equivalent if the identity map is a norm isomorphism from $(V, \| \ \|)$ to $(V, \| \ \|')$. We have already seen in Example 1.20 or 1.21 that two norms need not be equivalent. We should also point out the trivial case of $\mathbb{R}$ itself. Definition 1.1(b) implies that any norm on $\mathbb{R}$ is equivalent to the absolute value $| \ |$. We shall prove in Section 3 that if $\dim(V) < \infty$, then any two norms on V are equivalent.

The equivalence of norms is an important idea in analysis and topology alike. If two norms are equivalent on V, then they generate the same topology. By this we mean the open sets relative to the two norms are the same collection of sets. More precisely, we have the following theorem:

Theorem 2.2: Let $\| \ \|$ and $\| \ \|'$ be equivalent norms on V. A subset A of V is open with respect to $\| \ \|$ if and only if A is open with respect to $\| \ \|'$.

Proof: Suppose A is open with respect to the norm $\| \ \|$. Then for every $\alpha \in A$, there exists an $r(\alpha) > 0$ such that $B_{r(\alpha)}(\alpha) = \{\xi \mid \|\xi - \alpha\| < r(\alpha)\} \subseteq A$. Clearly, $A = \bigcup_{\alpha \in A} B_{r(\alpha)}(\alpha)$. Thus, to argue that A is an open set with respect to $\| \ \|'$, it suffices to show that any ball $B_r(\alpha) = \{\xi \mid \|\xi - \alpha\| < r\}$ is an open set in $(V, \| \ \|')$.

Let us suppose $\|\alpha\| \leqslant b\|\alpha\|'$ for every $\alpha \in V$. Let $\beta \in B_r(\alpha)$. Since $B_r(\alpha)$ is an open subset of $(V, \| \ \|)$ (see exercise 4 of Section 1], there exists an $s > 0$ such that $B_s(\beta) \subseteq B_r(\alpha)$. Set $D = \{\xi \mid \|\xi - \beta\|' < s/b\}$. D is just an open ball in the $\| \ \|'$-norm around β. If $\xi \in D$, then $\|\xi - \beta\| \leqslant b\|\xi - \beta\|' < b(s/b) = s$. Hence, $D \subseteq B_s(\beta) \subseteq B_r(\alpha)$. We have now shown that $B_r(\alpha)$ is an open set in $(V, \| \ \|')$.

If we reverse the roles of $\| \ \|$ and $\| \ \|'$, we get every ball in the $\| \ \|'$-norm is an open set in $(V, \| \ \|)$. This completes the proof of the theorem. $\square$

Theorem 2.2 implies that all topological notions remain the same when switching between equivalent norms. This being the case, one should try to choose an equivalent norm in which the given problem becomes easier computationally. For example, in $\mathbb{R}^n$ the Euclidean norm $\| \ \|$ is equivalent to the uniform norm $\| \ \|_\infty$ (see Exercise 1 at the end of this section). It often happens in specific problems that $\| \ \|_\infty$ is easier to handle than the Euclidean norm when doing arithmetic. Thus we often switch from $\| \ \|$ to $\| \ \|_\infty$ when dealing with problems in $\mathbb{R}^n$. Since these norms are equivalent, there is no loss in generality from a topological point of view in making this switch.

An immediate corollary to Theorem 2.2 is the following remark, whose proof we leave to the reader.

Corollary 2.3: Let V and W be normed linear vector spaces. Then the set $\mathscr{B}(V, W)$ in $\mathrm{Hom}_{\mathbb{R}}(V, W)$ remains the same if either the norm in V or W is replaced by an equivalent norm. Changing norms in V and W to equivalent norms results in equivalent uniform norms on $\mathscr{B}(V, W)$. $\square$

Now suppose we consider a product $V \times W$ of two normed linear vector spaces $(V, \| \ \|_1)$ and $(W, \| \ \|_2)$, We want to discuss what norms are available on $V \times W$. One norm which readily comes to mind is the so called sum norm $\|(\alpha, \beta)\|_s = \|\alpha\|_1 + \|\beta\|_2$. The reader can easily verify that $\| \ \|_s$ is a norm on $V \times W$. For example, the norm $\| \ \|_1$ given in equation 1.4(a) is clearly the sum norm on $\mathbb{R}^n$.

The sum norm on $V \times W$ has the property that the canonical injections θ_i and projections π_i associated with $V \times W$ are all bounded linear operators. Thus,

2.4:

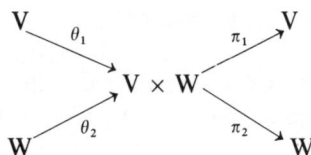

is a diagram of bounded linear operators when $V \times W$ is given the sum norm $\| \ \|_s$. To see this, we have $\|\pi_2(\alpha, \beta)\|_2 = \|\beta\|_2 \leqslant \|\alpha\|_1 + \|\beta\|_2 = \|(\alpha, \beta)\|_s$. Thus, π_2, and similarly π_1, is bounded. We also have $\|\theta_1(\alpha)\|_s = \|(\alpha, 0)\|_s = \|\alpha\|_1 + \|0\|_2 = \|\alpha\|_1$. Thus, θ_1, and similarly θ_2, is bounded. Note that 1 is a bound for all four maps in 2.4.

The norms on $V \times W$ which have some intrinsic relation with the given norms on V and W turn out to be the most useful norms to study on products. This prompts the following definition.

Definition 2.5: Let V and W be normed linear vector spaces. A norm on the product $V \times W$ is called a product norm if the canonical maps in Diagram 2.4 are all bounded linear operators.

Thus, a norm $\| \ \|$ on $V \times W$ is a product norm if there exist four positive constants a, b, c, and d such that the following inequalities are satisfied:

2.6: (1) $\|\theta_1(\alpha)\| \leqslant a\|\alpha\|_1$

(2) $\|\theta_2(\beta)\| \leqslant b\|\beta\|_2$ for all $\alpha \in V, \ \beta \in W$

(3) $\|\pi_1(\alpha, \beta)\|_1 \leqslant c\|(\alpha, \beta)\|$

(4) $\|\pi_2(\alpha, \beta)\|_2 \leqslant d\|(\alpha, \beta)\|$

The sum norm $\|(\alpha, \beta)\|_s = \|\alpha\|_1 + \|\beta\|_2$ is a typical example of a product norm on $V \times W$. In fact, up to equivalence, the sum norm is the only product norm on $V \times W$.

Theorem 2.7: Let $(V, \| \ \|_1)$ and $(W, \| \ \|_2)$ be normed linear vector spaces. Then any product norm $\| \ \|$ on $V \times W$ is equivalent to the sum norm $\|(\alpha, \beta)\|_s = \|\alpha\|_1 + \|\beta\|_2$.

Proof: Let $\| \ \|$ be any product norm on $V \times W$. Since θ_i and π_i are bounded, there exists constants a, b, c, and d satisfying the inequalities in 2.6. Let $e = \max\{a, b\}$. Then $\|(\alpha, \beta)\| = \|(\alpha, 0) + (0, \beta)\| \leqslant \|(\alpha, 0)\| + \|(0, \beta)\| = \|\theta_1(\alpha)\| + \|\theta_2(\beta)\| \leqslant a\|\alpha\|_1 + b\|\beta\|_2 \leqslant e(\|\alpha\|_1 + \|\beta\|_2) = e\|(\alpha, \beta)\|_s$. Thus, $\|(\alpha, \beta)\| \leqslant e\|(\alpha, \beta)\|_s$.

Let $f = c + d$. Then $\|(\alpha, \beta)\|_s = \|\alpha\|_1 + \|\beta\|_2 = \|\pi_1(\alpha, \beta)\|_1 + \|\pi_2(\alpha, \beta)\|_2 \leqslant c\|(\alpha, \beta)\| + d\|(\alpha, \beta)\| = f\|(\alpha, \beta)\|$.

We conclude that $\| \ \|$ and $\| \ \|_s$ are equivalent. □

We can generalize Theorem 2.7 to n factors in the obvious way. We state the result and leave the proof for the reader.

Theorem 2.8: Let $(V_i, \| \ \|_i)$, $i = 1, \ldots, n$, be a finite number of normed linear vector spaces. Then any product norm on $V_1 \times \cdots \times V_n$ is equivalent to the sum norm $\| \ \|_s$ given in (a) below. The formulas in (b) and (c) also define product norms on $V_1 \times \cdots \times V_n$.

(a) $\|(\alpha_1, \ldots, \alpha_n)\|_s = \sum_{i=1}^n \|\alpha_i\|_i$.

(b) $\|(\alpha_1, \ldots, \alpha_n)\|_e = (\sum_{i=1}^n \|\alpha_i\|_i^2)^{1/2}$.

(c) $\|(\alpha_1, \ldots, \alpha_n)\|_\infty = \sup\{\|\alpha_1\|_1, \ldots, \|\alpha_n\|_n\}$.

In particular, these three norms are all equivalent.   □

In Theorem 2.8, $\| \ \|_s$ is the sum norm on $V_1 \times \cdots \times V_n \cdot \| \ \|_e$ is called the Euclidean norm and $\| \ \|_\infty$ the uniform norm.

The definitions in this section allow us to make some comments about addition and scalar multiplication in a normed linear vector space $(V, \| \ \|)$. We can think of vector addition in V as a linear transformation $T: V \times V \to V$ given by $T(\alpha, \beta) = \alpha + \beta$. We claim T is a continuous function with respect to any product norm on $V \times V$. By Theorem 2.7, it suffices to show T is a bounded linear operator with respect to the sum norm on $V \times V$. Since $\|T(\alpha, \beta)\| = \|\alpha + \beta\| \leqslant \|\alpha\| + \|\beta\| = \|(\alpha, \beta)\|_s$, we see T is bounded by 1. We have now proved the first part of the following lemma:

Lemma 2.9: Let V be a normed linear vector space. The operation of addition is a bounded linear operator from $V \times V$ to V. The operation of scalar multiplication is a continuous function from $\mathbb{R} \times V$ to V.   □

The map $f: \mathbb{R} \times V \to V$ given by $f(x, \alpha) = x\alpha$ is not a linear transformation. However, it is easy to show that f is a continuous map on $\mathbb{R} \times V$ with respect to any product norm. We leave this point as an exercise at the end of this section.

We have discussed what sorts of norms are available on a product $V_1 \times \cdots \times V_n$ of a finite number of normed linear vector spaces $(V_i, \| \ \|_i)$ $i = 1, \ldots, n$. If $V = V_1 \times \cdots \times V_n$ is given any product norm, then the projections $\pi_i: V \to V_i$ are all bounded linear operators. If we identify V_i with $\theta_i(V_i)$ (these spaces are norm isomorphic), then V is an internal direct sum of the V_i, and the projections of V onto V_i are all bounded. In the remainder of this section, we shall extend these remarks to internal direct sums in general.

Suppose $(V, \| \ \|)$ is a normed linear vector space, and let $V_1, \ldots, V_n$ be subspaces of V. We assume that $V = V_1 \oplus \cdots \oplus V_n$. Then every vector $\alpha \in V$ can be written uniquely in the form $\alpha = \alpha_1 + \cdots + \alpha_n$, where $\alpha_i \in V_i$ for each $i = 1, \ldots, n$. Recall that the map sending α to α_i is a well-defined linear transformation from V to V_i. We call this map the ith projection and denote it by P_i. Thus, $P_i \in \text{Hom}_{\mathbb{R}}(V, V_i)$.

Now we can restrict the norm $\| \ \|$ to each subspace V_i. Then each V_i is a normed linear vector space in its own right, and we can consider the sum norm on the product $V_1 \times \cdots \times V_n$. This leads to the following definition:

Definition 2.10: Let $(V, \| \ \|)$ be a normed linear vector space with subspaces $V_1, \ldots, V_n$. We say that V is a norm direct sum of the V_i if

  (a) $V = V_1 \oplus \cdots \oplus V_n$.
  (b) The map $S: V_1 \times \cdots \times V_n \to V$ given by $S(\alpha_1, \ldots, \alpha_n) = \alpha_1 + \cdots + \alpha_n$ is a norm isomorphism.

In this definition, each V_i is normed with $\| \ \|$ restricted to V_i. The norm on $V_1 \times \cdots \times V_n$ is the sum norm $\|(\alpha_1, \ldots, \alpha_n)\|_s = \sum_{i=1}^n \|\alpha_i\|$. Since V is the direct sum of the V_i, the map S in (b) is an isomorphism. Thus, $V = V_1 \oplus \cdots \oplus V_n$ is a norm direct sum if and only if S and S^{-1} are bounded linear operators. A typical example to keep in mind is an external direct product $V_1 \times \cdots \times V_n = V$ (with sum norm) of normed linear vector spaces $(V_i, \| \ \|_i)$ and subspaces $\theta_i(V_i)$.
  Suppose $V = V_1 \oplus \cdots \oplus V_n$ is any (internal) direct sum. Then $\|S(\alpha_1, \ldots, \alpha_n)\| = \|\alpha_1 + \cdots + \alpha_n\| \leqslant \sum_{i=1}^n \|\alpha_i\| = \|(\alpha_1, \ldots, \alpha_n)\|_s$. Thus, the map S given in definition 2.10(b) is always a bounded linear operator. In particular, V is a norm direct sum of the V_i if S^{-1} is a bounded linear operator. This last remark can be said in a slightly different way.

Theorem 2.11: Suppose V is a normed linear vector space and an internal direct sum $V = V_1 \oplus \cdots \oplus V_n$ of subspaces $V_1, \ldots, V_n$. Let P_i denote the projection of V onto V_i. Then V is a norm direct sum of the V_i if and only if each P_i is a bounded linear operator.

Proof: From our discussion above, we know V is a norm direct sum of the V_i if and only if S^{-1} is a bounded linear operator. S^{-1} is bounded if there exists a $c > 0$ such that $\|S^{-1}(\alpha)\|_s \leqslant c\|\alpha\|$ for all $\alpha \in V$. This last inequality means $\sum_{i=1}^n \|\alpha_i\| \leqslant c\|\alpha\|$ whenever $\alpha = \alpha_1 + \cdots + \alpha_n$ with $\alpha_i \in V_i$ for $i = 1, \ldots, n$.
  Now suppose V is the norm direct sum of the V_i. Let $\alpha \in V$ and write $\alpha = \alpha_1 + \cdots + \alpha_n$ with $\alpha_j \in V_j$ for $j = 1, \ldots, n$. For any fixed i, we have $\|P_i(\alpha)\| = \|\alpha_i\| \leqslant \sum_{j=1}^n \|\alpha_j\| = \|S^{-1}(\alpha)\|_s \leqslant c\|\alpha\|$. Thus, P_i is a bounded linear operator.
  Conversely, suppose each P_i is bounded. Then there exists a $k_i > 0$ such that $\|P_i(\alpha)\| \leqslant k_i\|\alpha\|$ for all $\alpha \in V$. Let $c = \sum_{i=1}^n k_i$. Then $\|S^{-1}(\alpha)\|_s = \sum_{i-1}^n \|\alpha_i\| = \sum_{i=1}^n \|P_i(\alpha)\| \leqslant c\|\alpha\|$. Thus, S^{-1} is bounded. Consequently, V is a norm direct sum of the V_i. $\square$

EXERCISES FOR SECTION 2

  (1) Show that the norms $\| \ \|_1$, $\| \ \|$, and $\| \ \|_\infty$ given in 1.4 are all equivalent (and hence product norms) on $\mathbb{R}^n$.

  (2) Is Exercise 1 true for $V = \bigoplus_{i=1}^\infty \mathbb{R}$?

  (3) Give a detailed proof of Corollary 2.3.

(4) Let V and W denote normed linear vector spaces, and consider the projection $\pi_1 \colon V \times W \to V$. Show that relative to any product norm on $V \times W$, π_1 is an open map [i.e., U open in $V \times W \Rightarrow \pi_1(U)$ open in V].

(5) Prove Theorem 2.8.

(6) Let V be a normed linear vector space and define $f \colon \mathbb{R} \times V \to V$ by $f(x, \alpha) = x\alpha$. Show that f is continuous relative to any product norm on $\mathbb{R} \times V$.

(7) Let V_i, $i = 1, 2, 3$, be normed linear vector spaces. Show that the map $X \colon \mathscr{B}(V_1, V_2) \times \mathscr{B}(V_2, V_3) \to \mathscr{B}(V_1, V_3)$ given by $X(S, T) = TS$ is a continuous map. Here the norm on $\mathscr{B}(V_i, V_j)$ is the uniform norm given in equation 1.23, and the norm on the product is any product norm.

(8) Give an example of a normed linear space V and two subspaces V_1 and V_2 of V such that $V = V_1 \oplus V_2$, but V is not the norm direct sum of V_1 and V_2.

(9) Suppose $V = V_1 \oplus V_2$ is a norm direct sum. Show that each V_i is a closed subset of V.

(10) Let $f \colon V \to \mathbb{R}$ be a seminorm on V (Exercise 2 of Section 1). Prove the following assertions:
 (a) $N = \{\alpha \in V \mid f(\alpha) = 0\}$ is a subspace of V.
 (b) f is constant on each coset $\alpha + N$ of N.
 (c) f induces a norm $\| \, \|$ on the quotient space V/N [given by $\|\alpha + N\| = f(\alpha)$] such that the following diagram is commutative:

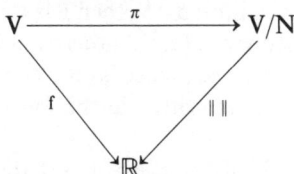

$$V \xrightarrow{\ \ \pi\ \ } V/N$$

f $\| \, \|$

$$\searrow \quad \swarrow$$

$$\mathbb{R}$$

Here π is the natural map $\pi(\alpha) = \alpha + N$.

(11) Show that $\mathscr{B}(\mathbb{R}^n, \mathbb{R}^n)$ and $M_{n \times n}(\mathbb{R})$ are norm isomorphic.

(12) Let V and W be normed linear vector spaces, and suppose $\omega \colon V \times W \to \mathbb{R}$ is a bounded bilinear map. This means there exists a positive constant b such that $|\omega(\alpha, \beta)| \leqslant b\|\alpha\| \, \|\beta\|$ for all $\alpha \in V$, and $\beta \in W$. Let $T \colon V \to W^*$ be defined by $T(\alpha)(\beta) = \omega(\alpha, \beta)$. Show that T is a bounded linear map from V to $\mathscr{B}(W, \mathbb{R})$ and compute $\|T\|$.

(13) Let N be a closed subspace of a normed linear vector space $(V, \| \, \|)$. Define a real valued function $\| \, \|' \colon V/N \to \mathbb{R}$ by $\|\alpha + N\|' = \inf\{\|\alpha + \beta\| \mid \beta \in N\}$. Show that $(V/N, \| \, \|')$ is a normed linear vector space. Prove that the natural map $\alpha \to \alpha + N$ is a bounded linear map from V to V/N.

(14) Suppose V and W are normed linear vector spaces, and $T \in \mathscr{B}(V, W)$. Let N be a closed subspace of V contained in ker(T). Let $\bar{T}$ be the induced map on V/N given by $\bar{T}(\alpha + N) = T(\alpha)$. If we norm V/N as in Exercise 13, show $\bar{T}$ is a bounded linear map from V/N to W such that $\|\bar{T}\| = \|T\|$.

(15) Let V and W be normed linear vector spaces, and let $T \in \mathscr{B}(V, W)$. T is called an *isometry* if $\|T(\alpha)\| = \|\alpha\|$ for all $\alpha \in V$. Examine the bounded maps in this section and decide which are isometries.

3. SEQUENTIAL COMPACTNESS AND THE EQUIVALENCE OF NORMS

In this section, we shall prove that any two norms on a finite-dimensional vector space V are equivalent. In order to do this, we need to develop a certain number of topological facts that are true in any metric space. However, we shall state all the results we need in the language of normed linear vector spaces. Throughout this section, V will denote a normed linear vector space with some fixed norm $\| \ \|$.

We begin with the notion of a sequence of vectors in V.

Definition 3.1: (a) A sequence $\{\alpha_n\}$ in V is a function $f: \mathbb{N} \to V$ such that $f(n) = \alpha_n$ for all $n \in \mathbb{N}$.

(b) A sequence $\{\alpha_n\}$ is said to have a limit $\beta \in V$ if for every $r > 0$ there exists an $m \in \mathbb{N}$ such that $k \geqslant m \Rightarrow \|\alpha_k - \beta\| < r$.

If a sequence $\{\alpha_n\}$ has a limit $\beta \in V$, then β is clearly unique. Thus, we can refer to β as *the* limit of the sequence $\{\alpha_n\}$. In this case, we shall say $\{\alpha_n\}$ *converges to* β and write $\{\alpha_n\} \to \beta$. We say a sequence is convergent if the sequence has a limit in V. Note that 3.1(b) can be rewritten in the following form:

3.2: $\{\alpha_n\} \to \beta$ if and only if for every $r > 0$ there exists an $m \in \mathbb{N}$ such that $k \geqslant m \Rightarrow \alpha_k \in B_r(\beta)$.

Having introduced the notion of sequences, we now explore the relationships between these functions and some of the other ideas we have been discussing. Our first lemma says that addition and scalar multiplication preserves limits.

Lemma 3.3: Suppose $\{\alpha_n\} \to \alpha$, and $\{\beta_n\} \to \beta$. Then for any $x, y \in \mathbb{R}$, $\{x\alpha_n + y\beta_n\} \to x\alpha + y\beta$.

Proof: We first show that $\{\alpha_n + \beta_n\} \to \alpha + \beta$. Let $r > 0$. Then there exists natural numbers m_1 and m_2 such that $k \geqslant m_1 \Rightarrow \|\alpha_k - \alpha\| < r/2$, and $k \geqslant m_2 \Rightarrow \|\beta_k - \beta\| < r/2$. Let $m = \max\{m_1, m_2\}$. If $k \geqslant m$, then $\|(\alpha_k + \beta_k) - (\alpha + \beta)\| \leqslant \|\alpha_k - \alpha\| + \|\beta_k - \beta\| < r/2 + r/2 = r$. Thus, $\{\alpha_n + \beta_n\} \to \alpha + \beta$.

To prove the lemma, it now suffices to show that $\{x\alpha_n\} \to x\alpha$. This is clear if $x = 0$. Thus, we assume $x \neq 0$. Let $r > 0$. Since $\{\alpha_n\} \to \alpha$, there exists an $m \in \mathbb{N}$ such that $k \geqslant m \Rightarrow \|\alpha_k - \alpha\| < r/|x|$. Then for $k \geqslant m$, we have $\|x\alpha_k - x\alpha\| = |x| \|\alpha_k - \alpha\| < r$. Thus, $\{x\alpha_n\} \to x\alpha$, and the proof of the lemma is complete. $\square$

Our next lemma gives us a sequential characterization of the closure of a set in V.

Lemma 3.4: Let A be a subset of V, and denote the closure of A by $\bar{A}$. Then $\alpha \in \bar{A}$ if and only if there exists a sequence $\{\alpha_n\} \subseteq A$ such that $\{\alpha_n\} \to \alpha$.

Proof: Recall from Exercise 7 of Section 1 that $\alpha \in \bar{A}$ if and only if $B_r(\alpha) \cap A \neq \varnothing$ for all $r > 0$. So, let us first suppose $\alpha \in \bar{A}$. Then for every $n \in \mathbb{N}$, there exists an $\alpha_n \in B_{1/n}(\alpha) \cap A$. Clearly, $\{\alpha_n\} \to \alpha$.

Conversely, suppose a sequence $\{\alpha_n\}$ in A converges to α. Let $r > 0$. Then there exists a natural number m such that $k \geqslant m \Rightarrow \alpha_k \in B_r(\alpha)$. In particular, $B_r(\alpha) \cap A \neq \varnothing$. Thus, $\alpha \in \bar{A}$. $\square$

Lemma 3.4 says that the smallest closed set in V containing A is precisely the set of all limits from sequences in A. For instance, the closure of $B_1(0)$ in the three norms given in Example 1.13 are the following sets:

3.5:

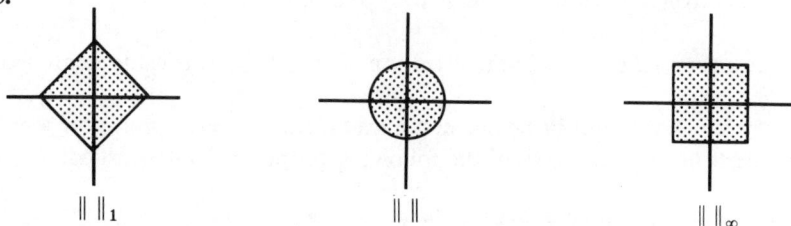

$\| \ \|_1$ $\| \ \|$ $\| \ \|_\infty$

The boundary, $(B_1(0))^\partial = \overline{B_1(0)} - B_1(0)$, of $B_1(0)$ in the three norms are the three curves pictured below:

3.6:

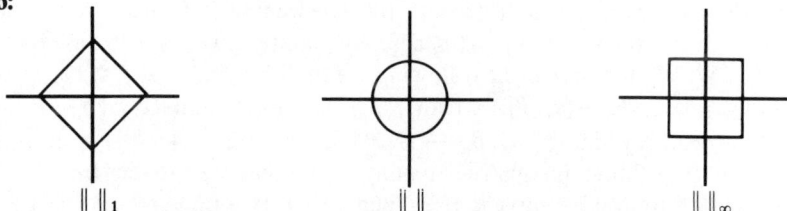

$\| \ \|_1$ $\| \ \|$ $\| \ \|_\infty$

We can characterize continuous functions on normed linear vector spaces using sequences.

Lemma 3.7: Let V and W be normed linear vector spaces, and let A be a subset of V. Suppose $f: A \to W$ is a function. Let $\alpha \in A$. Then f is continuous at α if and only if for every sequence $\{\alpha_n\}$ in A converging to α, $\{f(\alpha_n)\} \to f(\alpha)$.

Proof: Suppose f is continuous at α, and let $\{\alpha_n\}$ be a sequence in A that converges to α. Let $r > 0$. Since f is continuous at α, there exists an $s > 0$ such that $\xi \in A$ and $\|\xi - \alpha\| < s \Rightarrow \|f(\xi) - f(\alpha)\| < r$. Since $\{\alpha_n\} \to \alpha$, there exists an $m \in \mathbb{N}$ such that $k \geqslant m \Rightarrow \|\alpha_k - \alpha\| < s$. In particular, $k \geqslant m \Rightarrow \|f(\alpha_k) - f(\alpha)\| < r$. Thus, $\{f(\alpha_n)\} \to f(\alpha)$.

For the converse, suppose f fails to be continuous at α. Then there exists an $r > 0$ such that for every $s > 0$ there is a $\beta \in A$ such that $\|\beta - \alpha\| < s$, but $\|f(\beta) - f(\alpha)\| \geqslant r$. In particular, for every $n \in \mathbb{N}$, there exists an $\alpha_n \in A$ such that $\|\alpha_n - \alpha\| < 1/n$, but $\|f(\alpha_n) - f(\alpha)\| \geqslant r$. We now have a contradiction. $\{\alpha_n\} \to \alpha$ in A, but $\{f(\alpha_n)\}$ does not converge to $f(\alpha)$. We conclude that f must be continuous at α. $\square$

The reader will recall that two norms $\| \ \|$ and $\| \ \|'$ are equivalent on V if the identity map $(V, \| \ \|) \to (V, \| \ \|')$ is a norm isomorphism, that is, continuous in both directions. Using the sequential characterization of continuity given in Lemma 3.7, we get the following corollary:

Corollary 3.8: Two norms on V are equivalent if and only if the set of convergent sequences in V relative to one of the norms is precisely the same as the set of convergent sequences relative to the other norm. $\square$

We also get a sequential characterization of product norms using Lemma 3.7.

Corollary 3.9: Let V and W be normed linear vector spaces. A norm on $V \times W$ is a product norm if and only if the following property ($\mathscr{P}$) is satisfied:

($\mathscr{P}$): $\{(\alpha_n, \beta_n)\} \to (\alpha, \beta)$ in $V \times W \Leftrightarrow \{\alpha_n\} \to \alpha$ in V, and $\{\beta_n\} \to \beta$ in W

Proof: Suppose $\| \ \|_1$ and $\| \ \|_2$ are the norms on V and W, respectively. Let $\| \ \|$ denote a norm on $V \times W$. We first note that the sum norm $\|(\alpha, \beta)\|_s = \|\alpha\|_1 + \|\beta\|_2$ clearly satisfies property ($\mathscr{P}$). If $\| \ \|$ is a product norm on $V \times W$, then $\| \ \|$ is equivalent to $\| \ \|_s$ by Theorem 2.7. Hence, there exist constants $a, b > 0$ such that $\|(\xi, \eta)\| \leqslant a\|(\xi, \eta)\|_s$, and $\|(\xi, \eta)\|_s \leqslant b\|(\xi, \eta)\|$ for all $(\xi, \eta) \in V \times W$. We then have the following inequalities: $\|\alpha_n - \alpha\|_1 \leqslant \|\alpha_n - \alpha\|_1 + \|\beta_n - \beta\|_2 = \|(\alpha_n, \beta_n) - (\alpha, \beta)\|_s \leqslant b\|(\alpha_n, \beta_n) - (\alpha, \beta)\|$. Similarly, $\|\beta_n - \beta\|_2 \leqslant b\|(\alpha_n, \beta_n) - (\alpha, \beta)\|$. Finally, $\|(\alpha_n, \beta_n) - (\alpha, \beta)\| \leqslant a\|(\alpha_n, \beta_n) - (\alpha, \beta)\|_s = a(\|\alpha_n - \alpha\|_1 + \|\beta_n - \beta\|_2)$. These inequalities readily imply that $\| \ \|$ satisfies ($\mathscr{P}$).

Conversely, suppose $\| \ \|$ satisfies ($\mathscr{P}$). Then $\{(\alpha_n, \beta_n)\} \to (\alpha, \beta)$ relative to $\| \ \|$ in $V \times W$ if and only if $\{\alpha_n\} \to \alpha$ relative to $\| \ \|_1$ in V, and $\{\beta_n\} \to \beta$ relative to $\| \ \|_2$ in W. The same statement is true for the sum norm $\| \ \|_s$. We thus conclude that the sets of convergent sequences relative to $\| \ \|$ and $\| \ \|_s$ in $V \times W$ are precisely the same. Hence, $\| \ \|$ is equivalent to $\| \ \|_s$ by Corollary 3.8. In particular, $\| \ \|$ is a product norm on $V \times W$. $\square$

We can now introduce the central ideas of this section. In order to discuss sequential compactness, we need the notion of a subsequence of $\{\alpha_n\}$.

Definition 3.10: A sequence $\{\beta_n\}$ is a subsequence of $\{\alpha_n\}$ if there exists a strictly increasing function $f: \mathbb{N} \to \mathbb{N}$ such that $\beta_n = \alpha_{f(n)}$ for all $n \in \mathbb{N}$.

If we set $f(k) = n_k$ in Definition 3.10, then $n_1 < n_2 < n_3 < \ldots$, and $\beta_k = \alpha_{n_k}$ for all $k \in \mathbb{N}$. For this reason, we often use the notation $\{\alpha_{n_k}\}$ to indicate a subsequence of $\{\alpha_n\}$. We shall also use the notation $\{\beta_n\} \hookrightarrow \{\alpha_n\}$ to indicate that $\{\beta_n\}$ is a subsequence of $\{\alpha_n\}$.

There are a few elementary remarks concerning subsequences that we shall use implicitly throughout the rest of this section. We gather these remarks together in the following lemma:

Lemma 3.11: (a) If $\{\alpha_n\} \to \alpha$, and $\{\beta_n\} \hookrightarrow \{\alpha_n\}$, then $\{\beta_n\} \to \alpha$.

   (b) If a sequence $\{\alpha_n\}$ does not converge to some vector α, then there exists a subsequence $\{\beta_n\} \hookrightarrow \{\alpha_n\}$ such that no subsequence of $\{\beta_n\}$ converges to α.

   (c) Suppose $\{\alpha_n\}$ is a sequence in V, and $\alpha \in V$. If every subsequence of $\{\alpha_n\}$ has a subsequence that converges to α, then $\{\alpha_n\} \to \alpha$.

Proof: (a) This proof is trivial. We leave it to the reader.

   (b) Suppose $\{\alpha_n\}$ does not converge to α. Then there exists an $r > 0$ such that for any $m \in \mathbb{N}$ there exists an $n \geq m$ with $\|\alpha_n - \alpha\| \geq r$. By letting m get large, we can construct a strictly increasing sequence $n_1 < n_2 < \cdots$ such that $\|\alpha_{n_k} - \alpha\| \geq r$. Clearly, the subsequence $\{\alpha_{n_k}\}$ has no subsequence converging to α since every term of $\{\alpha_{n_k}\}$ is a distance at least r from α.

   (c) This statement follows immediately from (b). If a sequence $\{\alpha_n\}$ does not converge to α, then by (b), $\{\alpha_n\}$ has a subsequence $\{\beta_n\}$ such that no subsequence of $\{\beta_n\}$ converges to α. But, this is contrary to our assumptions in (c). Hence, $\{\alpha_n\} \to \alpha$. $\square$

Definition 3.12: Let $(V, \| \ \|)$ be a normed linear vector space. Suppose A is a subset of V. We say A is *sequentially compact* if every sequence in A has a subsequence that converges to a vector in A.

Thus, if A is sequentially compact, and $\{\alpha_n\}$ is a sequence in A, then there exists an $\alpha \in A$ and a subsequence $\{\alpha_{n_k}\}$ of $\{\alpha_n\}$ such that $\{\alpha_{n_k}\} \to \alpha$. The notion of sequential compactness of course depends on the specific norm in V. However,

Corollary 3.8 implies that if two norms are equivalent, then a set A is sequentially compact with respect to one norm if and only if A is sequentially compact with respect to the other norm. Let us consider some simple examples in $\mathbb{R}$ before continuing.

Example 3.13: Let $V = \mathbb{R}$, with $\| \| = | |$. Any finite set A in $\mathbb{R}$ must be sequentially compact since any sequence in A necessarily contains a constant subsequence.

An infinite set in $\mathbb{R}$ need not be sequentially compact. If $A = \mathbb{Z}$, for example, then the sequence $\{n\}$ contains no convergent subsequence. A more interesting example is the set $C = \{1/n \mid n \in \mathbb{N}\}$. Now any sequence in C has a convergent subsequence. The sequence $\{1/n\}$ is contained in C and has the property that any subsequence of $\{1/n\}$ converges to 0. Since 0 is not in C, we conclude that C is not sequentially compact. □

Our first property of a sequentially compact subset of V is perhaps the most important one of all.

Theorem 3.14: Let A be a subset of a normed linear vector space $(V, \| \|)$. If A is sequentially compact, then A is closed and bounded.

Proof: We have seen in Exercise 7 of Section 1, that the closure $\bar{A}$ of A is the smallest closed set in V containing A. Thus, to show that A is closed, it suffices to show that $\bar{A} \subseteq A$. Let $\alpha \in \bar{A}$. Then Lemma 3.4 implies that there exists a sequence $\{\alpha_n\}$ in A such that $\{\alpha_n\} \to \alpha$. Since A is sequentially compact, there exists a subsequence $\{\alpha_{n_k}\}$ of $\{\alpha_n\}$ and an element $\beta \in A$ such that $\{\alpha_{n_k}\} \to \beta$. But Lemma 3.11(a) implies $\{\alpha_{n_k}\} \to \alpha$. Thus, $\alpha = \beta$. In particular, $\alpha \in A$. Hence, $\bar{A} \subseteq A$, and A is closed.

Recall that A is bounded if $A \subseteq B_r(0)$ for some $r > 0$. Suppose A is not bounded. Then for every $n \in \mathbb{N}$, A is not contained in $B_n(0)$. Thus, for every $n \in \mathbb{N}$, there exists an $\alpha_n \in A$ such that $\|\alpha_n\| \geq n$. Since A is sequentially compact, there exists a subsequence of $\{\alpha_n\}$, say $\{\alpha_{n_k}\}$, and a vector $\beta \in A$ such that $\{\alpha_{n_k}\} \to \beta$. Now we have noted in Lemma 1.16 that the norm is a continuous function from V to $\mathbb{R}$. Lemma 3.7 then implies $\{\|\alpha_{n_k}\|\} \to \|\beta\|$ in $\mathbb{R}$. But from our construction of the sequence $\{\alpha_n\}$, $\{\|\alpha_{n_k}\|\} \to +\infty$. Thus, $\|\beta\| = +\infty$ which of course is impossible. We conclude that A is a bounded subset of V. □

At this point, it may be a good idea to say a few words about a possible converse of Theorem 3.14. We shall show later on in this section that if $\dim_\mathbb{R}(V) < \infty$, then the converse of 3.14 is true. For the time being, we merely note that neither hypothesis, closed nor bounded, alone implies sequential compactness. $\mathbb{Z}$ is a closed subset (but not bounded) of $\mathbb{R}$, and $\mathbb{Z}$ is not sequentially compact. $C = \{1/n \mid n \in \mathbb{N}\}$ is a bounded subset (but not closed) of $\mathbb{R}$, and C is not sequentially compact.

Another important property that a sequentially compact subset has is that a continuous, real valued function on such a set obtains both a maximum and minimum value somewhere on the set. This remark easily follows from our next theorem.

Theorem 3.15: Let V and W be normed linear vector spaces, and let A be a subset of V. Suppose f: A → W is a continuous function. If A is a sequentially compact subset of V, then f(A) is a sequentially compact subset of W.

Proof: Let $\{\gamma_n\}$ be a sequence in f(A). Then for every n, there exists an $\alpha_n \in A$ with $f(\alpha_n) = \gamma_n$. Since A is sequentially compact, $\{\alpha_n\}$ has a convergent subsequence $\{\beta_n\} \to \beta \in A$. Since f is continuous, $\{f(\beta_n)\} \to f(\beta)$ by Lemma 3.7. Since $\beta \in A$, $f(\beta) \in f(A)$. Thus, the sequence $\{\gamma_n\}$ has a convergent subsequence $\{f(\beta_n)\} \to f(\beta)$ in f(A). We conclude that f(A) is sequentially compact.   □

Corollary 3.16: Let A be a sequentially compact subset of V and suppose f: A → ℝ is a continuous function. Then f is bounded on A. Furthermore, f assumes both a maximum and minimum value on A.

Proof: Before proving the corollary, let us discuss its meaning. We say f is bounded on A if there exists a positive number b such that $|f(\alpha)| < b$ for all $\alpha \in A$. To say that f assumes a maximum value on A means there exists an $\alpha \in A$ such that $f(\xi) \leqslant f(\alpha)$ for all $\xi \in A$. Similarly, f assumes a minimum value on A if there exists a $\beta \in A$ such that $f(\xi) \geqslant f(\beta)$ for all $\xi \in A$.

Now by Theorem 3.15, f(A) is a sequentially compact subset of ℝ. Thus, by Theorem 3.14, f(A) is a bounded subset of ℝ. This of course means f is bounded on A.

Let $x = \sup f(A) = \sup\{f(\xi) \,|\, \xi \in A\}$. Let $y = \inf f(A)$. Since f(A) is a bounded subset of ℝ, both x and y exist. Theorem 3.14 also implies that f(A) is a closed subset of ℝ. The reader can easily check that any closed (and bounded) subset of ℝ contains both its infimum and supremum. In particular, $x, y \in f(A)$. If $\alpha \in A$ such that $f(\alpha) = x$, then clearly f assumes a maximum value x on A at α. Similarly, if $\beta \in A$ such that $f(\beta) = y$, then f assumes a minimum value y on A at β.   □

We now turn our attention to ℝ. Since any norm on ℝ is equivalent to the absolute value $|\ |$, we can, with no loss in generality, state all our results relative to $|\ |$. We first remind the reader of some familiar definitions from the calculus.

Definition 3.17: Let $\{x_n\}$ be a sequence in ℝ.

(a) We say $\{x_n\}$ is increasing if $x_n \leqslant x_{n+1}$ for all n.
(b) $\{x_n\}$ is decreasing if $x_n \geqslant x_{n+1}$ for all n.
(c) $\{x_n\}$ is monotone if $\{x_n\}$ is either increasing or decreasing.
(d) $\{x_n\}$ is bounded if there exists a b > 0 such that $|x_n| \leqslant b$ for all n.

The important facts about monotone sequences are contained in the following two lemmas:

Lemma 3.18: Any bounded, monotone sequence in $\mathbb{R}$ converges.

Proof: Let $\{x_n\}$ be an increasing, bounded sequence. Since $\{x_n\}$ is bounded, $x = \sup\{x_n \mid n \in \mathbb{N}\}$ exists. We claim $\{x_n\} \to x$. To see this, let $r > 0$. Then $x - r$ is not an upper bound of the set $\{x_n \mid n \in \mathbb{N}\}$. Hence, there exists an m such that $x_m > x - r$. Since $\{x_n\}$ is increasing, $x - r < x_m \leqslant x_n \leqslant x$ for all $n \geqslant m$. In particular, $n \geqslant m \Rightarrow |x_n - x| < r$. Thus, $\{x_n\} \to x$.

If $\{x_n\}$ is decreasing and $y = \inf\{x_n \mid n \in \mathbb{N}\}$, then a similar proof shows $\{x_n\} \to y$. $\square$

Lemma 3.19: Every sequence in $\mathbb{R}$ has a monotone subsequence.

Proof: Let $\{x_n\}$ be an arbitrary sequence in $\mathbb{R}$. Let us call a term x_k of $\{x_n\}$ a *peak term* if $x_k \geqslant x_{k+j}$ for all $j \geqslant 1$. Suppose $\{x_n\}$ has infinitely many peak terms, say $x_{n_1}, x_{n_2}, \ldots$. Here we label those terms with $n_1 < n_2 < \cdots$. Clearly, $\{x_{n_k}\}$ is a decreasing subsequence of $\{x_n\}$.

Suppose $\{x_n\}$ has only finitely many peak terms (possibly none). Then there is a last peak term, say x_{n_0}. Every term x_k of $\{x_n\}$ after x_{n_0} is not a peak term and, so, is strictly less than some x_{k+j}. In particular, we can find a strictly increasing sequence $n_0 < n_1 < n_2 < \cdots$ such that $x_{n_1} < x_{n_2} < \cdots$. Thus, $\{x_n\}$ has a strictly increasing subsequence. $\square$

We can combine Lemmas 3.18 and 3.19 into the following important theorem:

Theorem 3.20: Every bounded sequence in $\mathbb{R}$ has a convergent subsequence.

Proof: Let $\{x_n\}$ be a bounded sequence in $\mathbb{R}$. By Lemma 3.19, $\{x_n\}$ has a monotone subsequence $\{x_{n_k}\}$. Clearly, $\{x_{n_k}\}$ is bounded since $\{x_n\}$ is. Hence, Lemma 3.18 implies $\{x_{n_k}\}$ converges. $\square$

We can generalize Theorem 3.20 to $\mathbb{R}^n$ and any product norm as follows:

Theorem 3.21: Let $\| \ \|$ be any product norm on $\mathbb{R}^n$. Then any bounded sequence in $(\mathbb{R}^n, \| \ \|)$ has a convergent subsequence.

Proof: From Theorem 2.8 and Corollary 3.8, we may assume that $\| \ \|$ is the sum norm $\|(x_1, \ldots, x_n)\|_1 = \sum_{i=1}^{n} |x_i|$. We proceed by induction on n, the case $n = 1$ having been proved in 3.20.

Suppose $\{\alpha_k\}$ is a bounded sequence in $\mathbb{R}^n$. Then there exists a constant $c > 0$ such that $\|\alpha_k\|_1 \leqslant c$ for all k. Let us write $\alpha_k = (x_{1k}, \ldots, x_{nk})$ for all $k \in \mathbb{N}$, and set $\beta_k = (x_{1k}, \ldots, x_{n-1k}) \in \mathbb{R}^{n-1}$. We can think of $\mathbb{R}^n$ as the product $\mathbb{R}^n = \mathbb{R}^{n-1} \times \mathbb{R}$.

Then $\alpha_k = (\beta_k, x_{nk})$. Since $\|\alpha_k\|_1 = \|\beta_k\|_1 + |x_{nk}| \geqslant \|\beta_k\|_1, |x_{nk}|$, both $\{\beta_k\}$ and $\{x_{nk}\}$ are bounded sequences in $\mathbb{R}^{n-1}$ and $\mathbb{R}$, respectively. By our induction hypothesis, $\{\beta_k\}$ contains a convergent subsequence $\{\beta_{k_i}\}$. Suppose $\{\beta_{k_i}\} \to \beta \in \mathbb{R}^{n-1}$.

Now consider the corresponding subsequence $\{x_{nk_i}\}$ of $\{x_{nk}\}$. Since $\{x_{nk}\}$ is bounded, $\{x_{nk_i}\}$ is bounded. Thus, Theorem 3.20 implies $\{x_{nk_i}\}$ has a convergent subsequence $\{y_j\}$. Recall that this means there exists a strictly increasing function $f: \mathbb{N} \to \{k_1, k_2, k_3, \ldots \}$ such that $y_j = x_{nf(j)}$ for all $j \in \mathbb{N}$. Suppose $\{y_j\} \to y$. For each $j \in \mathbb{N}$, set $\gamma_j = \beta_{f(j)}$. Then $\{\gamma_j\}$ is a subsequence of $\{\beta_{k_i}\}$. At this point, a diagram of the sequences we have constructed may be helpful.

3.22:
$$\{y_j\} \hookrightarrow \{x_{nk_i}\} \hookrightarrow \{x_{nk}\}$$
$$\{\gamma_j\} \hookrightarrow \{\beta_{k_i}\} \hookrightarrow \{\beta_k\}$$

Since $\{\beta_{k_i}\} \to \beta$, $\{\gamma_j\} \to \beta$. We now claim $\{(\gamma_j, y_j)\}$ converges to (β, y). We first observe that $\|(\gamma_j, y_j) - (\beta, y)\|_1 = \|\gamma_j - \beta\|_1 + |y_j - y|$. Since $\{\gamma_j\} \to \beta$, and $\{y_j\} \to y$, this last equation implies $\{(\gamma_j, y_j)\} \to (\beta, y)$. The sequence $\{(\gamma_j, y_j)\}$ is clearly a subsequence of $\{\alpha_k\}$, and, consequently, $\{\alpha_k\}$ contains a convergent subsequence. This completes the proof of 3.21. $\square$

One important corollary to Theorem 3.21 is the converse of Theorem 3.14 for product norms on $\mathbb{R}^n$.

Theorem 3.23: Let $\| \ \|$ be any product norm on $\mathbb{R}^n$. A set A in $(\mathbb{R}^n, \| \ \|)$ is sequentially compact if and only if A is closed and bounded.

Proof: If A is sequentially compact, then we have seen in Theorem 3.14 that A is closed and bounded.

Suppose, conversely, that A is closed and bounded. Let $\{\alpha_k\}$ be a sequence in A. Since A is bounded, clearly $\{\alpha_k\}$ is a bounded sequence in $\mathbb{R}^n$. Hence, Theorem 3.21 implies $\{\alpha_k\}$ has a convergent subsequence $\{\beta_k\}$. Suppose $\{\beta_k\} \to \beta$. Since each vector β_k lies in A, and $\{\beta_k\} \to \beta$, we conclude $B_r(\beta) \cap A \neq \varnothing$ for any $r > 0$. Thus, $\beta \in \bar{A}$. But, $\bar{A} = A$ since A is closed. Therefore, $\beta \in A$. We have now shown that every sequence in A has a subsequence that converges to a vector in A. Thus, A is sequentially compact. $\square$

At this point, we have developed enough material to be able to state and prove the principal result of this section. Namely, that any two norms on $\mathbb{R}^n$ are equivalent. In proving this assertion, it clearly suffices to show that an arbitrary norm $\| \ \|$ is equivalent to the sum norm $\| \ \|_1$.

Theorem 3.24: Let $\| \ \|$ be an arbitrary norm on $\mathbb{R}^n$. Then $\| \ \|$ is equivalent to the sum norm $\| \ \|_1$.

Proof: Let $\underline{\delta} = \{\delta_1, \ldots, \delta_n\}$ be the canonical basis of $\mathbb{R}^n$. Set $a = \max\{\|\delta_1\|, \ldots, \|\delta_n\|\}$. Then $a > 0$, and for every $\alpha = (x_1, \ldots, x_n) \in \mathbb{R}^n$, we have $\|\alpha\| = \|\sum_{i=1}^n x_i \delta_i\| \leqslant \sum_{i=1}^n |x_i| \|\delta_i\| \leqslant a \sum_{i=1}^n |x_i| = a \|\alpha\|_1$. Thus, we have established one of the two inequalities we need in order to show $\| \ \|$ is equivalent to $\| \ \|_1$.

For any $\alpha, \beta \in \mathbb{R}^n$, we have $\big| \|\alpha\| - \|\beta\| \big| \leqslant \|\alpha - \beta\| \leqslant a \|\alpha - \beta\|_1$. If we think of $\| \ \|$ as a real valued function from $(\mathbb{R}^n, \| \ \|_1)$ to $(\mathbb{R}, | \ |)$, then this last inequality implies $\| \ \|$ is a continuous function on the normed linear vector space $(\mathbb{R}^n, \| \ \|_1)$.

Set $S = \{\alpha \in \mathbb{R}^n \mid \|\alpha\|_1 = 1\}$. The reader can easily check that S is a closed and bounded subset of $(\mathbb{R}^n, \| \ \|_1)$. In particular, Theorem 3.23 implies that S is a sequentially compact subset of $(\mathbb{R}^n, \| \ \|_1)$.

Since $\| \ \|$ is a continuous function on $\mathbb{R}^n$, certainly $\| \ \|$ is a continuous function on S. We can now apply Corollary 3.16 to the continuous map $\| \ \|: S \to \mathbb{R}$. We conclude that $\| \ \|$ assumes a minimum value m on S. Thus, there exists a $\gamma \in S$ such that $\|\gamma\| = m$, and $\|\alpha\| \geqslant m$ for all $\alpha \in S$. Note that $m > 0$. For if $m \leqslant 0$, then $\|\gamma\| \leqslant 0$. Since $\| \ \|$ is a norm on $\mathbb{R}^n$, we would then conclude that $\gamma = 0$. This is impossible since 0 is not in S.

We have now constructed a positive constant m, such that $\|\alpha\| \geqslant m$ for all $\alpha \in S$. We can rewrite this last inequality as $\|\alpha\| \geqslant m\|\alpha\|_1$ for all $\alpha \in S$. Let $\beta \in \mathbb{R}^n - \{0\}$. Then $\beta/\|\beta\|_1 \in S$. Consequently, $\|\beta/\|\beta\|_1\| \geqslant m\|\beta/\|\beta\|_1\|_1$ Thus, $\|\beta\| \geqslant m\|\beta\|_1$. This last inequality also holds when $\beta = 0$. Thus, setting $b = 1/m$, we have shown $\|\alpha\|_1 \leqslant b\|\alpha\|$ for all $\alpha \in \mathbb{R}^n$. Since we had previously argued that $\|\alpha\| \leqslant a\|\alpha\|_1$, we conclude that $\| \ \|$ is equivalent to $\| \ \|_1$. $\square$

In the rest of this section, we shall develop the important corollaries that come from Theorem 3.24. We have already mentioned our first corollary.

Corollary 3.25: Any two norms on $\mathbb{R}^n$ are equivalent. $\square$

Notice then that any norm on $\mathbb{R}^n$ being equivalent to $\| \ \|_1$ is automatically a product norm. Hence, we can drop the adjective "product" when dealing with norms on $\mathbb{R}^n$. Any norm on $\mathbb{R}^n$ is a product norm.

Corollary 3.26: Let V be a finite-dimensional vector space over $\mathbb{R}$. Then any two norms on V are equivalent.

Proof: Suppose $\dim(V) = n$. Then any coordinate map gives us an isomorphism $T: \mathbb{R}^n \cong V$. Suppose $\| \ \|$ and $\| \ \|'$ are two norms on V. Then $f(\alpha) = \|T(\alpha)\|$ and $g(\alpha) = \|T(\alpha)\|'$ define two norms on $\mathbb{R}^n$. By Corollary 3.25, f and g are equivalent. Since T is surjective, the equivalence of f and g immediately implies $\| \ \|$ and $\| \ \|'$ are equivalent on V. $\square$

There is an important application of Corollary 3.26, which we list as another corollary.

Corollary 3.27: Let $(V, \| \ \|)$ be a finite-dimensional, normed linear vector space. If $\dim_{\mathbb{R}}(V) = n$, then $(V, \| \ \|)$ is norm isomorphic to $(\mathbb{R}^n, \| \ \|_1)$.

Proof: Suppose $\underline{\alpha} = \{\alpha_1, \ldots, \alpha_n\}$ is a basis of V over $\mathbb{R}$. Then we have an isomorphism $S: V \cong \mathbb{R}^n$ given by $S(\xi) = (x_1, \ldots, x_n)$ where $\sum_{i=1}^n x_i\alpha_i = \xi$. We can define a new norm $\| \ \|'$ on V by the equation $\|\xi\|' = \|S(\xi)\|_1$. It is a simple matter to check that S is now a norm isomorphism between $(V, \| \ \|')$ and $(\mathbb{R}^n, \| \ \|_1)$. By Corollary 3.26, $\| \ \|$ and $\| \ \|'$ are equivalent. This means that the identity map from $(V, \| \ \|)$ to $(V, \| \ \|')$ is a norm isomorphism. Composing these two norm isomorphisms, we get $(V, \| \ \|)$ and $(\mathbb{R}^n, \| \ \|_1)$ are norm isomorphic. $\square$

Thus, for finite-dimensional, normed linear vector spaces, the theory is particularly easy. We can always assume that our space is $(\mathbb{R}^n, \| \ \|_1)$ up to norm isomorphism.

Returning to a remark made earlier in this section, we can now prove the following generalization of Theorem 3.23:

Corollary 3.28: Let $(V, \| \ \|)$ be a finite-dimensional, normed linear vector space. Then a subset A of V is sequentially compact if and only if A is closed and bounded.

Proof: $(V, \| \ \|)$ is norm isomorphic to $(\mathbb{R}^n, \| \ \|_1)$ for some n. So the result follows from Theorem 3.23. $\square$

We should point out here that Corollary 3.28 is really a theorem about $\mathbb{R}^n$. It is not true in general. If $(V, \| \ \|)$ is an infinite-dimensional, normed linear vector space, and we set $B = \{\alpha \in V \mid \|\alpha\| \leqslant 1\}$, then B is closed and bounded in V. However, B is never sequentially compact. We ask the reader to provide a proof of this assertion in Exercise 10 at the end of this section.

We can also use Corollary 3.26 to show that all linear transformations on finite-dimensional spaces are bounded.

Corollary 3.29: Let V and W denote finite-dimensional, normed linear vector spaces. Then $\mathscr{B}(V, W) = \mathrm{Hom}_{\mathbb{R}}(V, W)$.

Proof: Let $T \in \mathrm{Hom}_{\mathbb{R}}(V, W)$. Since any two norms on V (as well as W) are equivalent, it suffices to argue that T is bounded with respect to a specific choice of norms on V and W.

Suppose $\dim(V) = n$ and $\dim(W) = m$. Let $\underline{\alpha} = \{\alpha_1, \ldots, \alpha_n\}$ be a basis of V and $\underline{\beta} = \{\beta_1, \ldots, \beta_m\}$ a basis of W. Then we have the following commutative diagram:

3.30:

The vertical maps in 3.30 are the usual coordinate isomorphisms: $f(\alpha) = (x_1, \ldots, x_n)$, where $\sum_{i=1}^{n} x_i \alpha_i = \alpha$, and $g(\beta) = (y_1, \ldots, y_m)$, where $\sum_{i=1}^{m} y_i \beta_i = \beta$. If $\Gamma(\underline{\alpha}, \underline{\beta})(T) = (t_{ij})$, then S in 3.30 is the linear transformation given by

$$S((x_1, \ldots, x_n)) = \left(\sum_{j=1}^{n} t_{1j} x_j, \ldots, \sum_{j=1}^{n} t_{mj} x_j \right)$$

Now let us norm $\mathbb{R}^n$ with the usual sum norm $\| \ \|_1$. Then $\|\alpha\| = \|f(\alpha)\|_1$ defines a norm on V. Let us norm $\mathbb{R}^m$ with the uniform norm $\| \ \|_\infty$ (notation as in Example 1.3). Then $\|\beta\|' = \|g(\beta)\|_\infty$ is a norm on W. We now have the following commutative diagram of normed linear vector spaces:

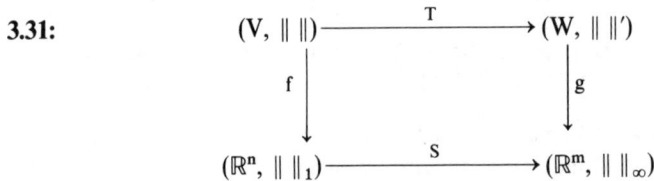

3.31:

$$
\begin{array}{ccc}
(V, \| \ \|) & \xrightarrow{\ \ T\ \ } & (W, \| \ \|') \\
\Big\downarrow {\scriptstyle f} & & \Big\downarrow {\scriptstyle g} \\
(\mathbb{R}^n, \| \ \|_1) & \xrightarrow{\ \ S\ \ } & (\mathbb{R}^m, \| \ \|_\infty)
\end{array}
$$

It suffices to argue that T is bounded with respect to the norms $\| \ \|$ and $\| \ \|'$.

Set $b = \max\{|t_{ij}| \ | \ i = 1, \ldots, m; \ j = 1, \ldots, n\}$. For any vector $\gamma = (x_1, \ldots, x_n) \in \mathbb{R}^n$, we have the following inequalities:

3.32:
$$
\begin{aligned}
\|S(\gamma)\|_\infty &= \left\| \left(\sum_{j=1}^{n} t_{1j} x_j, \ldots, \sum_{j=1}^{n} t_{mj} x_j \right) \right\|_\infty \\
&= \max\left\{ \left| \sum_{j=1}^{n} t_{ij} x_j \right| \ \middle| \ i = 1, \ldots, m \right\} \\
&\leqslant b\|\gamma\|_1
\end{aligned}
$$

Thus, the map S is a bounded linear operator. Since Diagram 3.31 is commutative, we have

3.33: $\|T(\alpha)\|' = \|gT(\alpha)\|_\infty = \|Sf(\alpha)\|_\infty \leqslant b\|f(\alpha)\|_1 = b\|\alpha\|$

This is precisely the statement that T is a bounded linear operator from $(V, \| \ \|)$ to $(W, \| \ \|')$. $\square$

The conclusion we can draw from Corollaries 3.27 and 3.29 is that when dealing with finite-dimensional, normed linear vector spaces V and W (and the linear transformations between them), we can make the following assumptions up to equivalence:

(a) $V = \mathbb{R}^n$ and $W = \mathbb{R}^m$.
(b) The norms on V and W can be any we choose for computational convenience.

(c) The linear operator between V and W is given by multiplication on the right by an n × m matrix.

Our last topic in this section is another corollary that has important applications in least-squares problems.

Corollary 3.34: Let $(V, \| \ \|)$ be a normed linear vector space, and suppose W is a finite-dimensional subspace of V. Then for any vector $\beta \in V$, there exists an $\alpha \in W$ such that $d(\beta, W) = \|\alpha - \beta\|$.

Proof: Recall that $d(\beta, W) = \inf\{\|\xi - \beta\| \mid \xi \in W\}$ is called the distance (in the $\| \ \|$-norm) between β and W. Since $d = d(\beta, W)$ is the infimum of the set $\{\|\xi - \beta\| \mid \xi \in W\}$, there exists a sequence $\{\alpha_n\}$ in W such that $\{\|\alpha_n - \beta\|\} \to d$. We restrict the norm to W and claim that $\{\alpha_n\}$ is a bounded sequence in W. To see this, suppose $\{\alpha_n\}$ is not bounded in W. Then there exists a subsequence $\{\alpha_{n_k}\}$ of $\{\alpha_n\}$ such that $\{\|\alpha_{n_k}\|\} \to +\infty$. Now $\{\|\alpha_{n_k} - \beta\|\}$ is a subsequence of $\{\|\alpha_n - \beta\|\}$. Therefore, $\{\|\alpha_{n_k} - \beta\|\} \to d$. On the other hand, $\| \|\alpha_{n_k}\| - \|\beta\| \| \leq \|\alpha_{n_k} - \beta\|$. These last two facts imply that $\|\alpha_{n_k}\|$ cannot be approaching $+\infty$. This is a contradiction. Thus, we conclude that $\{\alpha_n\}$ is a bounded sequence in W.

Now W is finite dimensional. Hence, it follows from Corollary 3.27 that the normed linear space $(W, \| \ \|)$ is norm isomorphic to $(\mathbb{R}^n, \| \ \|_1)$ for some n. In particular, Theorem 3.21 implies $\{\alpha_n\}$ has a subsequence $\{\alpha_{n_k}\}$ that converges to some vector $\alpha \in W$. Clearly, $\{\alpha_{n_k} - \beta\} \to \alpha - \beta$ in V. Since the norm is a continuous function, we have $d = \lim\{\|\alpha_{n_k} - \beta\|\} = \|\alpha - \beta\|$. This completes the proof of 3.34. □

A few comments about Corollary 3.34 are in order here. Suppose W is an arbitrary subspace of some normed linear vector space $(V, \| \ \|)$. Let $\beta \in V$. One of the central problems of linear algebra is how to find a vector in W (if such a vector exists) that is closest to β. Thus, we seek a vector $\alpha \in W$ such that $\|\alpha - \beta\| = d(\beta, W)$. In the case where the norm is induced from an inner product on V (see Section 1 of Chapter V), the search for α usually amounts to minimizing a sum of squares. Hence, these types of problems are called least-squares problems.

Corollary 3.34 guarantees that the least-squares problem is always solvable if W is finite dimensional. In this case, we can always find an $\alpha \in W$ such that $\|\alpha - \beta\| = d(\beta, W)$. In particular, if V itself is finite dimensional, then a vector in W closest to β always exists.

If W is not finite dimensional, then a vector $\alpha \in W$ such that $\|\alpha - \beta\| = d(\beta, W)$ may not exist. We complete this section with an example that illustrates this point.

Example 3.35: Let $V = \{f \in C([0, 1]) \mid f(0) = 0\}$. Clearly, V is a subspace of $C([0, 1])$. We norm V with the uniform norm $\| \ \|_\infty$ given in equation 1.6(c). Let

$W = \{f \in V \mid \int_0^1 f(t)\, dt = 0\}$. Since the integral is a bounded linear operator on V, W is a closed subspace of V. W is not finite dimensional over $\mathbb{R}$.

Let β be any function in V such that $\int_0^1 \beta(t)\, dt = 1$. If $\xi \in W$, then $\|\xi - \beta\|_\infty \geqslant \int_0^1 |\xi - \beta| \geqslant |\int_0^1 \xi - \beta| = \int_0^1 \beta = 1$. Thus, $d(\beta, W) \geqslant 1$.

We next claim that we can find a vector $\gamma \in W$ such that $\|\beta - \gamma\|_\infty$ is as close to 1 as we please. To see this, let h be any function in V such that $\int_0^1 h(t)\, dt \neq 0$. Set $c = 1/(\int_0^1 h(t)\, dt)$. Then $\beta - ch = \gamma$ is a vector in W. $\|\beta - \gamma\|_\infty = \|ch\|_\infty = \sup\{|h(t)| \mid t \in [0, 1]\}/|\int_0^1 h(t)\, dt|$. We can certainly choose h such that this last quotient is as close to 1 as we want. For example,

3.36:

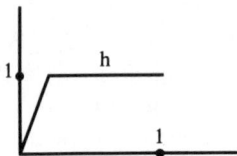

We conclude that $d(\beta, W) = 1$. To complete the example, we now argue that $\|\alpha - \beta\|_\infty > 1$ for any $\alpha \in W$. Consequently, $d(\beta, W) \neq \|\alpha - \beta\|_\infty$ for any α in W.

Suppose there exists an $\alpha \in W$ such that $\|\alpha - \beta\|_\infty = 1$. Set $h(t) = \beta - \alpha$. Then $h(t)$ is a continuous function on $[0, 1]$, $h(0) = 0$, $\int_0^1 h(t)\, dt = 1$, and $\sup\{|h(t)| \mid t \in [0, 1]\} = 1$. If you try to draw the graph of h, you will immediately see that no such function can exist. Hence, there is no $\alpha \in W$ such that $\|\alpha - \beta\|_\infty = 1$. $\square$

EXERCISES FOR SECTION 3

(1) If a sequence $\{\alpha_n\}$ in V converges to two vectors β and β', show that $\beta = \beta'$.

(2) Show that the sum norm $\|(\alpha, \beta)\|_s = \|\alpha\|_1 + \|\beta\|_2$ satisfies property ($\mathscr{P}$) in Corollary 3.9.

(3) Let A be a closed and bounded subset of $\mathbb{R}$. Show that inf(A) and sup(A) are elements in A.

(4) Let V and W be normed linear vector spaces, and suppose f: A → W is a continuous function on a sequentially compact subset A of V. If f is a bijection from A to f(A), show f^{-1}: f(A) → A is continuous.

(5) Construct a sequence $\{x_n\}$ in $[0, 1]$ such that every $y \in [0, 1]$ is the limit of some subsequence of $\{x_n\}$.

(6) If A and B are sequentially compact subsets of a normed linear vector space V, show A + B is also sequentially compact.

(7) Unlike sequential compactness, the sum of two closed sets in V need not be closed. Exhibit an example of this fact.

(8) Let V be a normed linear vector space. If W is a subspace of V, show that the closure $\bar{W}$ of W is also a subspace of V.

(9) Modify the argument in Corollary 3.34 to show that any finite-dimensional subspace of V is a closed set in V.

(10) Suppose V is an infinite-dimensional, normed linear vector space. Show that the subset $B = \{\alpha \in V \mid \|\alpha\| \leqslant 1\}$ is closed and bounded, but not sequentially compact. (*Hint*: Show that B cannot be covered by a finite number of open balls of radius $\frac{1}{3}$.)

(11) Suppose W is a proper closed subspace of a normed linear vector space V. Let $0 < r < 1$. Show there exists a vector $\beta \in V$ such that $\|\beta\| = 1$, and $d(\beta, W) > 1 - r$.

(12) In Exercise 11, suppose we assume $\dim(V) < \infty$. Show there exists a vector $\beta \in V$ such that $\|\beta\| = d(\beta, W) = 1$.

(13) Suppose V and W are normed linear vector spaces, and let $T \in \mathscr{B}(V, W)$. Show that $\ker(T)$ is a closed subspace of V. Is $\text{Im}(T)$ a closed subspace of W?

(14) Consider the function $f: \mathbb{R}^2 \to \mathbb{R}$ defined as follows: $f(x, y) = xy/(x^2 + y^2)$ if $(x, y) \neq (0, 0)$, and $f(0, 0) = 0$. Use Lemma 3.7 to prove the following assertions:

(a) For all $x \in \mathbb{R}$, $f(x, \ ): \mathbb{R} \to \mathbb{R}$ is continuous at 0.

(b) For all $y \in \mathbb{R}$, $f(\ , y): \mathbb{R} \to \mathbb{R}$ is continuous at 0.

(c) f is not continuous at $(0, 0)$.

(15) Suppose V and W are normed linear vector spaces, and let A and B be sequentially compact subsets of V and W, respectively. Prove that $A \times B$ is a sequentially compact subset of $V \times W$ relative to any product norm on $V \times W$.

(16) Let V be a normed linear vector space. A subset A of V is said to be dense in V if $\bar{A} = V$. Give an example of a proper subset of $\mathbb{R}^n$ that is dense in $\mathbb{R}^n$. Suppose W is a second normed linear vector space and f and g two continuous functions from V to W. Suppose $f = g$ on some dense subset of V. Prove $f = g$.

(17) Let V and W be normed linear vector spaces. Let $f: A \to W$ be a function from a subset A of V to W. We say f is *uniformly continuous* on A if for every $r > 0$, there exists an $s > 0$ such that for all α, $\beta \in A$, $\|\alpha - \beta\| < s \Rightarrow \|f(\alpha) - f(\beta)\| < r$.

(a) If f is uniformly continuous on A, prove that f is continuous on A. Show that the converse is false.

(b) If A is sequentially compact, and f continuous on A, prove that f is uniformly continuous.

4. BANACH SPACES

In this section, we take a brief look at Banach spaces. Our goal is to introduce the terminology most frequently used in the literature and prove that any normed linear vector space is contained in some Banach space. As usual, $(V, \| \, \|)$ will denote some normed linear vector space over $\mathbb{R}$. We first remind the reader about a definition familiar from the calculus.

Definition 4.1: A sequence $\{\alpha_n\}$ in V is said to be *Cauchy* (or is called a *Cauchy sequence*) if for every $r > 0$, there exists an $m \in \mathbb{N}$ such that $p, q \geqslant m \Rightarrow \|\alpha_p - \alpha_q\| < r$.

There are several easy facts about Cauchy sequences that we shall need in the sequel. We gather these facts together in our first lemma. We leave the proof for the exercises at the end of this section.

Lemma 4.2: (a) If $\{\alpha_n\} \to \alpha$, then $\{\alpha_n\}$ is Cauchy.

(b) Any Cauchy sequence is bounded.

(c) If $\{\alpha_n\}$ is Cauchy and contains a subsequence converging to say α, then $\{\alpha_n\} \to \alpha$.

(d) A sequence $\{\alpha_n\}$ is Cauchy if and only if $\lim_{n,m \to \infty} \|\alpha_n - \alpha_m\| = 0$.

(e) Let V and W be normed linear vector spaces, and suppose A is a subset of V. Let f: $A \to W$ be a function. If f is Lipschitz on A, and $\{\alpha_n\}$ is a Cauchy sequence in A, then $\{f(\alpha_n)\}$ is a Cauchy sequence in W.

(f) With the same notation as in (e), suppose f is uniformly continuous on A. If $\{\alpha_n\}$ is a Cauchy sequence in A, then $\{f(\alpha_n)\}$ is a Cauchy sequence in W. $\square$

We note that 4.2(f) is not true in general for continuous functions. See Exercise 2 at the end of this section. We can now introduce the central definition of this section.

Definition 4.3: A normed linear vector space V is said to be *complete* if every Cauchy sequence in V converges.

Again we remind the reader that this definition depends on the particular norm on V. It is more precise to say $(V, \| \, \|)$ is complete. If $\| \, \|$ and $\| \, \|'$ are two equivalent norms on V, then clearly a sequence $\{\alpha_n\}$ is Cauchy in the $\| \, \|$-norm if and only if $\{\alpha_n\}$ is Cauchy in the $\| \, \|'$-norm. In particular, $(V, \| \, \|)$ is complete if and only if $(V, \| \, \|')$ is complete.

A complete, normed linear vector space is called a *Banach space* in honor of the great analyst Stefan Banach. One of the most important examples of a Banach space is $\mathbb{R}$ itself.

Theorem 4.4: $\mathbb{R}$ is a Banach space.

Proof: Let $\{x_n\}$ be a Cauchy sequence in $\mathbb{R}$. Then $\{x_n\}$ is bounded by Lemma 4.2(b). Then Theorem 3.20 implies $\{x_n\}$ contains a convergent subsequence $\{x_{n_k}\}$. But then 4.2(c) implies $\{x_n\}$ converges. $\square$

We can follow the same sort of argument given in Theorem 3.21, and get the following corollary to Theorem 4.4:

Corollary 4.5: $\mathbb{R}^n$ is a Banach space.

Proof: Any norm on $\mathbb{R}^n$ is equivalent to the sum norm $\| \ \|_1$. Hence, it suffices to show $(\mathbb{R}^n, \| \ \|_1)$ is complete. If $\{\alpha_k = (x_{1k}, \ldots, x_{nk})\}$ is a Cauchy sequence in $\mathbb{R}^n$, then $\{\beta_k = (x_{1k}, \ldots, x_{n-1k})\}$ is a Cauchy sequence in $\mathbb{R}^{n-1}$, and $\{x_{nk}\}$ is a Cauchy sequence in $\mathbb{R}$. Induction then implies that $\{\beta_k\} \to \beta$ for some $\beta \in \mathbb{R}^{n-1}$, and $\{x_{nk}\} \to x$ for some $x \in \mathbb{R}$. Then the reader can easily check that $\{\alpha_k\} \to (\beta, x)$. $\square$

Corollary 4.6: If V is any finite-dimensional, normed linear vector space, then V is a Banach space.

Proof: We have seen in Corollary 3.27 that V is norm isomorphic to $(\mathbb{R}^n, \| \ \|_1)$ for some integer n. The result now follows from 4.5. $\square$

If V is not finite dimensional, then V need not be complete. Consider the following example:

Example 4.7: Let $V = \bigoplus_{i=1}^{\infty} \mathbb{R}$, and consider the norm $\| \ \|_\infty$ given in equation 1.8(c). We claim that V is not complete in the $\| \ \|_\infty$-norm. To see this, let $\alpha_n = (1, 1/2, 1/3, \ldots, 1/n, 0, 0, 0, \ldots)$. If $n > m$, then $\|\alpha_n - \alpha_m\|_\infty = 1/(m+1)$. Thus, Lemma 4.2(d) implies $\{\alpha_n\}$ is a Cauchy sequence in V. Clearly, $\{\alpha_n\}$ has no limit in V. Hence, V is not complete. $\square$

We do not intend to embark on a full-scale study of Banach spaces. We shall present only one last theorem, which says that every normed linear vector space can be imbedded in a Banach space. In order to prove this assertion, we need one preliminary result.

Lemma 4.8: Let $(V, \| \ \|)$ be a normed linear vector space, and suppose A is a subset of V such that $\bar{A} = V$. If every Cauchy sequence in A has a limit in V, then V is complete.

Proof: Let $\{\alpha_n\}$ be a Cauchy sequence in V. We must argue $\{\alpha_n\}$ has a limit in V. Since $\bar{A} = V$, each $\alpha_n \in \bar{A}$. In particular, for each $n \in \mathbb{N}$, there exists a $\beta_n \in A$ such that $\|\beta_n - \alpha_n\| < 1/n$. Then $\|\beta_n - \beta_m\| \leqslant \|\beta_n - \alpha_n\| + \|\alpha_n - \alpha_m\| + \|\alpha_m - \beta_m\|$

$< 1/n + 1/m + \|\alpha_n - \alpha_m\|$. Since $\{\alpha_n\}$ is Cauchy, we conclude from these inequalities that $\{\beta_n\}$ is Cauchy. Our hypotheses now imply that there exists a $\beta \in V$ such that $\{\beta_n\} \to \beta$. But then $\|\alpha_n - \beta\| \leqslant \|\alpha_n - \beta_n\| + \|\beta_n - \beta\| \leqslant 1/n + \|\beta_n - \beta\|$ implies $\{\alpha_n\} \to \beta$. Thus, V is complete. □

We can now state our main result.

Theorem 4.9: Let $(V, \|\ \|)$ be a normed linear vector space. Then there exists a Banach space $(V', \|\ \|')$ and a monomorphism $\theta: V \to V'$ such that the following properties are satisfied:

(a) $\|\theta(\alpha)\|' = \|\alpha\|$ for all $\alpha \in V$.
(b) The image of θ is dense in V', i.e. $\overline{\theta(V)} = V'$.

Furthermore, if $(V'', \|\ \|'')$ is another Banach space admitting a monomorphism $\psi: V \to V''$ satisfying (a) and (b), then there exists a norm isomorphism $\chi: V' \cong V''$ such that the following diagram is commutative:

4.10:

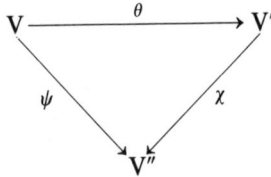

Let us say a few words concerning Theorem 4.9 before giving its proof. A Banach space $(V', \|\ \|')$ satisfying the conditions (a) and (b) in 4.9 is called a completion of $(V, \|\ \|)$. The theorem guarantees that every normed linear vector space has a completion. The second half of the theorem says that any completion of V is unique up to norm isomorphism. Hence, we may refer to *the* completion of $(V, \|\ \|)$.

A linear transformation between normed linear vector spaces which preserves distances is called an isometry. Thus, 4.9(a) says V is isometrically imbedded in its completion. 4.9(b) says via θ, V sits in its completion as a dense subset.

Proof of 4.9: Consider the vector space $V^{\mathbb{N}}$ given in Example 1.6 of Chapter 1. $V^{\mathbb{N}}$ is nothing but the set of all sequences in V with addition and scalar multiplication defined pointwise. Let $S = \{\{\alpha_n\} \in V^{\mathbb{N}} \mid \{\alpha_n\}$ is Cauchy$\}$. If $\{\alpha_n\}$ and $\{\beta_n\}$ are Cauchy sequences in V, and x, $y \in \mathbb{R}$, then clearly $\{x\alpha_n + y\beta_n\}$ is a Cauchy sequence in V. Thus, S is a subspace of $V^{\mathbb{N}}$.

Let $\{\alpha_n\} \in S$. The inequality $\big| \|\alpha_n\| - \|\alpha_m\| \big| \leqslant \|\alpha_n - \alpha_m\|$ implies that $\{\|\alpha_n\|\}$ is a Cauchy sequence in $\mathbb{R}$. Since $\mathbb{R}$ is complete, the sequence $\{\|\alpha_n\|\}$ has a limit in $\mathbb{R}$. In particular, it makes sense to talk about the limit, $\lim\{\|\alpha_n\|\}$, of the sequence $\{\|\alpha_n\|\}$ for any vector $\{\alpha_n\} \in S$. We can then define a function p: $S \to \mathbb{R}$ by $p(\{\alpha_n\}) = \lim\{\|\alpha_n\|\}$. The reader can easily verify that the function p satisfies the following properties:

4.11: (1) $p(\{\alpha_n\}) \geqslant 0$

(2) $p(x\{\alpha_n\}) = |x|p(\{\alpha_n\})$

(3) $p(\{\alpha_n\} + \{\beta_n\}) \leqslant p(\{\alpha_n\}) + p(\{\beta_n\})$

These inequalities hold for all $\{\alpha_n\}$, $\{\beta_n\} \in S$, and all $x \in \mathbb{R}$. In the language of Exercise 2 of Section 1, p is a seminorm on S. Note that $p(\{\alpha_n\}) = 0$ does not mean $\{\alpha_n\} = 0$. So, p is not a norm on S.

We now follow the ideas laid out in Exercise 10 of Section 2. Set $N = \{\{\alpha_n\} \in S \mid p(\{\alpha_n\}) = 0\}$. N is precisely those Cauchy sequences in V whose norms $\|\alpha_n\|$ have limit zero. By Exercise 10, N is a subspace of S, and S/N is a normed linear vector space with norm given by $\|\{\alpha_n\} + N\|' = p(\{\alpha_n\}) = \lim\{\|\alpha_n\|\}$. Set $V' = S/N$. We claim $(V', \|\ \|')$ is a Banach space satisfying (a) and (b).

We first define a map $\theta : V \to V'$ by the equation $\theta(\alpha) = \{\alpha_n\} + N$, where $\{\alpha_n\}$ is the constant sequence $\alpha_n = \alpha$ for all $n \in \mathbb{N}$. We shall need some special notation for constant sequences in this proof. We shall let $\{\alpha\}$ denote the constant sequence in V every term of which is α. Clearly, any constant sequence $\{\alpha\}$ is Cauchy, and, thus, $\{\alpha\} \in S$. The map θ is now given by $\theta(\alpha) = \{\alpha\} + N$. Clearly, θ is a linear transformation from V to V'.

Suppose $\theta(\alpha) = 0$ for some $\alpha \in V$. Then $\{\alpha\} + N = N$ in V'. Thus, $\{\alpha\} \in N$. But then $\|\alpha\| = \lim \|\alpha\| = 0$. Therefore, $\alpha = 0$ since $\|\ \|$ is a norm on V. We conclude that θ is a monomorphism from V to V'. Since $\|\theta(\alpha)\|' = \|\{\alpha\} + N\|' = \lim \|\alpha\| = \|\alpha\|$, we see θ is an isometry. Thus, we have established (a).

In order to establish (b), let us first observe that we have the following commutative diagram of seminormed vector spaces and linear maps:

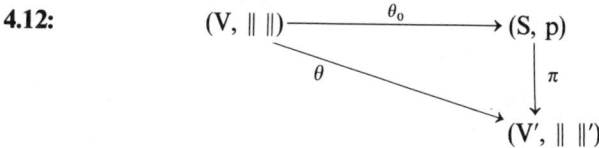

4.12:

$$ (V, \|\ \|) \xrightarrow{\ \theta_0\ } (S, p) $$

with θ from $(V, \|\ \|)$ to $(V', \|\ \|')$ and π from (S, p) to $(V', \|\ \|')$.

In this diagram, θ_0 is the constant sequence map given by $\theta_0(\alpha) = \{\alpha\}$. π is the natural projection given by $\pi(\{\gamma_n\}) = \{\gamma_n\} + N$. Note that these maps are "isometries" of the seminorm spaces in question. We have $p(\theta_0(\alpha)) = \|\alpha\|$ for all $\alpha \in V$. Also, $\|\pi(\{\gamma_n\})\|' = p(\{\gamma_n\})$ for any $\{\gamma_n\} \in S$.

To argue 4.9(b), we must show that every ball $B_r(\{\alpha_n\} + N)$ around a point $\{\alpha_n\} + N$ in V' has a nontrivial intersection with $\text{Im}(\theta)$. We claim that it is enough to prove this assertion on the θ_0-level of diagram 4.12. Hence, we claim that 4.9(b) follows from our next assertion.

4.13: For every $\{\alpha_n\} \in S$, and for every $r > 0$, there exists a vector $\eta \in V$ such that $p(\theta_0(\eta) - \{\alpha_n\}) < r$.

To prove 4.13, fix $r > 0$ and $\{\alpha_n\} \in S$. Since $\{\alpha_n\}$ is a Cauchy sequence in V, there exists an $m \in \mathbb{N}$ such that $n \geq m \Rightarrow \|\alpha_n - \alpha_m\| < r$. Set $\eta = \alpha_m$. Then $p(\theta_0(\eta) - \{\alpha_n\}) = p(\{\alpha_m\}_{n=1}^{\infty} - \{\alpha_n\}_{n=1}^{\infty}) = \lim \|\alpha_m - \alpha_n\| < r$. (The limit here is as $n \to \infty$.) This proves 4.13 and completes the proof of 4.9(b).

It now remains to show that $(V', \| \ \|')$ is complete. For this, it suffices by Lemma 4.8 to show that every Cauchy sequence in $\theta(V)$ has a limit in V'. Let $\{\theta(\xi_n)\} = \{\theta_0(\xi_n) + N\}$ be a Cauchy sequence in $\theta(V)$. Thus, $\xi_1, \xi_2, \ldots$, are vectors in V. Now $\|\theta(\xi_n) - \theta(\xi_m)\|' = \|\theta(\xi_n - \xi_m)\|' = \|\pi\theta_0(\xi_n - \xi_m)\|' = p(\theta_0(\xi_n) - \theta_0(\xi_m))$. Thus, $\{\theta_0(\xi_n)\}$ is a Cauchy sequence in the seminorm space S. [For the seminorm space (S, p), a sequence $\{\eta_n\}$ in S is said to be Cauchy if for every $r > 0$, there exists an m such that, $s, t \geq m \Rightarrow p(\eta_s - \eta_t) < r$.] Each $\theta_0(\xi_n)$ is the constant sequence $\{\xi_n\}_{m=1}^{\infty}$ in V. Thus, $\{\theta_0(\xi_n)\}$ is a Cauchy sequence of constant sequences in S. We need the following fact:

4.14: Every Cauchy sequence $\{\theta_0(\alpha_n)\}$ in $\theta_0(V)$ has a limit in S, that is, there exists a vector $\{\beta_k\} \in S$ such that $\lim_{n \to \infty} p(\theta_0(\alpha_n) - \{\beta_k\}) = 0$.

To prove 4.14, we first note that $\{\theta_0(\alpha_n)\}$ being Cauchy in S means $0 = \lim_{n,m \to \infty} p(\theta_0(\alpha_n) - \theta_0(\alpha_m)) = \lim_{n,m \to \infty} \|\alpha_n - \alpha_m\|$. In particular, $\{\alpha_n\}$ is a Cauchy sequence in V. Thus, $\{\alpha_n\} = \gamma \in S$. We claim the sequence $\{\theta_0(\alpha_n)\}$ converges to γ. For any fixed $m \in \mathbb{N}$, we have $p(\theta_0(\alpha_m) - \gamma) = p(\{\alpha_m\}_{n=1}^{\infty} - \{\alpha_n\}_{n=1}^{\infty}) = \lim_{n \to \infty} \|\alpha_m - \alpha_n\|$. Since $\{\alpha_n\}$ is a Cauchy sequence in V, this last limit can be made as small as we please by choosing m large. Therefore, $\{\theta_0(\alpha_n)\} \to \{\alpha_n\}$. This completes the proof of 4.14.

If we now apply 4.14 to the sequence $\{\theta_0(\xi_n)\}$, we see $\{\theta_0(\xi_n)\} \to \{\xi_n\}$ in S. Using diagram 4.12, it easily follows that $\{\theta(\xi_n)\} \to \{\xi_n\} + N$ in V'. Thus, the hypotheses of Lemma 4.8 are satisfied. We conclude that $(V', \| \ \|')$ is complete.

The fact that a completion of V is unique up to a norm isomorphism χ satisfying 4.10 is straightforward. We leave the details to the exercises at the end of this section. $\square$

EXERCISES FOR SECTION 4

(1) Prove the six assertions in Lemma 4.2.

(2) Give an example in $\mathbb{R}$ showing 4.2(f) is not true in general for continuous functions.

(3) Fill out the details in the proof of Corollary 4.5.

(4) Show that $C([0, 1])$ is a Banach space with respect to the uniform norm $\|f\|_{\infty} = \sup\{|f(t)| \,|\, t \in [0, 1]\}$.

(5) Suppose $(V_1, \| \ \|_1), \ldots, (V_n, \| \ \|_n)$ are a finite number of Banach spaces. Show that the product $V_1 \times \cdots \times V_n$ is a Banach space relative to any product norm.

(6) Suppose V is a normed linear vector space and W is a Banach space. Show that $\mathscr{B}(V, W)$ is a Banach space with respect to the uniform norm (see Definition 1.23).

(7) In the proof of Theorem 4.9, show that assertion 4.13 indeed implies the closure of $\theta(V)$ in V' is all of V'.

(8) Complete the proof of Theorem 4.9 by showing that the completion $(V', \| \ \|')$ of V is unique up to a norm isomorphism satisfying 4.10.

(9) Suppose $(V, \| \ \|)$ is a Banach space, and V itself is also an algebra over $\mathbb{R}$. We say $(V, \| \ \|)$ is a Banach algebra if the following two properties are satisfied:

(i) $\|\alpha\beta\| \leqslant \|\alpha\| \|\beta\|$ for all $\alpha, \beta \in V$.

(ii) $\|1\| = 1$.

If V is a Banach space, show that $\mathscr{B}(V, V)$ is a Banach algebra with respect to the uniform norm.

(10) Suppose $(V, \| \ \|)$ is a Banach algebra.

(a) If $\alpha \in V$ has $\|\alpha\| < 1$, then show $1 - \alpha$ is invertible in V. More precisely, show $(1 - \alpha)^{-1} = \sum_{n=0}^{\infty} \alpha^n$. (Recall that an element α in an algebra V is invertible if there exists an element $\beta \in V$ such that $\alpha\beta = \beta\alpha = 1$.)

(b) Let $U = \{\alpha \in V \mid \alpha$ is invertible$\}$. Show that U is a nonempty, open subset of V. Is U a subspace of V?

(c) With U as above, show the map $\alpha \to \alpha^{-1}$ is a continuous map on U.

(d) Deduce the following theorem: if V is a Banach space, then the invertible transformations in $\mathscr{B}(V, V)$ form a nonempty, open subset. The map $T \to T^{-1}$ is continuous on this open set.

(11) Suppose $(V, \| \ \|)$ is a Banach algebra. Show that the multiplication map $(\alpha, \beta) \to \alpha\beta$ is a bounded bilinear map from $V \times V \to V$.

(12) Show that $(C([0, 1]), \| \ \|)$ is not a Banach space. ($\| \ \|$ as in 1.6(b)).

(13) Let $(V, \| \ \|)$ be a Banach space. Suppose W is a closed subspace of V. Prove that $(W, \| \ \|)$ is a Banach space.

(14) Suppose $(V, \| \ \|)$ is a normed linear vector space. If V is sequentially compact, prove that $(V, \| \ \|)$ is a Banach space. Is the converse true?

(15) Let $(V, \| \ \|)$ be a normed linear vector space. Let $\{\alpha_n\}$ be a sequence in V. We say the infinite series $\sum_{n=1}^{\infty} \alpha_n$ converges (to say $\beta \in V$) if $\{\sigma_n\} \to \beta$. Here $\{\sigma_n\}$ is the usual sequence of partial sums given by $\sigma_n = \sum_{k=1}^{n} \alpha_k$. We say $\sum_{n=1}^{\infty} \alpha_n$ is absolutely convergent if $\sum_{n=1}^{\infty} \|\alpha_n\|$ converges. If $(V, \| \ \|)$ is a Banach space, prove that every absolutely convergent series converges.

(16) Prove the converse of Exercise 15: If every absolutely convergent series in V is convergent, then V is a Banach space.

(17) Use Exercise 16 to show that if N is a closed subspace of a Banach space V, then V/N is a Banach space. The norm on V/N is given in Exercise 13 of Section 2.

Chapter V

Inner Product Spaces

---

1. REAL INNER PRODUCT SPACES

In this chapter, we return to the material of Section 7 of chapter I. We want to study inner products for both real and complex vector spaces. In this section and the next, we shall concentrate solely on real inner product spaces. Later, we shall modify our definitions and results for complex spaces. Throughout this section then, V will denote a vector space over the real numbers $\mathbb{R}$. We do not assume V is finite dimensional over $\mathbb{R}$.

The reader will recall that an inner product on V is a bilinear form $\omega: V \times V \to \mathbb{R}$, which is symmetric and whose associated quadratic form is positive definite. If ω is an inner product on V, then we shall shorten our notation and write $\omega(\alpha, \beta) = \langle \alpha, \beta \rangle$. Thus, we can rewrite Definition 7.12 of Chapter I using this new notation as follows:

Definition 1.1: Let V be a real vector space. An inner product on V is a function $\langle \ , \ \rangle: V \times V \to \mathbb{R}$ satisfying the following conditions:

(a) $\langle x\alpha + y\beta, \gamma \rangle = x\langle \alpha, \gamma \rangle + y\langle \beta, \gamma \rangle$.

(b) $\langle \gamma, x\alpha + y\beta \rangle = x\langle \gamma, \alpha \rangle + y\langle \gamma, \beta \rangle$.

(c) $\langle \alpha, \beta \rangle = \langle \beta, \alpha \rangle$.

(d) $\langle \alpha, \alpha \rangle$ is a positive real number for any $\alpha \neq 0$.

Conditions (a)–(d) in 1.1 are to hold for all vectors α, β, γ in V and for all x, $y \in \mathbb{R}$. Note that (a) and (d) imply that $\langle \alpha, \alpha \rangle \geqslant 0$ with equality if and only if $\alpha = 0$.

A vector space V together with some inner product $\langle \ , \ \rangle$ on V will be called an *inner product space*. Of course, the same space V may be regarded as an inner product space in many different ways. More precisely then, an inner product space is an ordered pair $(V, \langle \ , \ \rangle)$ consisting of a real vector space V and a real valued function $\langle \ , \ \rangle : V \times V \rightarrow \mathbb{R}$ satisfying the conditions in 1.1. Let us review some of the examples from Chapter I.

Example 1.2: Let $V = \mathbb{R}^n$, and set $\langle \alpha, \beta \rangle = \sum_{i=1}^{n} x_i y_i$, where $\alpha = (x_1, \ldots, x_n)$ and $\beta = (y_1, \ldots, y_n)$. It is a simple matter to check that $\langle \ , \ \rangle$ satisfies conditions (a)–(d) in 1.1. We shall refer to this particular inner product as the *standard inner product* on $\mathbb{R}^n$. $\square$

Example 1.3: Let $V = \bigoplus_{i=1}^{\infty} \mathbb{R}$. Define an inner product on V by setting $\langle \alpha, \beta \rangle = \sum_{i=1}^{\infty} x_i y_i$. Here $\alpha = (x_1, x_2, \ldots)$ and $\beta = (y_1, y_2, \ldots)$. Since any vector in V has only finitely many nonzero components, $\langle \ , \ \rangle$ is clearly an inner product on V. Thus, V is an example of an infinite-dimensional, inner product space. $\square$

Example 1.4: Let $V = C([a, b])$. Set $\langle f, g \rangle = \int_a^b f(x)g(x)\,dx$. An easy computation shows $\langle \ , \ \rangle$ is an inner product on V. $\square$

A less familiar example is the normed linear vector space mentioned in Example 1.9 of Chapter IV.

Example 1.5: Let $V = \{\{x_n\} \in \mathbb{R}^{\mathbb{N}} \mid \sum_{n=1}^{\infty} x_n^2 < \infty\}$. We can define an inner product on V by setting $\langle \{x_n\}, \{y_n\} \rangle = \sum_{n=1}^{\infty} x_n y_n$. We ask the reader to verify that with this definition of $\langle \ , \ \rangle$, $(V, \langle \ , \ \rangle)$ is an inner product space. This space is usually denoted by l^2 in the literature. $\square$

Let $(V, \langle \ , \ \rangle)$ be an inner product space. If $T: W \rightarrow V$ is an injective linear transformation, then we can define an inner product on W by setting $\langle \alpha, \beta \rangle' = \langle T(\alpha), T(\beta) \rangle$ for all $\alpha, \beta \in W$. In this way, we can produce many new examples from the examples we already have. A special case of this procedure is that in which W is a subspace of V. If we restrict $\langle \ , \ \rangle$ to $W \times W$, then $(W, \langle \ , \ \rangle)$ becomes an inner product space in its own right. For instance, $\mathbb{R}^n$ is a subspace of V in Example 1.3. When we restrict $\langle \ , \ \rangle$ to $\mathbb{R}^n \times \mathbb{R}^n$, we get the standard inner product on $\mathbb{R}^n$.

Our first general fact about an inner product space $(V, \langle , \rangle)$ is that V has a natural, normed linear vector space structure that is intimately related to the inner product $\langle \ , \ \rangle$. To see this, we need an inequality, known as Schwarz's inequality.

Lemma 1.6: Let V be an inner product space. Then $|\langle \alpha, \beta \rangle| \leqslant \langle \alpha, \alpha \rangle^{1/2} \langle \beta, \beta \rangle^{1/2}$ for all $\alpha, \beta \in V$.

Proof: Fix α and β in V. For each real number t, let $p(t) = \langle \alpha - t\beta, \alpha - t\beta \rangle$. Then 1.1 implies that $p(t)$ is a quadratic function on $\mathbb{R}$ such that $p(t) \geq 0$ for all $t \in \mathbb{R}$. It follows that the discriminant, $4\langle \alpha, \beta \rangle^2 - 4\langle \alpha, \alpha \rangle \langle \beta, \beta \rangle$, of $p(t)$ must be negative or zero. Thus, $\langle \alpha, \beta \rangle^2 \leq \langle \alpha, \alpha \rangle \langle \beta, \beta \rangle$. Taking square roots gives us the desired conclusion. $\square$

We take this opportunity to point out that although Schwarz's inequality is easy to prove, its conclusions in specific examples are not at all obvious. In Example 1.2, for instance, the inequality becomes

1.7:
$$\left(\sum_{i=1}^{n} x_i y_i \right)^2 \leq \left(\sum_{i=1}^{n} x_i^2 \right) \left(\sum_{i=1}^{n} y_i^2 \right).$$

In 1.7, $x_1, y_1, \ldots, x_n, y_n$ are arbitrary real numbers.

In Example 1.4, Schwarz's inequality becomes

1.8:
$$\left| \int_a^b f(x)g(x)\,dx \right| \leq \left(\int_a^b f(x)^2\,dx \right)^{1/2} \left(\int_a^b g(x)^2\,dx \right)^{1/2}$$

Our most important application of Lemma 1.6 is the following corollary:

Corollary 1.9: Let $(V, \langle \ , \ \rangle)$ be an inner product space. Then $\|\alpha\| = \langle \alpha, \alpha \rangle^{1/2}$ is a norm on V.

Proof: We must verify that $\| \ \|$ satisfies the conditions in Definition 1.1 of Chapter IV. Of these conditions, only the triangle inequality $\|\alpha + \beta\| \leq \|\alpha\| + \|\beta\|$, requires any proof. Fix α and β in V. Using Schwarz's inequality, we have $\|\alpha + \beta\|^2 = \langle \alpha + \beta, \ \alpha + \beta \rangle = \langle \alpha, \alpha \rangle + 2\langle \alpha, \beta \rangle + \langle \beta, \beta \rangle \leq \|\alpha\|^2 + 2|\langle \alpha, \beta \rangle| + \|\beta\|^2 \leq \|\alpha\|^2 + 2\|\alpha\| \|\beta\| + \|\beta\|^2 = (\|\alpha\| + \|\beta\|)^2$. If we now take the square root of both sides of this inequality, we get $\|\alpha + \beta\| \leq \|\alpha\| + \|\beta\|$. $\square$

Corollary 1.9 implies that every inner product space $(V, \langle \ , \ \rangle)$ is a normed linear vector space via $\|\alpha\| = \langle \alpha, \alpha \rangle^{1/2}$. We shall call the norm given in 1.9 the norm associated with the inner product $\langle \ , \ \rangle$. In Example 1.2, for instance, the norm associated with the standard inner product is just the ordinary Euclidean norm given in equation 1.4(b) of Chapter IV. In Example 1.3, the norm associated with $\langle \ , \ \rangle$ is the natural extension of the Euclidean norm to $\bigoplus_{i=1}^{\infty} \mathbb{R}$ [1.8(b) of Chapter IV]. In Example 1.4, the norm associated with the inner product there is given by 1.6(b) of Chapter IV.

Since any inner product space $(V, \langle \ , \ \rangle)$ is a normed linear vector space with respect to the norm associated with $\langle \ , \ \rangle$, we have all the topological machinery from Chapter IV at our disposal. In particular, we can talk about the distance between two vectors, open and closed sets, continuity, limits, completeness, and so on. It will be understood that these notions are all relative to the norm associated with the inner product $\langle\ ,\ \rangle$. Thus, when we speak of the distance between two vectors α and β, for instance, we mean $\|\alpha - \beta\| = \langle\alpha - \beta, \alpha - \beta\rangle^{1/2}$.

When dealing with inner product spaces, we shall always use the symbol $\|\ \|$ to represent the norm associated with the inner product, that is, $\|\alpha\| = \langle\alpha, \alpha\rangle^{1/2}$. In terms of the associated norm, Schwarz's inequality can be rewritten as follows:

Theorem 1.10 (Schwarz's Inequality): Let $(V, \langle\ ,\ \rangle)$ be an inner product space with associated norm $\|\ \|$. Then $|\langle\alpha, \beta\rangle| \leqslant \|\alpha\|\,\|\beta\|$ for all $\alpha, \beta \in V$. $\square$

There is an important corollary that follows from Theorem 1.10.

Corollary 1.11: Let $(V, \langle\ ,\ \rangle)$ be an inner product space with associated norm $\|\ \|$. Then the inner product $\langle\ ,\ \rangle : V \times V \to \mathbb{R}$ is a continuous function with respect to any product norm on $V \times V$.

Proof: By Theorem 2.7 of Chapter IV, any two product norms on $V \times V$ are equivalent. So, we might as well prove the assertion for the sum norm $\|(\alpha, \beta)\|_s = \|\alpha\| + \|\beta\|$ on $V \times V$. In order to show $\langle\ ,\ \rangle$ is continuous, it suffices by Lemma 3.7 of Chapter IV to show that if $\{(\alpha_n, \beta_n)\} \to (\alpha, \beta)$ in $V \times V$, then $\{\langle\alpha_n, \beta_n\rangle\} \to \langle\alpha, \beta\rangle$ in $\mathbb{R}$.

So, suppose $\{(\alpha_n, \beta_n)\} \to (\alpha, \beta)$ in $V \times V$. The convergence here is relative to the sum norm $\|\ \|_s$. Consequently, $\{\alpha_n\} \to \alpha$, and $\{\beta_n\} \to \beta$ in V. These statements follow from $(\mathscr{P})$ in Corollary 3.9 of Chapter IV. Since $\{\alpha_n\} \to \alpha$, we know the sequence $\{\alpha_n\}$ is bounded. Suppose $\|\alpha_n\| \leqslant c$ for all $n \in \mathbb{N}$. Applying Schwarz's inequality, we have

$$|\langle\alpha_n, \beta_n\rangle - \langle\alpha, \beta\rangle| \leqslant |\langle\alpha_n, \beta_n\rangle - \langle\alpha_n, \beta\rangle| + |\langle\alpha_n, \beta\rangle - \langle\alpha, \beta\rangle|$$

$$= |\langle\alpha_n, \beta_n - \beta\rangle| + |\langle\alpha_n - \alpha, \beta\rangle|$$

$$\leqslant \|\alpha_n\|\,\|\beta_n - \beta\| + \|\beta\|\,\|\alpha_n - \alpha\|$$

$$\leqslant c\|\beta_n - \beta\| + \|\beta\|\,\|\alpha_n - \alpha\|$$

From this inequality, it is now obvious that $\{\langle\alpha_n, \beta_n\rangle\} \to \langle\alpha, \beta\rangle$ in $\mathbb{R}$. $\square$

In the sequel, we shall often use the following special case of 1.11. Suppose in some inner product space $(V, \langle\ ,\ \rangle)$, $\{\alpha_n\} \to \alpha$. Here it is understood that the convergence is relative to the norm $\|\ \|$ associated to $\langle\ ,\ \rangle$. Then for any $\beta \in V$, $\{\langle\alpha_n, \beta\rangle\} \to \langle\alpha, \beta\rangle$ in $\mathbb{R}$. This conclusion follows immediately from applying the continuous function $\langle\ ,\ \rangle$ to the convergent sequence $\{(\alpha_n, \beta)\} \to (\alpha, \beta)$ in $V \times V$.

We have seen that any inner product space is a normed linear vector space. Spaces of this type are called *pre-Hilbert spaces*. Let us introduce the following more precise definition:

Definition 1.12: A normed linear vector space $(V, \| \|)$ is called a *pre-Hilbert space* if there exists an inner product $\langle \, , \, \rangle$ on V such that $\|\alpha\| = \langle \alpha, \alpha \rangle^{1/2}$ for all $\alpha \in V$.

Thus, $(\mathbb{R}^n, \| \|)$ (notation as in 1.4 of Chapter IV) is a finite-dimensional pre-Hilbert space. An example of an infinite-dimensional pre-Hilbert space is given by $(C([a, b]), \| \|)$ (notation as in equation 1.6 of Chapter IV). Note that the notion of a space being pre-Hilbert depends on the particular norm being discussed. Unlike most of the ideas discussed in Chapter IV, the property of being pre-Hilbert is not invariant with respect to equivalence of norms. For example, the sum norm $\| \|_1$ on $\mathbb{R}^n$ is equivalent to the Euclidean norm $\| \|$. However, $(\mathbb{R}^n, \| \|_1)$ is not a pre-Hilbert space. To see this, suppose $\|\alpha\|_1 = \langle \alpha, \alpha \rangle^{1/2}$ for some inner product $\langle \, , \, \rangle$ on $\mathbb{R}^n$. As usual, let $\underline{\delta} = \{\delta_1, \ldots, \delta_n\}$ denote the canonical basis of $\mathbb{R}^n$. Set $\langle \delta_i, \delta_j \rangle = a_{ij}$, and let S denote the boundary of the unit ball in $(\mathbb{R}^n, \| \|_1)$. Then $S = \{\alpha = (x_1, \ldots, x_n) \mid \|\alpha\|_1 = 1\} = \{(x_1, \ldots, x_n) \mid \sum_{i=1}^n |x_i| = 1\}$. On the other hand, $\|\alpha\|_1^2 = \langle \alpha, \alpha \rangle = \sum_{i,j=1}^n a_{ij} x_i x_j$. Thus, S is the set of zeros in $\mathbb{R}^n$ of the quadratic polynomial $(\sum_{i,j=1}^n a_{ij} X_i X_j) - 1$. This is clearly impossible. S has too many corners to be the set of zeros of any polynomial. Thus, the sum norm $\| \|_1$ cannot be the associated norm of any inner product on $\mathbb{R}^n$. In particular, the normed linear vector space $(V, \| \|_1)$ is not a pre-Hilbert space.

Definition 1.13: A pre-Hilbert space $(V, \| \|)$ is called a *Hilbert space* if V is complete with respect to $\| \|$.

Thus, a Hilbert space is a Banach space whose norm is given by an inner product. For example, $(\mathbb{R}^n, \| \|)$ is a Hilbert space. More generally, Corollaries 3.27 and 4.5 of Chapter IV imply that any finite-dimensional pre-Hilbert space is in fact a Hilbert space. For an example of an infinite-dimensional Hilbert space, we return to Example 1.5. We ask the reader to confirm that l^2 is a Hilbert space, infinite dimensional over $\mathbb{R}$. (See Exercise 2 at the end of this section.) An important point here when dealing with pre-Hilbert spaces is the analog of Theorem 4.9 of Chapter IV.

Theorem 1.14: Let $(V, \| \|)$ be a pre-Hilbert space. Then there exists a Hilbert space $(V', \| \|')$ and an isometry $\theta: V \to V'$ such that the closure of $\theta(V)$ in V' is all of V'. Furthermore, if $(V'', \| \|'')$ is a second Hilbert space admitting an isometry $\psi: V \to V''$ such that $\psi(V)$ is dense in V'', then there exists a norm isomorphism $\chi: V' \cong V''$ such that $\chi\theta = \psi$.

Proof: We shall not use this theorem in the rest of this text. Hence, we define V' and leave the rest of the details to the exercises. Let $(V', \| \|')$ denote the completion of V constructed in Theorem 4.9 of Chapter IV. We define an inner product on $V' = S/N$ by the following formula:

1.15:
$$\langle \{\alpha_n\} + N, \{\beta_n\} + N \rangle = \lim_{n \to \infty} \langle \alpha_n, \beta_n \rangle$$

The reader can easily argue that this formula is well defined and gives an inner product on V' whose associated norm is $\| \ \|'$. $\square$

Throughout the rest of this section, $(V, \| \ \|)$ will denote a pre-Hilbert space. Let us recall a few familiar definitions from the calculus.

Definition 1.16: (a) Two vectors α and β in V are said to be *orthogonal* if $\langle \alpha, \beta \rangle = 0$. If α and β are orthogonal, we shall indicate this by writing $\alpha \perp \beta$.

(b) Two subsets A and B of V are orthogonal if $\alpha \perp \beta$ for all $\alpha \in A$ and $\beta \in B$. In this case, we shall write $A \perp B$.

(c) If A is a subset of V, then $A^{\perp} = \{ \alpha \in V \mid \langle \alpha, \beta \rangle = 0$ for all $\beta \in A \}$.

(d) A set of vectors $\{ \alpha_i \mid i \in I \}$ is said to be *pairwise orthogonal* if $\langle \alpha_i, \alpha_j \rangle = 0$ whenever $i \neq j$.

(e) A collection of subsets $\{ A_i \mid i \in I \}$ is said to be pairwise orthogonal if $A_i \perp A_j$ whenever $i \neq j$.

Note that $A^{\perp}$ is a subspace of V such that $A^{\perp} \cap L(A) = 0$. In fact, we even have $A^{\perp} \cap \overline{L(A)} = 0$. For suppose $\alpha \in \overline{L(A)}$, and $\beta \in A^{\perp}$. By Lemma 3.4 of Chapter IV, there exists a sequence $\{ \alpha_n \}$ in L(A) such that $\{ \alpha_n \} \to \alpha$. Using the continuity of the inner product, we have $\{ \langle \alpha_n, \beta \rangle \} \to \langle \alpha, \beta \rangle$. But, $\langle \alpha_n, \beta \rangle = 0$ for all $n \in \mathbb{N}$. Hence, $\langle \alpha, \beta \rangle = 0$. In particular, if $\alpha \in A^{\perp} \cap \overline{L(A)}$, then $\langle \alpha, \alpha \rangle = 0$. Thus $\alpha = 0$.

Vectors that are orthogonal behave nicely with respect to length formulas.

Lemma 1.17 (Parallelogram Law): Let $\alpha, \beta \in V$. Then

(a) $\| \alpha + \beta \|^2 + \| \alpha - \beta \|^2 = 2(\| \alpha \|^2 + \| \beta \|^2)$.

(b) $\alpha \perp \beta$ if and only if $\| \alpha + \beta \|^2 = \| \alpha \|^2 + \| \beta \|^2$.

(c) If $\{ \alpha_i \mid i = 1, \ldots, n \}$ is pairwise orthogonal, then $\| \sum_{i=1}^{n} \alpha_i \|^2 = \sum_{i=1}^{n} \| \alpha_i \|^2$.

Proof: (a) $\| \alpha + \beta \|^2 + \| \alpha - \beta \|^2 = \langle \alpha + \beta, \alpha + \beta \rangle + \langle \alpha - \beta, \alpha - \beta \rangle = 2 \langle \alpha, \alpha \rangle + 2 \langle \beta, \beta \rangle = 2(\| \alpha \|^2 + \| \beta \|^2)$.

(b) $\| \alpha + \beta \|^2 = \langle \alpha + \beta, \alpha + \beta \rangle = \| \alpha \|^2 + 2 \langle \alpha, \beta \rangle + \| \beta \|^2$. Thus, $\| \alpha + \beta \|^2 = \| \alpha \|^2 + \| \beta \|^2$ if and only if $\langle \alpha, \beta \rangle = 0$.

(c) This assertion follows trivially from (b). $\square$

The Parallelogram Law has an interesting corollary:

Corollary 1.18: Suppose $A = \{ \alpha_i \mid i \in I \}$ is a set of pairwise orthogonal, nonzero vectors in V. Then A is linearly independent over $\mathbb{R}$.

Proof: Suppose $c_1 \alpha_{i_1} + \cdots + c_n \alpha_{i_n} = 0$ is a linear combination of vectors from A. By Lemma 1.17(c), we have $0 = \| \sum_{j=1}^{n} c_j \alpha_{i_j} \|^2 = \sum_{j=1}^{n} |c_j|^2 \| \alpha_{i_j} \|^2$. Since no α_{i_j} is zero, we conclude that every $c_j = 0$. Thus, $\alpha_{i_1}, \ldots, \alpha_{i_n}$ are linearly independent. In particular, A is linearly independent over $\mathbb{R}$. $\square$

At this point, we return to the study of least-squares problems in the context of a pre-Hilbert space V. Suppose W is a subspace of V. Let $\beta \in V$. We want to decide when the distance, $d(\beta, W)$, between β and W is given by $\| \alpha - \beta \|$ for some $\alpha \in W$. We first note that $d(\beta, W)$ may not equal $\| \alpha - \beta \|$ for any vector $\alpha \in W$. We had seen an instance of this in Example 3.35 of Chapter IV for the normed linear vector space $(V = \{ f \in C([0, 1]) \mid f(0) = 0 \}, \| \ \|_\infty)$. Unfortunately, $(V, \| \ \|_\infty)$ is not a pre-Hilbert space. The reader can easily argue that $\| \ \|_\infty$ is not the norm associated with any inner product on V. To produce an example that fits our present context, we can return to Example 1.5. If we set $W = \{ \{ x_n \} \in l^2 \mid x_n = 0$ for all n sufficiently large$\}$, then W is a subspace of l^2. Let $\beta = \{ 1/n \}$. Then $\beta \in l^2 - W$. The reader can easily check that $d(\beta, W) = 0$. (In fact, $\overline{W} = l^2$.) Thus, $d(\beta, W) = 0$, but $0 \neq \| \alpha - \beta \|$ for any $\alpha \in W$. We ask the reader to verify these remarks in Exercise 7 at the end of this section.

Thus, in a pre-Hilbert space V, a given subspace W may contain no vector α that is closest to β in the sense that $d(\beta, W) = \| \alpha - \beta \|$. However, if W does contain a vector α such that $\| \alpha - \beta \| = d(\beta, W)$, then we can give a nice geometric characterization of α.

Theorem 1.19: Let $(V, \| \ \|)$ be a pre-Hilbert space. Suppose W is a subspace of V, and let $\beta \in V$. Let $\alpha \in W$. Then $\| \alpha - \beta \| = d(\beta, W)$ if and only if $(\alpha - \beta) \perp W$.

Proof: Suppose $(\alpha - \beta) \perp W$. Let $\xi \in W - \{\alpha\}$. Then $\| \xi - \beta \|^2 = \| (\xi - \alpha) + (\alpha - \beta) \|^2 = \| \xi - \alpha \|^2 + \| \alpha - \beta \|^2$. Since $\alpha - \beta$ is orthogonal to W, this last equality comes from 1.17(b). Taking square roots, we see $\| \xi - \beta \| > \| \alpha - \beta \|$. In particular, $d(\beta, W) = \inf \{ \| \gamma - \beta \| \mid \gamma \in W \} = \| \alpha - \beta \|$.

Conversely, suppose $\| \alpha - \beta \| = d(\beta, W)$. Fix a vector $\xi \in W - \{0\}$. Then for any real number t, we have $\alpha + t\xi \in W$. Thus, $\| \alpha - \beta \|^2 \leq \| \alpha + t\xi - \beta \|^2 = \langle \alpha - \beta + t\xi, \ \alpha - \beta + t\xi \rangle = \| \alpha - \beta \|^2 + 2t \langle \alpha - \beta, \ \xi \rangle + t^2 \| \xi \|^2$. Thus, the quadratic form $q(t) = 2t \langle \alpha - \beta, \ \xi \rangle + t^2 \| \xi \|^2$ is nonnegative on $\mathbb{R}$. This can only happen if the discriminant of $q(t)$ is not positive. Thus, $4 \langle \alpha - \beta, \ \xi \rangle^2 - 4 \| \xi \|^2 (0) \leq 0$. Hence, $\langle \alpha - \beta, \ \xi \rangle = 0$. If $\xi = 0$, then clearly $\langle \alpha - \beta, \ \xi \rangle = 0$. We conclude that $\alpha - \beta$ is orthogonal to W. $\square$

Let us make a couple of observations about the proof of Theorem 1.19. If $\beta \in W$, then of course $\alpha = \beta$, and the result is trivial. Suppose β is not in W. If W contains a vector α such that $\| \alpha - \beta \| = d(\beta, W)$, then α is the unique vector in W with this property. For we have seen in the proof of 1.19, that if $\xi \in W - \{\alpha\}$, then $\| \xi - \beta \| > \| \alpha - \beta \|$. Hence, if W contains a vector closest to β, then that vector is unique. This point is a characteristic feature of pre-Hilbert space theory. If W is a subspace of an arbitrary normed linear vector space V, then W may contain several vectors that are closest to a given vector β. In pre-Hilbert spaces, if W contains a vector α closest to β in the $\| \ \|$-norm, then α is unique. We want to give a special name to the vector α when it exists.

Definition 1.20: Let W be a subspace of the pre-Hilbert space V. Let $\beta \in V$. If W contains a vector α such that $(\alpha - \beta) \perp W$, then α will be called the *orthogonal projection* of β onto W. In this case, we shall use the notation $\alpha = \text{Proj}_W(\beta)$ to indicate that α is the orthogonal projection of β onto W.

We caution the reader that $\text{Proj}_W(\beta)$ does not always exist. By Theorem 1.19, $\text{Proj}_W(\beta)$ [when it exists] is the unique vector in W closest to β in the $\| \ \|$-norm. We have seen an example (Exercise 7 at the end of this section) that shows that in general there is no vector in W closest to β. If $\text{Proj}_W(\beta)$ does exist, then $\text{Proj}_W(\beta) - \beta$ is orthogonal to W. Notice also that $\text{Proj}_W(\beta) = \beta$ if and only if $\beta \in W$.

There is one important case in which $\text{Proj}_W(\beta)$ always exists.

Theorem 1.21: Let W be a subspace of the pre-Hilbert space $(V, \| \ \|)$. Suppose $(W, \| \ \|)$ is a Banach space. Then $\text{Proj}_W(\beta)$ exists for every $\beta \in V$. In this case, $V = W \oplus W^{\perp}$.

Proof: Let $\beta \in V$. If $\beta \in W$, then $\text{Proj}_W(\beta) = \beta$, and there is nothing to prove. Suppose β is not in W. Set $d = d(\beta, W)$. Then there exists a sequence $\{\alpha_n\}$ in W such that $\{\|\alpha_n - \beta\|\} \to d$. We claim that $\{\alpha_n\}$ is a Cauchy sequence in W. To see this, we first apply the Parallelogram Law. We have $\|\alpha_n - \alpha_m\|^2 = \|(\beta - \alpha_n) - (\beta - \alpha_m)\|^2 = 2(\|\beta - \alpha_n\|^2 + \|\beta - \alpha_m\|^2) - \|2\beta - (\alpha_n + \alpha_m)\|^2$. Since $\|2\beta - (\alpha_n + \alpha_m)\|^2 = 4\|\beta - (\alpha_n + \alpha_m)/2\|^2 \geqslant 4d^2$, we have $\|\alpha_n - \alpha_m\|^2 \leqslant 2(\|\beta - \alpha_n\|^2 + \|\beta - \alpha_m\|^2) - 4d^2$. The limit of this last expression is zero as m, n go to infinity. We conclude that $\lim_{n,m \to \infty} \|\alpha_n - \alpha_m\| = 0$. This proves that $\{\alpha_n\}$ is a Cauchy sequence in W.

Since W is complete, there exists a vector $\alpha \in W$ such that $\{\alpha_n\} \to \alpha$. Then continuity of the norm implies $\{\|\alpha_n - \beta\|\} \to \|\alpha - \beta\|$. Thus, $d = \|\alpha - \beta\|$. Theorem 1.19 now implies that $\text{Proj}_W(\beta) = \alpha$.

As for the second assertion, we always have $W \cap W^{\perp} = 0$. We need to argue that $V = W + W^{\perp}$. Let $\beta \in V$. From the first part of this argument, we know $\alpha = \text{Proj}_W(\beta)$ exists. The vector $\alpha - \beta$ is an element of $W^{\perp}$. Thus, $\beta - \alpha \in W^{\perp}$. Since $\beta = \alpha + (\beta - \alpha) \in W + W^{\perp}$, we conclude that $V = W \oplus W^{\perp}$. $\square$

Note that Theorem 1.21 is a generalization of Corollary 3.34 of Chapter IV when $(V, \| \ \|)$ is a pre-Hilbert space. For if W is a finite-dimensional subspace of V, then $(W, \| \ \|)$ is norm isomorphic to $(\mathbb{R}^n, \| \ \|_1)$ for some n. Thus, W is complete by Corollary 4.5 of Chapter IV. Some of the most important applications of Theorem 1.21 arise when V itself is finite dimensional. Then every subspace W of V is a Banach space, and, consequently, $\text{Proj}_W(\beta)$ exists for every vector $\beta \in V$.

Let us discuss a well-known example of the above results. Suppose V is the

Hilbert space $(\mathbb{R}^n, \|\ \|)$ given in Example 1.2. Let W be a subspace of V, and let $\beta \in V$. In this discussion, it will be convenient to identify $\mathbb{R}^n$ with the space of all $n \times 1$ matrices $M_{n \times 1}(\mathbb{R})$. The standard inner product on $\mathbb{R}^n$ is then given by the following formula: $\langle \xi, \eta \rangle = \xi^t \eta$. Here ξ^t is the transpose of the $n \times 1$ matrix ξ, and $\xi^t \eta$ is the matrix product of ξ^t with η.

Suppose $\{\alpha_1, \ldots, \alpha_m\}$ is a basis of W. Let us write each column vector α_i as $\alpha_i = (a_{1i}, \ldots, a_{ni})^t$ and form the $n \times m$ matrix $A = (a_{ij}) = (\alpha_1 | \cdots | \alpha_m)$. Then W is just the column space of A.

Let $\beta = (b_1, \ldots, b_n)^t$. Then finding the orthogonal projection, $\text{Proj}_W(\beta)$, of β onto W is equivalent to determining the least squares solution to the following system of linear equations:

1.22:
$$AX = \beta$$

In 1.22, $X = (x_1, \ldots, x_m)^t$ is a column vector in $M_{m \times 1}(\mathbb{R})$. If $\beta \in W$, then the linear system in 1.22 is consistent. In this case, equation 1.22 has a unique solution Z since $\text{rk}(A) = m$. If β is not in W, then the linear system in 1.22 is inconsistent. In either case, the words "least-squares solution" means a Z in $M_{m \times 1}(\mathbb{R})$ for which $\|AZ - \beta\|$ is as small as possible.

Now $\inf\{\|AX - \beta\| \mid X \in M_{m \times 1}(\mathbb{R})\} = \inf\{\|\gamma - \beta\| \mid \gamma \in W\} = d(\beta, W)$. Thus, by Theorem 1.19, the least-squares solution to 1.22 is a vector $Z \in M_{m \times 1}(\mathbb{R})$ such that $AZ = \text{Proj}_W(\beta)$. Since W is finite dimensional, we know from Theorem 1.21 that $\text{Proj}_W(\beta)$ exists. Since the rank of A is m, there exists a unique $Z \in M_{m \times 1}(\mathbb{R})$ such that $AZ = \text{Proj}_W(\beta)$. It is an easy matter to find a formula for Z. Since $AZ - \beta$ is orthogonal to W, we must have $(AX)^t(AZ - \beta) = 0$ for every $X \in M_{m \times 1}(\mathbb{R})$. Thus, $X^t(A^tAZ - A^t\beta) = 0$ for every X. This implies that $A^tAZ - A^t\beta = 0$. At this point, we need the following fact:

1.23: Let $A \in M_{n \times m}(\mathbb{R})$. If the rank of A is m, then A^tA is a nonsingular, symmetric $m \times m$ matrix.

A proof of 1.23 is easy. We leave this as an exercise at the end of this section. Returning to our computation, we see $Z = (A^tA)^{-1}A^t\beta$ and $\text{Proj}_W(\beta) = A(A^tA)^{-1}A^t\beta$. Let us summarize our results in the following theorem:

Theorem 1.24: Let $A \in M_{n \times m}(\mathbb{R})$ with $\text{rk}(A) = m$. Let $\beta \in M_{n \times 1}(\mathbb{R})$. Then the least-squares solution to the linear system $AX = \beta$ is given by

1.25:
$$Z = (A^tA)^{-1}A^t\beta$$

The orthogonal projection of β onto the column space W of A is given by

1.26:
$$\text{Proj}_W(\beta) = A(A^tA)^{-1}A^t\beta \qquad \square$$

If we look back at our discussion preceding Theorem 1.24, we see that the hypothesis rk(A) = m was used in 1.23 to conclude that A'A was invertible. If A is an arbitrary n × m matrix, then the linear system $AX = \beta$ still has a least-squares solution Z in $M_{m \times 1}(\mathbb{R})$. Z is not necessarily unique, but the same analysis as before shows that Z must satisfy the following equation:

1.27: $$A'AZ = A'\beta$$

Equation 1.27 is known as the "normal equation" of the least-squares solution to $AX = \beta$. If the rank of A is m, then equation 1.27 specializes to equation 1.25. If the rank of A is less than m, then we need the theory of the pseudoinverse of A to construct Z.

Let us now return to our general setting of an arbitrary pre-Hilbert space $(V, \| \ \|)$. Suppose W is a subspace of V for which $\mathrm{Proj}_W(\beta)$ exists for every $\beta \in V$. For example, W could be a finite-dimensional subspace of V. Then Theorem 1.21 implies that $V = W \oplus W^\perp$. Thus, the map $\mathrm{Proj}_W(\ ): V \to W$ is just the natural projection of V onto W determined by the direct sum decomposition $V = W \oplus W^\perp$ (see Section 4 of Chapter I). In particular, $\mathrm{Proj}_W(\ )$ satisfies the usual properties of a projection map:

1.28: (a) $\mathrm{Proj}_W(\ ) \in \mathrm{Hom}_{\mathbb{R}}(V, W)$.

   (b) $\mathrm{Proj}_W(\beta) = \beta$ if and only if $\beta \in W$.

   (c) $\mathrm{Proj}_W(\ )$ is an idempotent endomorphism of V.

Thus, if $\mathrm{Proj}_W(\beta)$ exists for every $\beta \in V$, then $W^\perp$ is a complement of W in V. The converse of this statement is true also. If $W^\perp$ is a complement of W, that is, $V = W \oplus W^\perp$, then $\mathrm{Proj}_W(\beta)$ exists for every $\beta \in V$. To see this, let $\beta \in V$. Write $\beta = \alpha + \delta$ with $\alpha \in W$, and $\delta \in W^\perp$. Then $\alpha - \beta = -\delta \in W^\perp$. Thus, $\alpha - \beta$ is orthogonal to W. Therefore, $\mathrm{Proj}_W(\beta) = \alpha$ by 1.20.

A careful examination of the coefficients of $\mathrm{Proj}_W(\beta)$ relative to some basis in W leads to the theory of Fourier coefficients. We need two preliminary results.

Lemma 1.29: Let $(V, \| \ \|)$ be a pre-Hilbert space. If $W_1, \ldots, W_n$ are pairwise orthogonal subspaces of V, then $W_1 + \cdots + W_n = W_1 \oplus \cdots \oplus W_n$. In addition, suppose $\mathrm{Proj}_{W_i}(\beta)$ exists for every $\beta \in V$ and every $i = 1, \ldots, n$. Set $W = W_1 + \cdots + W_n$. Then $\mathrm{Proj}_W(\beta)$ exists and is given by the following formula:

1.30: $$\mathrm{Proj}_W(\beta) = \sum_{i=1}^{n} \mathrm{Proj}_{W_i}(\beta)$$

Proof: In order to show $W_1 + \cdots + W_n = W_1 \oplus \cdots \oplus W_n$, we must argue that $W_i \cap \{\sum_{j \neq i} W_j\} = 0$ for every $i = 1, \ldots, n$. This will follow from the fact that the W_k are pairwise orthogonal. Fix i, and let $\gamma \in W_i \cap \{\sum_{j \neq i} W_j\}$. Then $\gamma = \gamma_i = \sum_{j \neq i} \gamma_j$. Here $\gamma_i \in W_i$, and $\gamma_j \in W_j$ for all $j \neq i$. Then $\langle \gamma, \gamma \rangle = \langle \gamma_i, \sum_{j \neq i} \gamma_j \rangle = \sum_{j \neq i} \langle \gamma_i, \gamma_j \rangle = 0$. Therefore, $\gamma = 0$.

To prove 1.30, let $\beta \in V$. For each $i = 1, \ldots, n$, set $\alpha_i = \text{Proj}_{W_i}(\beta)$. Then $(\alpha_i - \beta) \perp W_i$. Set $\alpha = \alpha_1 + \cdots + \alpha_n$. We claim $(\alpha - \beta) \perp W$. To see this, let $\gamma \in W$. Then $\gamma = \gamma_1 + \cdots + \gamma_n$, where $\gamma_i \in W_i$ for all $i = 1, \ldots, n$. So, $\langle \alpha - \beta, \gamma \rangle = \sum_{i=1}^n \langle \alpha - \beta, \gamma_i \rangle = \sum_{i=1}^n (\langle \alpha_i - \beta, \gamma_i \rangle + \sum_{j \neq i} \langle \alpha_j, \gamma_i \rangle) = \sum_{i=1}^n \langle \alpha_i - \beta, \gamma_i \rangle = 0$. Here $\langle \alpha_j, \gamma_i \rangle = 0$ whenever $j \neq i$ because the subspaces W_k are pairwise orthogonal. $\langle \alpha_i - \beta, \gamma_i \rangle = 0$ because $(\alpha_i - \beta)$ is orthogonal to W_i for all $i = 1, \ldots, n$. Thus, $(\alpha - \beta) \perp W$. This means that $\alpha = \text{Proj}_W(\beta)$. Therefore, $\sum_{i=1}^n \text{Proj}_{W_i}(\beta) = \alpha = \text{Proj}_W(\beta)$. This established formula 1.30 and completes the proof of the lemma. $\square$

Lemma 1.31: Let α be a nonzero vector in a pre-Hilbert space V. Then the orthogonal projection of V onto $\mathbb{R}\alpha$ is given by the following formula:

1.32: $\text{Proj}_{\mathbb{R}\alpha}(\beta) = (\langle \beta, \alpha \rangle / \langle \alpha, \alpha \rangle)\alpha$

Proof: The projection of β onto $\mathbb{R}\alpha$ is $x\alpha$ for some $x \in \mathbb{R}$. We also know that $\langle x\alpha - \beta, \alpha \rangle = 0$. Solving this equation for x gives the desired result. $\square$

The scalar $\langle \beta, \alpha \rangle / \langle \alpha, \alpha \rangle$ appearing in 1.32 is called the Fourier coefficient of β with respect to α. We can combine the last two lemmas in the following theorem:

Theorem 1.33: Let $\{\alpha_i \mid i = 1, \ldots, n\}$ be a collection of pairwise orthogonal, nonzero vectors in the pre-Hilbert space V. Set $W = L(\{\alpha_1, \ldots, \alpha_n\})$. Then the orthogonal projection of V onto W is given by the following formula:

1.34: $$\text{Proj}_W(\beta) = \sum_{i=1}^n (\langle \beta, \alpha_i \rangle / \langle \alpha_i, \alpha_i \rangle)\alpha_i$$

Proof: Since $W = \mathbb{R}\alpha_1 \oplus \cdots \oplus \mathbb{R}\alpha_n$, we can apply the last two lemmas and get the result. $\square$

Note then that if W has a basis $\{\alpha_1, \ldots, \alpha_n\}$ consisting of pairwise orthogonal vectors, then the vector in W closest to a given vector β is obtained by using the Fourier coefficients of β with respect to $\alpha_1, \ldots, \alpha_n$. The formula for $\text{Proj}_W(\beta)$ given in 1.34 makes it clear that orthogonal bases of W are very useful. Let us introduce the following definition:

Definition 1.35: A set of pairwise orthogonal vectors $\{\alpha_i \mid i \in I\}$ is said to orthonormal if $\|\alpha_i\| = 1$ for every $i \in I$.

A basis $\underline{\alpha} = \{\alpha_i \mid i \in I\}$ of some subspace W is said to be an orthonormal basis if the set $\underline{\alpha}$ is orthonormal. Thus, an orthonormal basis of W is a basis consisting of pairwise orthogonal vectors whose lengths are one. Any finite-dimensional subspace of V has an orthonormal basis. In fact, we have the following slightly stronger result, which is called the Gram–Schmidt theorem:

Theorem 1.36: Let $\{\alpha_i\}$ be a finite or infinite sequence of linearly independent vectors in the pre-Hilbert space V. Then there exists an orthonormal set $\{\varphi_i\}$ such that $L(\{\alpha_1, \ldots, \alpha_n\}) = L(\{\varphi_1, \ldots, \varphi_n\})$ for every n.

Proof: The finite sequence case is included in the infinite argument. Hence, we assume $\{\alpha_i\}$ is an infinite sequence of linearly independent vectors in V. We shall construct a sequence of pairwise orthogonal, nonzero vectors $\{\eta_i\}$ such that $L(\{\alpha_1, \ldots, \alpha_n\}) = L(\{\eta_1, \ldots, \eta_n\})$ for every $n \in \mathbb{N}$. Having done this, then $\{\varphi_i = \eta_i / \|\eta_i\|\}$ is the required orthonormal set in the theorem.

To construct the sequence $\{\eta_i\}$, we proceed as follows: Fix $n \in \mathbb{N}$. Define n vectors $\eta_1, \ldots, \eta_n$ in V inductively by the formulas below.

1.37:
$$\eta_1 = \alpha_1$$
$$\eta_2 = \alpha_2 - (\langle \alpha_2, \eta_1 \rangle / \langle \eta_1, \eta_1 \rangle)\eta_1$$
$$\vdots$$
$$\eta_n = \alpha_n - \sum_{i=1}^{n-1} (\langle \alpha_n, \eta_j \rangle / \langle \eta_j, \eta_j \rangle)\eta_j$$

It is obvious from the nature of the equations in 1.37 that $L(\{\alpha_1, \ldots, \alpha_i\}) = L(\{\eta_1, \ldots, \eta_i\})$ for all $i = 1, \ldots, n$. In particular, the vectors $\eta_1, \ldots, \eta_i$ are nonzero and linearly independent for all $i = 1, \ldots, n$. It is also obvious that each η_i is orthogonal to the η_k's defined before η_i. This implies that $\eta_1, \ldots, \eta_n$ are pairwise orthogonal.

Now in the last paragraph, n was a fixed natural number. Strictly speaking, the vectors $\eta_1, \ldots, \eta_n$ constructed in equation 1.37 depend on n and should be labeled more explicitly as $\eta_1^{(n)}, \ldots, \eta_n^{(n)}$. Thus, for each $n \in \mathbb{N}$, we have constructed a function $g_n: \{1, \ldots, n\} \to V$ given by $g_n(i) = \eta_i^{(n)}$ for $i = 1, \ldots, n$. The important point to note here is that when $n < m$, the vectors $\eta_1^{(n)}, \ldots, \eta_n^{(n)}$ are precisely the same as $\eta_1^{(m)}, \ldots, \eta_n^{(m)}$. Thus, the functions g_n and g_m agree on their common domain. Hence, it makes sense to define a function f: $\mathbb{N} \to V$ by $f(i) = g_n(i)$ for any $n \geqslant i$. Let the sequence $\{\eta_i\}$ be the function f. Thus, $\eta_i = f(i)$ for every $i \in \mathbb{N}$. Then the first n vectors of $\{\eta_i\}$ are precisely the vectors listed in equation 1.37. So, for every $n \in \mathbb{N}$, we have $L(\{\alpha_1, \ldots, \alpha_n\}) = L(\{\eta_1, \ldots, \eta_n\})$. This completes the proof of the theorem.   □

Of course, Theorem 1.36 implies that every finite-dimensional subspace W of a pre-Hilbert space V has an orthonormal basis $\underline{\alpha} = \{\alpha_1, \ldots, \alpha_n\}$. In terms of this basis, the orthogonal projection of V onto W is given by the following formula:

1.38:
$$\text{Proj}_W(\beta) = \sum_{i=1}^{n} (\langle \beta, \alpha_i \rangle)\alpha_i$$

We shall finish this section with a brief description of what analysts call a "basis" when dealing with Hilbert spaces. We need a formal name for the type of sequence constructed in Theorem 1.36.

Definition 1.39: A sequence $\{\varphi_i\}$ in a pre-Hilbert space V is called an *orthonormal sequence* if the set $\{\varphi_i \mid i \in \mathbb{N}\}$ is an orthonormal set.

Thus, a sequence $\{\varphi_i\}$ is orthonormal if $\|\varphi_i\| = 1$ for every $i \in \mathbb{N}$, and $\{\varphi_i \mid i \in \mathbb{N}\}$ is a set of pairwise orthogonal vectors in V. In particular, if $\{\varphi_i\}$ is an orthonormal sequence in V, then no φ_i is zero, and the set $\{\varphi_i \mid i \in \mathbb{N}\}$ is linearly independent over $\mathbb{R}$. Hence, V contains an orthonormal sequence if and only if $\dim_{\mathbb{R}}(V) = \infty$. There are two famous results about orthonormal sequences, which are used far more in analysis than in algebra.

Theorem 1.40: Let $\{\varphi_i\}$ be an orthonormal sequence in the pre-Hilbert space V. Let $\xi \in V$, and set $x_i = \langle \xi, \varphi_i \rangle$ for every $i \in \mathbb{N}$. Then

(a) (Bessel's Inequality): $\sum_{i=1}^{\infty} x_i^2 \leqslant \|\xi\|^2$.
(b) (Parseval's Equation): $\xi = \sum_{i=1}^{\infty} x_i \varphi_i$ if and only if $\sum_{i=1}^{\infty} x_i^2 = \|\xi\|^2$.

Proof: (a) For each $n \in \mathbb{N}$, set $\sigma_n = \sum_{i=1}^{n} x_i \varphi_i$. Then by equation 1.38 $(\xi - \sigma_n) \perp L(\{\varphi_1, \ldots, \varphi_n\})$. In particular, $\xi - \sigma_n$ and σ_n are orthogonal. Hence, the parallelogram law implies $\|\xi\|^2 = \|\xi - \sigma_n + \sigma_n\|^2 = \|\xi - \sigma_n\|^2 + \|\sigma_n\|^2 = \|\xi - \sigma_n\|^2 + \sum_{i=1}^{n} x_i^2$. We conclude that $\sum_{i=1}^{n} x_i^2 \leqslant \|\xi\|^2$ for every n. Thus, the infinite series $\sum_{i=1}^{\infty} x_i^2$ is absolutely convergent with sum no bigger than $\|\xi\|^2$. In symbols, $\sum_{i=1}^{\infty} x_i^2 \leqslant \|\xi\|^2$.

(b) For Parseval's equation, we first note that $\xi = \sum_{i=1}^{\infty} x_i \varphi_i$ means the sequence $\{\sigma_n\}$ converges to ξ. Now we had established in the proof of (a) that $\|\xi\|^2 = \|\xi - \sigma_n\|^2 + \sum_{i=1}^{n} x_i^2$. Thus, $\{\sigma_n\} \to \xi$ if and only if $\sum_{i=1}^{\infty} x_i^2 = \lim_{n \to \infty} \{\sum_{i=1}^{n} x_i^2\} = \|\xi\|^2$. $\square$

The infinite series $\sum_{i=1}^{\infty} \langle \xi, \varphi_i \rangle \varphi_i$ in Theorem 1.40 is called the Fourier series of ξ (relative to the orthonormal sequence $\{\varphi_i\}$). Parseval's equation allows us to test when the Fourier series of ξ converges to ξ.

Definition 1.41: Suppose $\{\varphi_i\}$ is an orthonormal sequence in the pre-Hilbert space V. The set $\{\varphi_i \mid i \in \mathbb{N}\}$ is called a "basis" of V if $\xi = \sum_{i=1}^{\infty} \langle \xi, \varphi_i \rangle \varphi_i$ for every vector $\xi \in V$.

Thus, if every vector ξ in V is equal to its Fourier series (relative to $\{\varphi_i\}$), then the set $\{\varphi_i \mid i \in \mathbb{N}\}$ is called a "basis" of V. The reader will note that the word "basis" is included in quotation marks here. This is because an orthonormal sequence that is a "basis" of V is not in general a vector space basis in the sense of Chapter I. Consider the following example:

Example 1.42: We return to the Hilbert space l^2 given in Example 1.5. For each $i \in \mathbb{N}$, let φ_i denote the sequence that is zero except in the ith position where it is one. Thus, $\varphi_i = (0, \ldots, 0, 1, 0, 0, \ldots)$. Clearly, the set $\{\varphi_i \mid i \in \mathbb{N}\}$ is linearly independent over $\mathbb{R}$. This set is not a basis of l^2 since, for instance, $\eta = \{1/n\}$ is not a finite linear combination of the vectors in $\{\varphi_i \mid i \in \mathbb{N}\}$.

The sequence $\{\varphi_i\}$ is clearly an orthonormal sequence in l^2. If $\xi = \{x_n\} \in l^2$, then the Fourier coefficient of ξ relative to φ_i is $\langle \xi, \varphi_i \rangle = x_i$. Thus, the Fourier series of ξ is $\sum_{i=1}^{\infty} x_i \varphi_i$. The nth partial sum, σ_n, of this series is clearly $(x_1, x_2, \ldots, x_n, 0, 0, \ldots)$. Thus, $\lim_{n \to \infty} \|\xi - \sigma_n\| = 0$. Therefore, $\xi = \sum_{i=1}^{\infty} x_i \varphi_i$. Since ξ is arbitrary, we conclude that $\{\varphi_i \mid i \in \mathbb{N}\}$ is a "basis" of l^2. $\square$

When dealing with infinite-dimensional pre-Hilbert spaces V, analysts usually use the word basis to mean an orthonormal sequence of vectors satisfying 1.41 (i.e., a "basis"). A vector space basis of V is usually called a *Hamel basis*. Algebraists, in general, do not use these words, and, so, the reader is advised to use some caution when dealing with this terminology. In this text, we have used the word basis consistently to mean a collection of linearly independent vectors whose linear span is all of V. We shall use quotation marks (i.e., "basis") around the word basis when referring to an orthonormal sequence satisfying 1.41. Remember, a "basis" of V (if V has a "basis") is not in general a vector space basis of V.

It is an easy matter to decide when an orthonormal sequence gives us a "basis" of V.

Theorem 1.43: Let $\{\varphi_i\}$ be an orthonormal sequence in the pre-Hilbert space V. Then $\{\varphi_i \mid i \in \mathbb{N}\}$ is a "basis" of V if and only if the closure of the subspace $L(\{\varphi_i \mid i \in \mathbb{N}\})$ is all of V.

Proof: Suppose $\{\varphi_i\}$ is a "basis" of V. Then $\xi = \sum_{i=1}^{\infty} \langle \xi, \varphi_i \rangle \varphi_i$ for every $\xi \in V$. Set $\sigma_n = \sum_{i=1}^{n} \langle \xi, \varphi_i \rangle \varphi_i$. Then $\{\sigma_n\} \to \xi$. Each vector σ_n is in $L(\{\varphi_i \mid i \in \mathbb{N}\})$. It now follows from Lemma 3.4 [Chapter IV] that ξ is in the closure of $L(\{\varphi_i \mid i \in \mathbb{N}\})$. Thus, $V = \overline{L(\{\varphi_i \mid i \in \mathbb{N}\})}$.

Conversely, suppose $\overline{L(\{\varphi_i \mid i \in \mathbb{N}\})} = V$. Let $\xi \in V$. Set $x_i = \langle \xi, \varphi_i \rangle$, and $\sigma_n = \sum_{i=1}^{n} x_i \varphi_i$. We want to show the sequence $\{\sigma_n\}$ converges to ξ. Let $r > 0$. Since $\xi \in \overline{L(\{\varphi_i \mid i \in \mathbb{N}\})}$, $B_r(\xi) \cap L(\{\varphi_i \mid i \in \mathbb{N}\}) \neq \varnothing$. Hence, there exists scalars $y_1, \ldots, y_m$ in $\mathbb{R}$ such that $\|\sum_{i=1}^{m} y_i \varphi_i - \xi\| < r$. Now by Equation 1.38, $\|\sum_{i=1}^{n} x_i \varphi_i - \xi\|$ is the distance between ξ and $L(\{\varphi_1, \ldots, \varphi_n\})$ for any $n \in \mathbb{N}$. In particular, for any $n \geq m$, $\|\sigma_n - \xi\| \leq \|\sum_{i=1}^{m} y_i \varphi_i - \xi\| < r$. This proves that $\{\sigma_n\} \to \xi$, and completes the proof of the theorem. $\square$

If V is a Hilbert space, then the criterion in Theorem 1.43 can be simplified significantly.

Corollary 1.44: Let $\{\varphi_i\}$ be an orthonormal sequence in the Hilbert space V. Then $\{\varphi_i \mid i \in \mathbb{N}\}$ is a "basis" of V if and only if $\{\varphi_i \mid i \in \mathbb{N}\}^{\perp} = (0)$.

Proof: If $\{\varphi_i\}$ is a "basis" for V, then $\{\varphi_i \mid i \in \mathbb{N}\}^{\perp} = L(\{\varphi_i \mid i \in \mathbb{N}\})^{\perp} = \overline{L(\{\varphi_i \mid i \in \mathbb{N}\})}^{\perp} = V^{\perp} = (0)$.

Conversely, suppose $\{\varphi_i \mid i \in \mathbb{N}\}^\perp = (0)$. Set $W = \overline{L(\{\varphi_i \mid i \in \mathbb{N}\})}$. Then W is a closed subspace of V for which $W^\perp = (0)$. Since V is complete, and W is closed, W is complete. (See Exercise 13, Section 4, Chapter IV.) By Theorem 1.21, $V = W \oplus W^\perp = W$. Hence, Theorem 1.43 implies $\{\varphi_i \mid i \in \mathbb{N}\}$ is a "basis" of V. $\square$

EXERCISES FOR SECTION 1

(1) Verify that the definition given in 1.1 is the same as that given in 7.12 of Chapter I.

(2) Show that l^2 is a Hilbert space.

(3) Show that the example given in 1.4 is a pre-Hilbert space but not a Hilbert space.

(4) Occasionally it is convenient to relax condition (d) in Definition 1.1. A function $f: V \times V \to \mathbb{R}$ is called a *semiscalar product* if f satisfies the following conditions:

(a) f is bilinear.

(b) $f(\alpha, \beta) = f(\beta, \alpha)$ for all $\alpha, \beta \in V$.

(c) $f(\alpha, \alpha) \geqslant 0$ for all $\alpha \in V$.

Give an example of a semiscalar product that is not an inner product on V. Show that Schwarz's inequality remains valid for any semiscalar product f.

(5) Let $V = \{f \in C^1([a, b]) \mid f'$ is continuous$\}$. Define $\langle f, g \rangle = f(a)g(a) + \int_a^b f'(t)g'(t)\, dt$. Show that $\langle\ ,\ \rangle$ is an inner product on V.

(6) Provide the technical details for the proof of Theorem 1.14.

(7) Let $V = \{f \in C([0, 1]) \mid f(0) = 0\}$. Let $\|\ \|_\infty$ be the norm defined in equation 1.6 of Chapter IV. Let $W = \{\{x_n\} \in l^2 \mid x_n = 0$ for all $n \gg 0\}$.

(a) Show that the normed linear vector space $(V, \|\ \|_\infty)$ is not a pre-Hilbert space.

(b) Let $\beta = \{1/n\}$. Show that $\beta \in l^2$, $d(\beta, W) = 0$ and $0 \neq \|\alpha - \beta\|$ for any $\alpha \in W$. Thus, W contains no vector closest to β.

(8) Give an example of a normed linear vector space $(V, \|\ \|)$, a subspace W, and a vector β such that W contains more than one vector α such that $\|\alpha - \beta\| = d(\beta, W)$. There is such an example in Chapter IV.

(9) If $|\langle \alpha, \beta \rangle| = \|\alpha\|\, \|\beta\|$ in Theorem 1.10, prove that α and β are linearly dependent.

(10) If $(V, \langle\ ,\ \rangle)$ is a finite-dimensional, inner product space, and W is a subspace of V, show directly that $V = W \oplus W^\perp$.

(11) In Exercise 10, show that $W^{\perp\perp} = W$. Give an example of an infinite-dimensional inner product space V and a subspace W such that $W^{\perp\perp} \neq W$.

(12) Let $(V, \langle \ , \ \rangle)$ be a finite-dimensional inner product space. Let $\{\alpha_1, \ldots, \alpha_n\}$ be a basis of V. Let $c_1, \ldots, c_n \in \mathbb{R}$. Show there exists a unique vector $\alpha \in V$ such that $\langle \alpha, \alpha_i \rangle = c_i$ for all $i = 1, \ldots, n$.

(13) Let $V = M_{n \times n}(\mathbb{R})$. If $A, B \in V$, set $\langle A, B \rangle = \mathrm{Tr}(AB^t)$.

    (a) Show that $(V, \langle \ , \ \rangle)$ is an inner product space.

    (b) Find the orthogonal complement of the subspace of all diagonal matrices in V.

(14) Let W be a finite-dimensional subspace of an inner product space V. Show that $\|\mathrm{Proj}_W(\beta)\| \leqslant \|\beta\|$ for all $\beta \in V$.

(15) In Exercise 14, suppose $T \in \mathscr{E}(V)$ is an idempotent map with $\mathrm{Im}(T) = W$. If $\|T(\beta)\| \leqslant \|\beta\|$ for all $\beta \in V$, prove $T = \mathrm{Proj}_W(\ )$.

(16) Prove the assertion in 1.23.

(17) Find the linear equation $y = mx + b$ that "best" fits the data $(x_1, y_1), \ldots, (x_n, y_n)$ by solving the normal equation. Here $x_1, \ldots, x_n$ are assumed all distinct.

(18) Find the polynomial $p(X) = b_m X^m + b_{m-1} X^{m-1} + \cdots + b_0$ that "best" fits the data in Exercise 17. Here we assume $n \gg m$.

(19) Give an example of a pre-Hilbert space in which a "basis" turns out to be a vector space basis of V also.

(20) Here is a calculus problem that can be solved using Schwarz's inequality: Suppose a positive term series $\sum_{n=1}^{\infty} a_n$ converges. Show that the series $\sum_{n=1}^{\infty} \sqrt{a_n}/n$ converges.

(21) Consider the vector space V_n given in Exercise 1, Section 2 of Chapter I. Here we assume $F = \mathbb{R}$. Define an inner product on V_n by setting $\langle f, g \rangle = \int_0^1 f(X)g(X) \, dX$. Apply the Gram–Schmidt process to the vectors 1, X, X^2, X^3, and X^4 to produce the first five Legendre polynomials.

(22) Let $V = C([-1, 1])$. Define an inner product on V by setting $\langle f, g \rangle = (2/\pi)\int_{-1}^{1} f(x)g(x)/(1 - x^2)^{1/2} \, dx$. Repeat Exercise 21 in this setting. The polynomials thus formed are the first five Chebyshev polynomials of the first kind.

2. SELF-ADJOINT TRANSFORMATIONS

As in Section 1, we suppose $(V, \langle \ , \ \rangle)$ is a real inner product space. The reader will recall from Section 6 of Chapter I that the dual V^* of V is the vector space $\mathrm{Hom}_{\mathbb{R}}(V, \mathbb{R})$. If $T: V \to W$ is a linear transformation, then the adjoint of T is the linear transformation $T^*: W^* \to V^*$ given by $T^*(f) = fT$.

    If V is an infinite-dimensional pre-Hilbert space, then V^* is too large to be of any interest. Recall that $\dim_{\mathbb{R}}(V) = \infty$ implies that $\dim_{\mathbb{R}}(V^*) > \dim_{\mathbb{R}}(V)$. We confine our attention to the bounded linear maps, $\mathscr{B}(V, \mathbb{R})$, in V*. Recall that $T \in \mathscr{B}(V, \mathbb{R})$ if and only if there exists a positive constant c such that $|T(\xi)| \leqslant c\|\xi\|$ for all $\zeta \in V$. If V is finite dimensional, then we had seen in Chapter IV that $\mathscr{B}(V, \mathbb{R}) = V*$. If V is any Hilbert space, finite or infinite dimensional, we shall see that $\mathscr{B}(V, \mathbb{R})$ is isomorphic to V in a natural way.

Any pre-Hilbert space $(V, \|\ \|)$ admits a linear transformation $\theta: V \to \mathscr{B}(V, \mathbb{R})$, which we formally define as follows:

Definition 2.1: Let $(V, \|\ \|)$ be a pre-Hilbert space with inner product $\langle\ ,\ \rangle$. Let $\theta: V \to V*$ denote the linear transformation defined by $\theta(\beta)(\alpha) = \langle \alpha, \beta \rangle$.

Thus, for any $\beta \in V$, $\theta(\beta)$ is the real valued function on V whose value at α is $\langle \alpha, \beta \rangle$. We can easily check that $\theta(\beta)$ is a linear transformation from V to $\mathbb{R}$. If $\alpha, \alpha' \in V$, and $x, y \in \mathbb{R}$, then we have $\theta(\beta)(x\alpha + y\alpha') = \langle x\alpha + y\alpha', \beta \rangle = x\langle \alpha, \beta \rangle + y\langle \alpha', \beta \rangle = x\theta(\beta)(\alpha) + y\theta(\beta)(\alpha')$. Thus, θ is a well-defined map from V to V*. We next note that θ is a linear transformation from V to V*. We have $\theta(x\beta + y\beta')(\alpha) = \langle \alpha, x\beta + y\beta' \rangle = x\langle \alpha, \beta \rangle + y\langle \alpha, \beta' \rangle = x\theta(\beta)(\alpha) + y\theta(\beta')(\alpha) = [x\theta(\beta) + y\theta(\beta')](\alpha)$. Since α is arbitrary, we conclude that $\theta(x\beta + y\beta') = x\theta(\beta) + y\theta(\beta')$. Hence, θ is a linear map from V to V*. Let us also note that θ is an injective linear transformation, For suppose, $\theta(\beta) = 0$. Then, in particular, $0 = \theta(\beta)(\beta) = \langle \beta, \beta \rangle$. Thus, $\beta = 0$.

The linear transformation θ gives an imbedding of V into V*. We claim that the image of θ actually lies in $\mathscr{B}(V, \mathbb{R})$. As usual for statements of this kind, we regard $\mathbb{R}$ as a normed linear vector space via the absolute value $|\ |$, and V as a normed linear vector space via the norm $\|\ \|$ associated with $\langle\ ,\ \rangle$. We claim $\theta(\beta) \in \mathscr{B}(V, \mathbb{R})$ for every $\beta \in V$. This follows from Schwarz's inequality. If $\alpha \in V$, then $|\theta(\beta)(\alpha)| = |\langle \alpha, \beta \rangle| \leqslant \|\beta\|\ \|\alpha\|$. Thus, $\theta(\beta)$ is a bounded linear operator on V. A bound for $\theta(\beta)$ is $\|\beta\|$. We have now shown that θ is an injective linear map from V to $\mathscr{B}(V, \mathbb{R})$.

Now recall that $\mathscr{B}(V, \mathbb{R})$ is a normed linear vector space relative to the uniform norm $\|T\| = \inf\{c \mid c$ is a bound of $T\}$. Here $T \in \mathscr{B}(V, \mathbb{R})$. Thus, in the last paragraph, we showed that $\|\theta(\beta)\| \leqslant \|\beta\|$ for all $\beta \in V$. On the other hand, if $\beta \neq 0$, then $\alpha = \beta/\|\beta\|$ has length one, and $|\theta(\beta)(\alpha)| = |\langle \beta/\|\beta\|, \beta \rangle| = \|\beta\|$. In particular, $\|\theta(\beta)\| \geqslant \|\beta\|$ by Lemma 1.24(b) of Chapter IV. We conclude that $\|\theta(\beta)\| = \|\beta\|$ for every $\beta \in V$. The reader will recall that a bounded linear map between normed linear vector spaces that preserves lengths is called an *isometry*. Hence, θ is an isometry of V into $\mathscr{B}(V, \mathbb{R})$. We have now proved the first part of the following theorem:

Theorem 2.2: Let $(V, \|\ \|)$ be a pre-Hilbert space with inner product $\langle\ ,\ \rangle$. The map θ given by $\theta(\beta)(\alpha) = \langle \alpha, \beta \rangle$ is an isometry of V into the Banach space $\mathscr{B}(V, \mathbb{R})$.

Proof: We have already established the fact that $\theta: V \to \mathscr{B}(V, \mathbb{R})$ is an isometry. It remains to show that $\mathscr{B}(V, \mathbb{R})$ is a Banach space with respect to the uniform norm. This is a special case of Exercise 6, Section 4 of Chapter IV. We sketch a brief proof of this special case here.

Suppose $\{T_n\}$ is a Cauchy sequence in $\mathscr{B}(V, \mathbb{R})$. We want to find a T in $\mathscr{B}(V, \mathbb{R})$ such that $\{T_n\} \to T$. Fix $\alpha \in V$. Then $|T_n(\alpha) - T_m(\alpha)| = |(T_n - T_m)(\alpha)| \leqslant \|T_n - T_m\| \|\alpha\|$. Since, $\{T_n\}$ is Cauchy, $\lim_{n,m \to \infty} \|T_n - T_m\| = 0$. We conclude that $\{T_n(\alpha)\}$ is a Cauchy sequence in $\mathbb{R}$. Since $\mathbb{R}$ is complete, the sequence $\{T_n(\alpha)\}$ converges in $\mathbb{R}$. We can thus define a function $T: V \to \mathbb{R}$ by $T(\alpha) = \lim\{T_n(\alpha)\}$. It is an easy matter to show that $T \in \mathscr{B}(V, \mathbb{R})$ and that $\{T_n\} \to T$. $\square$

If $\dim_{\mathbb{R}}(V) < \infty$, then $\mathscr{B}(V, \mathbb{R}) = V^*$, and the isometry θ is surjective by Theorem 3.33(b) of Chapter I. Thus, when V is finite dimensional over $\mathbb{R}$, $\theta: V \to \mathscr{B}(V, \mathbb{R}) = V^*$ is an isomorphism. If V is infinite dimensional over $\mathbb{R}$, then, in general, $\theta: V \to \mathscr{B}(V, \mathbb{R})$ is not surjective. However, we can tell precisely when θ is surjective.

Theorem 2.3: Let $(V, \| \|)$ be a pre-Hilbert space. The isometry $\theta: V \to \mathscr{B}(V, \mathbb{R})$ is surjective if and only if $(V, \| \|)$ is a Hilbert space.

Proof: Suppose θ is surjective. Then V is isometric via θ to the Banach space $\mathscr{B}(V, \mathbb{R})$. This implies that $(V, \| \|)$ is a Banach space. Since $(V, \| \|)$ is a pre-Hilbert space, we conclude that $(V, \| \|)$ is a Hilbert space.

Conversely, suppose $(V, \| \|)$ is a Hilbert space. Let $f \in \mathscr{B}(V, \mathbb{R}) - \{0\}$. Set $W = \ker(f)$. Since f is bounded, f is a continuous map from V to $\mathbb{R}$. $\{0\}$ is a closed subset of $\mathbb{R}$. Therefore, $W = f^{-1}(0)$ is a closed subspace of V. We have seen that a closed subspace of a complete space is itself complete (Exercise 13 of Section 4 of Chapter IV). Thus, W is a Banach space. It now follows from Theorem 1.21 that $V = W \oplus W^{\perp}$. Since $f \neq 0$, $\text{Im}(f) = \mathbb{R}$. Therefore, $W^{\perp} \cong V/W \cong \text{Im}(f) = \mathbb{R}$. We conclude that there exists a nonzero vector $\alpha \in W^{\perp}$ such that $V = W \oplus \mathbb{R}\alpha$.

Since α is not in W, $f(\alpha) \neq 0$. Set $\beta = (f(\alpha)/\|\alpha\|^2)\alpha$. Then $\mathbb{R}\alpha = \mathbb{R}\beta$ and consequently, $V = W \oplus \mathbb{R}\beta$. We claim $\theta(\beta) = f$. To see this, let $\xi \in V$. Write $\xi = \eta + x\beta$ for some $\eta \in W$, and $x \in \mathbb{R}$. Then $f(\xi) = f(\eta + x\beta) = f(\eta) + xf(\beta) = xf(\beta) = xf(\alpha)^2/\|\alpha\|^2$. On the other hand, $\theta(\beta)(\xi) = \langle \xi, \beta \rangle = \langle \eta + x\beta, \beta \rangle = \langle \eta, \beta \rangle + x \langle \beta, \beta \rangle = x \langle \beta, \beta \rangle = (xf(\alpha)^2/\|\alpha\|^4)\|\alpha\|^2 = xf(\alpha)^2/\|\alpha\|^2$. Thus, $\theta(\beta)(\xi) = f(\xi)$ for all $\xi \in V$. We conclude that $\theta(\beta) = f$, and the map θ is surjective. $\square$

Now suppose T is a bounded endomorphism of the pre-Hilbert space V. Thus, $T \in \mathscr{B}(V, V)$. We shall refer to T as an operator on V. Let T^* denote the adjoint of T. Thus, $T^* \in \text{Hom}_{\mathbb{R}}(V^*, V^*)$. Then we can consider the restriction of T^* to the subspace $\mathscr{B}(V, \mathbb{R})$. Suppose $f \in \mathscr{B}(V, \mathbb{R})$. Then $T^*(f) = fT$ is a bounded linear map by Theorem 1.27 of Chapter IV. In fact, $\|T^*(f)\| \leqslant \|T\| \|f\|$. Thus, T^*, when restricted to $\mathscr{B}(V, \mathbb{R})$, is a bounded linear operator from $\mathscr{B}(V, \mathbb{R})$ to

$\mathscr{B}(V, \mathbb{R})$. We have the following diagram:

2.4:

$$
\begin{array}{ccc}
V^* & \xleftarrow{\quad T^* \quad} & V^* \\
\Big\uparrow{\scriptstyle i} & & \Big\uparrow{\scriptstyle i} \\
\mathscr{B}(V, \mathbb{R}) & \xleftarrow{\quad T^*| \quad} & \mathscr{B}(V, \mathbb{R}) \\
\Big\uparrow{\scriptstyle \theta} & & \Big\uparrow{\scriptstyle \theta} \\
V & \xrightarrow{\quad T \quad} & V
\end{array}
$$

In diagram 2.4, the map $T^*|$ denotes the restriction of T^* to the subspace $\mathscr{B}(V, \mathbb{R})$. The map i is the inclusion of $\mathscr{B}(V, \mathbb{R})$ in V^*.

When $(V, \langle \ , \ \rangle)$ is a Hilbert space, the map θ in 2.4 is an isomorphism by Theorem 2.3. In this case, the composite map $S = \theta^{-1}(T^*|)\theta$ is a bounded linear operator from V to V. There is a simple relationship between S and T.

2.5: $\qquad\qquad \langle T\alpha, \beta \rangle = \langle \alpha, S\beta \rangle \qquad$ for all $\quad \alpha, \beta \in V$

In equation 2.5 and much of the sequel, we take the parentheses off the symbols $T(\alpha)$ and $S(\beta)$ to simplify the notation. To prove 2.5, we compare both sides. On the left, we have $\langle T\alpha, \beta \rangle = \theta(\beta)(T\alpha)$. On the right, we have $\langle \alpha, S\beta \rangle = \theta(S\beta)(\alpha) = [\theta(\theta^{-1}T^*\theta)(\beta)](\alpha) = [T^*\theta(\beta)](\alpha) = \theta(\beta)(T\alpha)$. Thus, $\langle T\alpha, \beta \rangle = \langle \alpha, S\beta \rangle$.

We also note that S is the only bounded linear operator on V that satisfies equation 2.5. For suppose $S' \in \mathscr{B}(V, V)$, and $\langle T\alpha, \beta \rangle = \langle \alpha, S'\beta \rangle$ for all $\alpha, \beta \in V$. Then $\langle \alpha, (S' - S)\beta \rangle = \langle \alpha, S'\beta \rangle - \langle \alpha, S\beta \rangle = \langle T\alpha, \beta \rangle - \langle T\alpha, \beta \rangle = 0$. In particular, $\|(S' - S)\beta\|^2 = \langle (S' - S)\beta, (S' - S)\beta \rangle = 0$ for all $\beta \in V$. We conclude that $S' = S$.

We have now proved the following theorem:

Theorem 2.6: Let $(V, \langle \ , \ \rangle)$ be a Hilbert space. For every bounded linear operator $T \in \mathscr{B}(V, V)$, there exists a unique bounded linear operator $S \in \mathscr{B}(V, V)$ such that $\langle T\alpha, \beta \rangle = \langle \alpha, S\beta \rangle$ for all $\alpha, \beta \in V$. $\quad\square$

In Section 1, we saw an instance where the terminology used in Hilbert space theory diverges from what is normally used in algebra. Here is a second and more important instance of the same phenomenon. In Hilbert space theory, the map S given in Theorem 2.6 is called the *adjoint* of T and written T^*. We shall follow this convention also. Thus, when $(V, \langle \ , \ \rangle)$ is a Hilbert space and $T \in \mathscr{B}(V, V)$, then the adjoint of T will be the unique, bounded linear operator $T^* \in \mathscr{B}(V, V)$ such that

2.7: $\qquad\qquad \langle T\alpha, \beta \rangle = \langle \alpha, T^*\beta \rangle \qquad$ for all $\quad \alpha, \beta \in V$

The reader is warned that we have changed our definition of the adjoint when dealing with bounded linear operators on Hilbert spaces. When we need to refer back to our old usage of the word "adjoint" (Section 6 of Chapter I), we shall use the words "algebra adjoint." Thus, if V is a Hilbert space and $T \in \mathscr{B}(V, V)$, then the "algebra adjoint" is the induced map on V^* given by $f \to fT$. The adjoint of T is the operator $T^* \in \mathscr{B}(V, V)$ that satisfies equation 2.7. When $\dim_{\mathbb{R}}(V) < \infty$, then $\mathscr{B}(V, \mathbb{R}) = V^*$ by Corollary 3.29 of Chapter IV. Thus, in this case, the adjoint of T only differs from the "algebra adjoint" of T by the isometry θ.

Now let $(V, \langle \ , \ \rangle)$ be an arbitrary Hilbert space. There are two types of bounded linear operators on V that we want to study in the remainder of this section.

Definition 2.8: Let $(V, \langle \ , \ \rangle)$ be a Hilbert space. Let $T \in \mathscr{B}(V, V)$.

(a) We say T is *self-adjoint* if $T = T^*$.

(b) We say T is *orthogonal* if $T^*T = I_V$.

We can give alternative definitions of self-adjoint and orthogonal operators by using equation 2.7. We have the following equivalent definition to 2.8:

2.8': (a) T is self-adjoint if and only if $\langle T\alpha, \beta \rangle = \langle \alpha, T\beta \rangle$ for all $\alpha, \beta \in V$.

(b) T is orthogonal if and only if $\langle T\alpha, T\beta \rangle = \langle \alpha, \beta \rangle$ for all $\alpha, \beta \in V$.

If T is self-adjoint, then $\langle T\alpha, \beta \rangle = \langle \alpha, T^*\beta \rangle = \langle \alpha, T\beta \rangle$. Conversely, if $\langle T\alpha, \beta \rangle = \langle \alpha, T\beta \rangle$, for all $\alpha, \beta \in V$, then $\langle \alpha, T\beta \rangle = \langle \alpha, T^*\beta \rangle$. In particular, $\|(T - T^*)\beta\|^2 = \langle (T - T^*)\beta, (T - T^*)\beta \rangle = 0$. Thus, $T = T^*$, and T is self-adjoint. If T is orthogonal, then $\langle T\alpha, T\beta \rangle = \langle \alpha, T^*T\beta \rangle = \langle \alpha, \beta \rangle$ for all $\alpha, \beta \in V$. Conversely, if $\langle T\alpha, T\beta \rangle = \langle \alpha, \beta \rangle$ for all $\alpha, \beta \in V$, then $\langle \alpha, T^*T\beta \rangle = \langle \alpha, \beta \rangle$. This implies that $T^*T\beta = \beta$ for all $\beta \in V$. Consequently, $T^*T = I_V$, and T is orthogonal. In either case, we have shown that 2.8 and 2.8' are equivalent definitions.

Note that Definition 2.8 is for bounded operators on V. Thus, an endomorphism T of V is said to be self-adjoint or orthogonal only if T is a bounded map on V satisfying 2.8.

If T is orthogonal, then T is left invertible. In particular, T is a monomorphism of V. If $\dim_{\mathbb{R}}(V) < \infty$, then T is orthogonal if and only if $T^* = T^{-1}$. In any case, $\langle T\alpha, T\beta \rangle = \langle \alpha, \beta \rangle$ implies $\|T(\alpha)\| = \|\alpha\|$ for all $\alpha \in V$. Thus, an orthogonal map is always an isometry of V into V. Let us consider examples in $(\mathbb{R}^n, \| \ \|)$.

Example 2.9: Suppose $V = \mathbb{R}^n$ with the standard inner product. Let $\varphi = \{\varphi_1, \ldots, \varphi_n\}$ be an orthonormal basis of $\mathbb{R}^n$. Then a linear transformation $T: \mathbb{R}^n \to \mathbb{R}^n$ is an orthogonal transformation if and only if $\{T(\varphi_1), \ldots, T(\varphi_n)\}$ is an orthonormal basis of $\mathbb{R}^n$. This remark is easy to prove. We leave it to the exercises.

In terms of matrices, we have T is orthogonal if and only if $\Gamma(\varphi, \varphi)(T) = A$ is an $n \times n$ matrix such that $A^t A = I_n$. Matrices with this property are said to be *orthogonal*.

If $\gamma = \{\gamma_1, \ldots, \gamma_n\}$ is a second orthonormal basis of $\mathbb{R}^n$, then there exists a $T \in Hom_{\mathbb{R}}(\mathbb{R}^n, \mathbb{R}^n)$ such that $T(\varphi_i) = \gamma_i$, $i = 1, \ldots, n$. T then is orthogonal. In particular, $\Gamma(\varphi, \varphi)(T) = M(\gamma, \varphi)$ is an orthogonal matrix. Thus, a change of basis matrix between orthonormal bases of $\mathbb{R}^n$ is orthogonal. $\square$

To produce examples of self-adjoint operators, we look at symmetric matrices. We have the following general result:

Lemma 2.10: Suppose $(V, \langle \ , \ \rangle)$ is a finite-dimensional Hilbert space. Let $\varphi = \{\varphi_1, \ldots, \varphi_n\}$ be an orthonormal basis for V. A linear transformation $T: V \to V$ is self-adjoint if and only if the matrix $\Gamma(\varphi, \varphi)(T)$ is symmetric.

Proof: Set $A = \Gamma(\varphi, \varphi)(T) = (a_{ij})$. Then $T(\varphi_j) = \sum_{k=1}^n a_{kj} \varphi_k$ for all $j = 1, \ldots n$. Suppose T is self-adjoint. Then for all i and j, we have $a_{ij} = \langle \varphi_i, T\varphi_j \rangle = \langle T\varphi_i, \varphi_j \rangle = a_{ji}$. Thus, $A^t = A$, and A is symmetric.

Conversely, suppose A is symmetric. Then $A^t = A$. For each i and j, we have $\langle \varphi_i, T\varphi_j \rangle = a_{ij} = a_{ji} = \langle T\varphi_i, \varphi_j \rangle$. Since $\langle \ , \ \rangle$ is bilinear, it now readily follows that $\langle T\alpha, \beta \rangle = \langle \alpha, T\beta \rangle$ for all $\alpha, \beta \in V$. Hence, T is self-adjoint. $\square$

Since the canonical basis $\underline{\delta} = \{\delta_1, \ldots, \delta_n\}$ of $\mathbb{R}^n$ is orthonormal, Lemma 2.10 implies that a linear transformation $T: \mathbb{R}^n \to \mathbb{R}^n$ is self-adjoint if and only if $A = \Gamma(\underline{\delta}, \underline{\delta})(T)$ is symmetric, that is, $A^t = A$.

Self-adjoint and orthogonal transformations are special cases of what we shall call *normal operators* in Section 4. In the rest of this section, we shall prove a special case of the spectral theorem for self-adjoint operators on a real Hilbert space. In Section 4, we shall use different techniques to give a proof of the spectral theorem for normal operators on a complex inner product space.

Let $(V, \langle \ , \ \rangle)$ be a Hilbert space. Suppose $T \in \mathcal{B}(V, V)$ is a self-adjoint operator. We say T is *nonnegative* if $\langle T\alpha, \alpha \rangle \geqslant 0$ for every $\alpha \in V$. We shall need the following lemma:

Lemma 2.11: Let V be a Hilbert space, and $T \in \mathcal{B}(V, V)$. Suppose T is self-adjoint and nonnegative. Then

(a) $\|T(\alpha)\| \leqslant \|T\|^{1/2} \langle T\alpha, \alpha \rangle^{1/2}$ for all $\alpha \in V$.
(b) $\langle T\alpha, \alpha \rangle = 0$ if and only if $T(\alpha) = 0$.
(c) $\{\langle T\alpha_n, \alpha_n \rangle\} \to 0$ if and only if $\{T(\alpha_n)\} \to 0$.

Proof: (a) We define a semiscalar product $[\ , \ ]$ on V by setting $[\alpha, \beta] = \langle T\alpha, \beta \rangle$. Clearly, $[\ , \ ]$ is a bilinear function on $V \times V$. Since T is self-adjoint, we have $[\beta, \alpha] = \langle T\beta, \alpha \rangle = \langle \beta, T^*\alpha \rangle =$

$\langle \beta, T\alpha \rangle = \langle T\alpha, \beta \rangle = [\alpha, \beta]$. Thus, $[\ ,\ ]$ is symmetric. Since T is nonnegative, $[\alpha, \alpha] = \langle T\alpha, \alpha \rangle \geqslant 0$. Thus, $[\ ,\ ]$ is a semiscalar product on V.

We had seen in Exercise 4 of Section 1 that any semiscalar product satisfies Schwarz's inequality. Therefore, we have

2.12: $|\langle T\alpha, \beta \rangle| = |[\alpha, \beta]| \leqslant [\alpha, \alpha]^{1/2}[\beta, \beta]^{1/2} = \langle T\alpha, \alpha \rangle^{1/2}\langle T\beta, \beta \rangle^{1/2}$

If we set $\beta = T\alpha$ in the inequality 2.12, then we have

2.13: $\qquad |\langle T\alpha, T\alpha \rangle| \leqslant \langle T\alpha, \alpha \rangle^{1/2}\langle T^2\alpha, T\alpha \rangle^{1/2} \qquad$ for all $\quad \alpha \in V$

Now by Schwarz's inequality again, we have $\langle T^2\alpha, T\alpha \rangle^{1/2} \leqslant \langle T^2\alpha, T^2\alpha \rangle^{1/4}\langle T\alpha, T\alpha \rangle^{1/4} = \|T^2\alpha\|^{1/2}\|T\alpha\|^{1/2} \leqslant (\|T\|\|T\alpha\|)^{1/2}\|T\alpha\|^{1/2} = \|T\|^{1/2}\|T\alpha\|$. Substituting this inequality into 2.13 gives us $\|T\alpha\|^2 \leqslant \langle T\alpha, \alpha \rangle^{1/2}\|T\|^{1/2}\|T\alpha\|$. We can assume $T(\alpha) \neq 0$. Consequently, $\|T\alpha\| \leqslant \langle T\alpha, \alpha \rangle^{1/2}\|T\|^{1/2}$. This completes the proof of (a). The assertions in (b) and (c) follow trivially from (a). $\quad\square$

We can now state and prove the spectral theorem for a self-adjoint operator on a finite-dimensional, real Hilbert space.

Theorem 2.14: Let $(V, \langle\ ,\ \rangle)$ be a finite-dimensional Hilbert space. Let T be a self-adjoint linear transformation on V. Then V has an orthonormal basis consisting entirely of eigenvectors of T.

Proof: The first order of business is to argue that T has an eigenvector. To this end, consider the following function $f(\xi) = \langle T\xi, \xi \rangle$. As ξ varies over V, f gives us a real valued function on V. Since T is continuous on V and $\langle\ ,\ \rangle$ is continuous on $V \times V$, we conclude that f is continuous on V.

Set $S = \{\xi \in V \mid \|\xi\| = 1\}$. We had seen in Chapter IV that S is a closed and bounded subset of V. Hence, S is sequentially compact by Corollary 3.28 of Chapter IV. Note that f is a bounded function on S. For if $\xi \in S$, then $|f(\xi)| = |\langle T\xi, \xi \rangle| \leqslant \|T\xi\|\|\xi\| \leqslant \|T\|\|\xi\|^2 = \|T\|$.

Set $m = \sup\{f(\xi) \mid \xi \in S\}$. It follows from Corollary 3.16 of Chapter IV that there exists a vector $\alpha \in S$ such that $f(\alpha) = m$. Now set $T_1 = mI_V - T$. The map T_1 is the difference of two self-adjoint operators, and, consequently, is self-adjoint. If $\xi \in S$, then $\langle T_1\xi, \xi \rangle = \langle m\xi - T\xi, \xi \rangle = m\|\xi\|^2 - \langle T\xi, \xi \rangle = m - f(\xi) \geqslant 0$. If $\xi \in V - \{0\}$, then $\xi/\|\xi\| \in S$. Therefore, $\langle T_1\xi, \xi \rangle/\|\xi\|^2 = \langle T_1(\xi/\|\xi\|), \xi/\|\xi\| \rangle \geqslant 0$. Thus, $\langle T_1\xi, \xi \rangle \geqslant 0$. We can now conclude that $\langle T_1\xi, \xi \rangle \geqslant 0$ for all $\xi \in V$. In particular, T_1 is a nonnegative, self-adjoint operator on V.

Now $\langle T_1\alpha, \alpha \rangle = \langle m\alpha - T\alpha, \alpha \rangle = m\|\alpha\|^2 - f(\alpha) = m - m = 0$. But then, Lemma 2.11 implies that $T_1(\alpha) = 0$. Thus, $T(\alpha) = m\alpha$, and we have found a unit eigenvector $\alpha \in V$ with eigenvalue m.

If $\dim_{\mathbb{R}}(V) = 1$, then $\{\alpha\}$ is an orthonormal basis of V, and we are done. If

$\dim_{\mathbb{R}}(V) > 1$, we proceed by induction on the dimension of V. We know from Theorem 1.21 that $V = \mathbb{R}\alpha \oplus (\mathbb{R}\alpha)^{\perp}$. Set $V_2 = (\mathbb{R}\alpha)^{\perp}$. If $\xi \in V_2$, then $\langle T\xi, \alpha \rangle = \langle \xi, T\alpha \rangle = \langle \xi, m\alpha \rangle = m\langle \xi, \alpha \rangle = 0$. Therefore, $T(\xi) \in (\mathbb{R}\alpha)^{\perp} = V_2$. Thus, V_2 is a T-invariant subspace of V. If we restrict T to V_2, we get a self-adjoint operator on the Hilbert space V_2. Since $\dim_{\mathbb{R}}(V_2) < \dim_{\mathbb{R}}(V)$, our induction hypothesis implies that V_2 has an orthonormal basis, say $\{\alpha_2, \ldots, \alpha_n\}$, consisting of eigenvectors of T. Then $\{\alpha, \alpha_2, \ldots, \alpha_n\}$ is an orthonormal basis of V which consists of eigenvectors of T. $\square$

Let us say a few words about the construction given in the proof of Theorem 2.14. To find an eigenvalue of T, form the function $f(\xi) = \langle T\xi, \xi \rangle$. The real number $m = \sup\{f(\xi) \,|\, \xi \in S\}$ is an eigenvalue for T. A vector $\alpha \in S$ such that $f(\alpha) = m$ is an eigenvector for T associated with m. We then pass to $V_2 = (\mathbb{R}\alpha)^{\perp}$. The proof shows that $m_2 = \sup\{f(\xi) \,|\, \xi \in S \cap V_2\}$ is an eigenvalue of the restriction of T to V_2. Of course, then m_2 is an eigenvalue of T. Note that $m \geqslant m_2$. Suppose $\dim_{\mathbb{R}}(V) = n$. Then n applications of this argument produce a decreasing sequence of eigenvalues $m \geqslant m_2 \geqslant \cdots \geqslant m_n$, and an associated sequence $\alpha, \alpha_2, \ldots, \alpha_n$ of eigenvectors such that $\underline{\alpha} = \{\alpha, \alpha_2, \ldots, \alpha_n\}$ is an orthonormal basis of V. In particular, $\Gamma(\underline{\alpha}, \underline{\alpha})(T) = \mathrm{diag}(m, m_2, \ldots, m_n)$.

Now in general, the m_i are not all distinct. Suppose $c_1, \ldots, c_r$ are the distinct real numbers in the set $\{m, m_2, \ldots, m_n\}$. We can label the c_i so that $c_1 > c_2 > \cdots > c_r$. Then $\{c_1, c_2, \ldots, c_r\} = \mathscr{S}_{\mathbb{R}}(T)$. Furthermore, if n_i equals the number of vectors in $\underline{\alpha}$ that are eigenvectors for c_i, then each c_i is repeated precisely n_i times on the diagonal of $\Gamma(\underline{\alpha}, \underline{\alpha})(T)$. In particular, the characteristic polynomial of T is $c_T(X) = \prod_{i=1}^{r}(X - c_i)^{n_i}$. The minimal polynomial of T is given by $m_T(X) = \prod_{i=1}^{r}(X - c_i)$ (see 4.22 of Chapter III).

Let us set $Y_i = \ker(T - c_i)$ for $i = 1, \ldots, r$. Then each Y_i is spanned by n_i of the vectors in $\underline{\alpha}$, and T restricted to Y_i is just $c_i I_{Y_i}$. The Y_i are clearly pairwise orthogonal subspaces of V. Hence, another version of Theorem 2.14, is as follows:

Corollary 2.15: Suppose T is a self-adjoint, linear transformation on a finite-dimensional Hilbert space V. Then there exists real numbers $c_1 > c_2 > \cdots > c_r$ and pairwise orthogonal subspaces $Y_1, \ldots, Y_r$ such that $V = Y_1 \oplus \cdots \oplus Y_r$, and the restriction of T to Y_i is $c_i I_{Y_i}$. $\square$

The reader can argue that the subspaces $Y_1, \ldots, Y_r$ of V in 2.15 are unique. We leave this point as an exercise at the end of this section. We have seen in Example 2.9 that a change of basis matrix between orthonormal bases of $\mathbb{R}^n$ is an orthogonal matrix. Hence, the matrix version of Theorem 2.14 is as follows:

Corollary 2.16: Let $A \in M_{n \times n}(\mathbb{R})$ be symmetric. Then there exists an orthogonal matrix P such that PAP^{-1} is diagonal. $\square$

Corollary 2.16 has many applications in applied mathematics. There are many problems in which we need to compute $A^r\alpha$, where $A \in M_{n \times n}(\mathbb{R})$,

$\alpha = (x_1, \ldots, x_n)^t$, and r is a large positive integer. If the matrix A is diagonalizable (e.g., if A is symmetric), then we can easily compute $A^r\alpha$. There exists a matrix P such that $PAP^{-1} = \text{diag}(m_1, \ldots, m_n)$. Then $A^r\alpha = (P^{-1}\text{diag}(m_1, \ldots, m_n)P)^r\alpha = P^{-1}\text{diag}(m_1^r, \ldots, m_n^r)P\alpha$. Thus, a potentially difficult computation becomes easy. Let us consider a specific example of these ideas.

Example 2.17: Consider the following system of differential equations:

2.18:
$$x_n' = a_{11}x_1 + \cdots + a_{1n}x_n$$
$$\vdots \qquad\qquad a_{ij} \in \mathbb{R}$$
$$x_n' = a_{n1}x_1 + \cdots + a_{nn}x_n$$

We assume here that the matrix $A = (a_{ij})$ of this system is symmetric. Then Corollary 2.16 implies that there exists an orthogonal matrix P such that $PAP^{-1} = D = \text{diag}(m_1, \ldots, m_n)$ for some $m_1, \ldots, m_n \in \mathbb{R}$. We had seen in Section 5 of Chapter III that any solution of 2.18 has the form $x = e^{tA}C$, where $C = (x_1(0), \ldots, x_n(0))^t$. Since $A = P^{-1}DP$, we have $e^{tA} = e^{t(P^{-1}DP)} = P^{-1}e^{tD}P = P^{-1}\text{diag}(e^{m_1t}, \ldots, e^{m_nt})P$.

The orthogonal matrix P is constructed by finding an orthonormal basis $\underline{\alpha}$ of $\mathbb{R}^n$ consisting of eigenvectors of A. The columns of P^{-1} are then the vectors in $\underline{\alpha}$. Such a basis $\underline{\alpha}$ exists by Theorem 2.14. Thus, if $\underline{\alpha} = \{\alpha_1, \ldots, \alpha_n\}$, then $P^{-1} = M(\underline{\alpha}, \underline{\delta})$, and $PAP^{-1} = \text{diag}(m_1, \ldots, m_n)$. A complete solution to 2.18 is given by the following equation:

2.19:
$$x = M(\underline{\alpha}, \underline{\delta})\, \text{diag}(e^{m_1t}, \ldots, e^{m_nt})M(\underline{\delta}, \underline{\alpha})C$$

For instance, consider the following 2×2 system:

2.20:
$$x_1' = -2x_1 + x_2$$
$$x_2' = x_1 - 2x_2$$

Then

$$A = \begin{pmatrix} -2 & 1 \\ 1 & -2 \end{pmatrix}$$

is symmetric. The characteristic polynomial of A is given by $c_A(X) = X^2 + 4X + 3$. Thus, the eigenvalues of A are -1 and -3. An orthonormal basis of $\mathbb{R}^2$ consisting of eigenvectors of A is easily seen to be $\underline{\alpha} = \{\alpha_1, \alpha_2\}$ where $\alpha_1 = (1/\sqrt{2}, 1/\sqrt{2})^t$, and $\alpha_2 = (1/\sqrt{2}, -1/\sqrt{2})^t$. Thus,

$$P^{-1} = M(\underline{\alpha}, \underline{\delta}) = \frac{\sqrt{2}}{2}\begin{pmatrix} 1 & 1 \\ 1 & -1 \end{pmatrix} \quad \text{and} \quad P = M(\underline{\delta}, \underline{\alpha}) = \frac{\sqrt{2}}{2}\begin{pmatrix} 1 & 1 \\ 1 & -1 \end{pmatrix}$$

Equation 2.19 then becomes

2.21:
$$x = \frac{1}{2}\begin{pmatrix} 1 & 1 \\ 1 & -1 \end{pmatrix} \text{diag}(e^{-t}, e^{-3t}) \begin{pmatrix} 1 & 1 \\ 1 & -1 \end{pmatrix} C \quad \square$$

The computations in Example 2.17 can be carried out for any matrix A that is similar to a diagonal matrix. The main point here is that symmetric matrices are always diagonalizable, and they are easy to recognize.

There is a third corollary to Theorem 2.14 that is worth mentioning here.

Corollary 2.22: Let $(V, \langle \ , \ \rangle)$ be a finite-dimensional Hilbert space. Suppose $T \in \text{Hom}_{\mathbb{R}}(V, V)$ is an isomorphism. Then $T = RS$, where R is orthogonal and S is a positive, self-adjoint operator.

Proof: A self-adjoint operator S on V is said to be positive if all the eigenvalues of S are positive. To prove the corollary, consider the adjoint T^* of T. Since $(T^*T)^* = T^*T^{**} = T^*T$, we see that T^*T is self-adjoint. By Theorem 2.14, V has an orthonormal basis $\underline{\alpha} = \{\alpha_1, \ldots, \alpha_n\}$ consisting of eigenvectors of T^*T. Suppose $T^*T(\alpha_i) = m_i\alpha_i$ for $i = 1, \ldots, n$. Since T is an isomorphism, $T(\alpha_i) \neq 0$. Therefore, $0 < \|T(\alpha_i)\|^2 = \langle T\alpha_i, \ T\alpha_i \rangle = \langle \alpha_i, T^*T\alpha_i \rangle = \langle \alpha_i, m_i\alpha_i \rangle = m_i\|\alpha_i\|^2 = m_i$. Thus, each m_i is a positive real number. Set $p_i = \sqrt{m_i}$.

We can define a linear transformation $S: V \to V$ by setting $S(\alpha_i) = p_i\alpha_i$ for all $i = 1, \ldots, n$. Then $S^2 = T^*T$. The reader can easily check that S is self-adjoint. Hence, S is a positive self-adjoint operator whose square is T^*T.

Set $P = ST^{-1}$. Since $\langle P\alpha, P\beta \rangle = \langle ST^{-1}\alpha, ST^{-1}\beta \rangle = \langle T^{-1}\alpha, S^2T^{-1}\beta \rangle = \langle T^{-1}\alpha, T^*TT^{-1}\beta \rangle = \langle T^{-1}\alpha, T^*\beta \rangle = \langle TT^{-1}\alpha, \beta \rangle = \langle \alpha, \beta \rangle$, P is orthogonal. In particular, $P^{-1} = P^*$ is also orthogonal. Set $R = P^{-1}$. Then $T = RS$ with R orthogonal, and S a positive, self-adjoint operator. $\square$

If we combine Corollaries 2.22 and 2.16, we get what is known as the UDV-decomposition of a nonsingular matrix.

Corollary 2.23: Let $A \in M_{n \times n}(\mathbb{R})$ be nonsingular. Then $A = UDV$, where D is diagonal and U and V are orthogonal matrices. $\square$

In the last part of this section, we shall discuss a generalization of Theorem 2.14. If $(V, \langle \ , \ \rangle)$ is an infinite-dimensional Hilbert space and T a self-adjoint operator on V, then T may not have enough eigenvectors to span V even in the "basis" sense discussed in Section 1. Thus, the infinite analog of Theorem 2.14 is not true in general. However, if T is a compact operator, then we can recover much of 2.14. The theorem we shall present is true for any pre-Hilbert space. Since we only defined a self-adjoint operator for Hilbert spaces, we need the following definition:

Definition 2.24: Let $(V, \langle \ , \ \rangle)$ be a pre-Hilbert space, and suppose $T \in \mathscr{B}(V, V)$. We say T is self-adjoint if $\langle T\alpha, \beta \rangle = \langle \alpha, T\beta \rangle$ for all $\alpha, \beta \in V$.

Obviously, our new definition agrees with the old one when V is a Hilbert space. We had argued this point in 2.8'(a). Note that Lemma 2.11 is still valid for any pre-Hilbert space V, and any nonnegative, self-adjoint operator T. The proof is precisely the same. The definition of a compact operator is as follows:

Definition 2.25: Let $(V, \| \ \|)$ be a pre-Hilbert space, and $T \in \mathscr{B}(V, V)$. Set $S = \{\xi \in V \mid \|\xi\| = 1\}$. We say T is a *compact operator* if the closure of T(S) in V is sequentially compact.

If $\dim_{\mathbb{R}}(V) < \infty$, then every linear transformation T on V is a compact operator. This follows from Corollary 3.29 of Chapter IV. From this corollary, we know T is bounded. Hence, for all $\xi \in S$, we have $\|T(\xi)\| \leqslant \|T\| \, \|\xi\| = \|T\|$. Let $\overline{T(S)}$ denote the closure of T(S) in V. If $\alpha \in \overline{T(S)}$, then there exists a sequence $\{\alpha_n\}$ contained in T(S) such that $\{\alpha_n\} \to \alpha$ (Lemma 3.4 of Chapter IV). We have just seen that $\|\alpha_n\| \leqslant \|T\|$ for all $n \in \mathbb{N}$. Since the norm is continuous, it follows that $\|\alpha\| \leqslant \|T\|$. In particular, $\overline{T(S)}$ is a bounded subset of V. It now follows from Corollary 3.28 of Chapter IV that $\overline{T(S)}$ is sequentially compact. Thus, T is a compact operator.

In general, suppose $T \in \mathscr{B}(V, V)$. Then the same reasoning as above shows that $\overline{T(S)}$ is a closed and bounded subset of V. However, when $\dim_{\mathbb{R}}(V) = \infty$, closed and bounded subsets of V are not necessarily sequentially compact. (Recall exercise 10, Section 3 of Chapter IV.) Hence, we cannot conclude that $\overline{T(S)}$ is sequentially compact. Those operators for which $\overline{T(S)}$ is sequentially compact are called *compact operators*. The generalization of Theorem 2.14 to possibly infinite-dimensional pre-Hilbert spaces is the following statement about compact operators:

Theorem 2.26: Let V be a pre-Hilbert space, and $T \in \mathscr{B}(V, V)$. Suppose T is a self-adjoint, compact operator. Let $W = \text{Im}(T)$.

If $\dim_{\mathbb{R}}(W) < \infty$, then W has an orthonormal basis consisting of eigenvectors of T.

If $\dim_{\mathbb{R}}(W) = \infty$, then there exists an orthonormal sequence $\{\varphi_i\}$ in W consisting entirely of eigenvectors of T. Furthermore, the set $\{\varphi_i \mid i \in \mathbb{N}\}$ is a "basis" of W. The corresponding sequence of eigenvalues $\{c_i\}$ associated with $\{\varphi_i\}$ converges to 0 in $\mathbb{R}$.

Before we give the proof of Theorem 2.26, let us discuss why this theorem is a generalization of Theorem 2.14. Suppose V is finite dimensional. Then T is just a self-adjoint, linear transformation from V to V. We need the following lemma:

Lemma 2.27: Let V be a finite-dimensional Hilbert space. Let T be a self-adjoint operator on V. Then V is the orthogonal direct sum of ker(T) and Im(T).

Proof: We first show $\ker(T) \cap \mathrm{Im}(T) = (0)$. Let $\alpha \in \ker(T) \cap \mathrm{Im}(T)$. Then $\alpha = T(\gamma)$ for some $\gamma \in V$. Since $\alpha \in \ker(T)$, we have $0 = T(\alpha) = T^2(\gamma)$. But then, $0 = \langle T^2\gamma, \gamma \rangle = \langle T\gamma, T\gamma \rangle = \|T(\gamma)\|^2$. We conclude that $\alpha = T(\gamma) = 0$.

To show $\ker(T) + \mathrm{Im}(T) = V$, we can use the first isomorphism theorem and count dimensions. We have $V/\ker(T) \cong \mathrm{Im}(T)$. Therefore, $\dim(\ker(T)) + \dim(\mathrm{Im}(T)) = \dim(V)$. Since $\ker(T) \cap \mathrm{Im}(T) = (0)$, the union of a basis from $\ker(T)$ with a basis from $\mathrm{Im}(T)$ is a set of linearly independent vectors in V. Since the dimensions add up right, the union of bases from $\ker(T)$ and $\mathrm{Im}(T)$ is in fact a basis of V. Therefore, $\ker(T) + \mathrm{Im}(T) = V$.

We have now shown that $V = \ker(T) \oplus \mathrm{Im}(T)$. It remains to show that these two subspaces are orthogonal. To see this, let $\alpha \in \ker(T)$, and $\beta \in \mathrm{Im}(T)$. Then $\beta = T(\gamma)$ for some $\gamma \in V$. We have $\langle \alpha, \beta \rangle = \langle \alpha, T\gamma \rangle = \langle T\alpha, \gamma \rangle = \langle 0, \gamma \rangle = 0$. Thus, $\ker(T)$ and $\mathrm{Im}(T)$ are orthogonal, and the proof of the lemma is complete. $\square$

We can now apply Lemma 2.27 to our discussion. The subspace $\ker(T)$ has an orthonormal basis $\underline{\alpha}$ by the Gram–Schmidt theorem. The vectors in $\underline{\alpha}$ are all eigenvectors of T with eigenvalue 0. If $\underline{\alpha}'$ is an orthonormal basis of $\mathrm{Im}(T)$, then Lemma 2.27 implies that the union $\underline{\alpha} \cup \underline{\alpha}'$ is an orthonormal basis of V. Hence, V has an orthonormal basis consisting of eigenvectors of T if and only if $\mathrm{Im}(T)$ has an orthonormal basis consisting of eigenvectors of T. In particular, Theorem 2.26 implies Theorem 2.14.

Proof of Theorem 2.26: Set $r_1 = \|T\| = \sup\{\|T(\xi)\| \,|\, \xi \in S\}$. If $r_1 = 0$, then $T = 0$, and the theorem is trivial. Hence, we assume $r_1 \neq 0$. Then there exists a sequence $\{\alpha_n\}$ in S such that $\{\|T(\alpha_n)\|\} \to r_1$. $\langle (r_1^2 - T^2)\alpha_n, \alpha_n \rangle = r_1^2 \langle \alpha_n, \alpha_n \rangle - \langle T^2\alpha_n, \alpha_n \rangle = r_1^2 \|\alpha_n\|^2 - \langle T\alpha_n, T\alpha_n \rangle = r_1^2 - \|T(\alpha_n)\|^2$. Since $\lim_{n \to \infty}(r_1^2 - \|T(\alpha_n)\|^2) = 0$, $\lim_{n \to \infty} \langle (r_1^2 - T^2)\alpha_n, \alpha_n \rangle = 0$. Now the reader can easily check that $r_1^2 - T^2$ is a nonnegative, self-adjoint operator on V. The proof is the same as that given in Theorem 2.14. Hence, Lemma 2.11(c) implies that $\{(r_1^2 - T^2)\alpha_n\} \to 0$.

The operator T is compact, and, hence, the sequence $\{T(\alpha_n)\}$ has a convergent subsequence. Replacing $\{T(\alpha_n)\}$ by its convergent subsequence, we can assume $\{T(\alpha_n)\} \to \beta$ for some $\beta \in V$. Since T is continuous, we have $\{T^2(\alpha_n)\} \to T(\beta)$. We also have $\{r_1^2 \alpha_n\} \to T(\beta)$ since $\{(r_1^2 - T^2)\alpha_n\} \to 0$. In particular, $\{\alpha_n\} \to T(\beta)/r_1^2$, and $\beta = \lim T(\alpha_n) = T^2(\beta)/r_1^2$. Since $\|\beta\| = \lim \|T(\alpha_n)\| = r_1 > 0$, we have found a nonzero vector β such that $T^2(\beta) = r_1^2 \beta$. Set $\alpha = \beta/\|\beta\|$.

We have found a vector $\alpha \in S$ such that $(r_1^2 - T^2)(\alpha) = 0$. Thus, $[(r_1 - T)(r_1 + T)](\alpha) = 0$. If $(r_1 + T)(\alpha) = 0$, then $T(\alpha) = -r_1\alpha$. Suppose $(r_1 + T)(\alpha) = \gamma \neq 0$. Then $(r_1 - T)(\gamma) = 0$. Thus, $T(\gamma) = r_1\gamma$. We could then divide γ by its length and produce a vector $\gamma' \in S$ such that $T(\gamma') = r_1\gamma'$. In either case, we have found a vector $\varphi_1 \in V$ such that $\|\varphi_1\| = 1$ and $T(\varphi_1) = c_1\varphi_1$, where $|c_1| = r_1$. Since $r_1 \neq 0$, $c_1 \neq 0$. In particular, $\varphi_1 = T(\varphi_1)/c_1 \in W$.

We now proceed as in the proof of Theorem 2.14. Set $V_2 = (\mathbb{R}\varphi_1)^{\perp}$. If $\alpha \in V_2$, then $\langle T\alpha, \varphi_1 \rangle = \langle \alpha, T\varphi_1 \rangle = \langle \alpha, c_1\varphi_1 \rangle = c_1 \langle \alpha, \varphi_1 \rangle = 0$. Thus, V_2 is a T-invariant subspace of V. Let $T|_{V_2}$ denote the restriction of T to V_2. Clearly, $T|_{V_2}$ is a self-adjoint operator on V_2. We claim $T|_{V_2}$ is compact as well. To ease notation, let us call $T|_{V_2}$ just T. We must argue that the closure of $T(S \cap V_2)$ is sequentially compact in V_2. Let us denote the closure of $T(S \cap V_2)$ in V_2 by Y. Then the vectors in Y are the limits in V_2 of all sequences from $T(S \cap V_2)$. Suppose $\{\alpha_n\}$ is a sequence in Y. Since $Y \subseteq \overline{T(S)}$, the sequence $\{\alpha_n\}$ has a subsequence which converges to a vector $\beta \in \overline{T(S)}$. We can replace $\{\alpha_n\}$ by its subsequence and assume $\{\alpha_n\} \to \beta$. Since each $\alpha_n \in V_2 = (\mathbb{R}\varphi_1)^\perp$, Corollary 1.11 implies $\beta \in V_2$. Since Y is closed in V_2, and $\{\alpha_n\}$ is a sequence in Y converging to β (in V_2), $\beta \in Y$. We have shown that any sequence in Y has a subsequence which converges to a vector in Y. Thus, Y is sequentially compact, and $T|_{V_2}$ is compact.

Set $r_2 = \|T|_{V_2}\|$. There are two cases to consider here. Either $r_2 = 0$, or $r_2 \neq 0$. Before proceeding with these cases, we need to make the following remark: $\ker(T) = W^\perp$. To see this, let $\alpha \in \ker(T)$, and $\beta \in W$. Then $\beta = T(\gamma)$ for some $\gamma \in V$. So, $\langle \alpha, \beta \rangle = \langle \alpha, T\gamma \rangle = \langle T\alpha, \gamma \rangle = \langle 0, \gamma \rangle = 0$. Thus, $\ker(T) \subseteq W^\perp$. For the other inclusion, let $\alpha \in W^\perp$. If $\beta \in V$, then $0 = \langle \alpha, T\beta \rangle = \langle T\alpha, \beta \rangle$. Since β here is arbitrary, we conclude that $T(\alpha) = 0$. Hence, $\alpha \in \ker(T)$, and $\ker(T) = W^\perp$.

Suppose $r_2 = \|T|_{V_2}\| = 0$. Then $T|_{V_2} = 0$, and thus, $V_2 \subseteq \ker(T)$. Since $\varphi_1 \in W$, $\ker(T) = W^\perp \subseteq (\mathbb{R}\varphi_1)^\perp = V_2$. Thus, $(\mathbb{R}\varphi_1)^\perp = \ker(T)$. We now apply Theorem 1.21 to the Banach space $\mathbb{R}\varphi_1$. We get $V = (\mathbb{R}\varphi_1) \oplus (\mathbb{R}\varphi_1)^\perp = (\mathbb{R}\varphi_1) \oplus \ker(T)$. In particular, $W = \text{Im}(T) = T(V) = T(\mathbb{R}\varphi_1) = \mathbb{R}\varphi_1$. Thus, $\dim(W) = 1$, and $\{\varphi_1\}$ is an orthonormal basis of W consisting entirely of eigenvectors of T. Hence, if $r_2 = 0$, the proof of 2.26 is complete.

Suppose $r_2 \neq 0$. Then we can repeat the argument given in the first three paragraphs of this proof for the map $T|_{V_2}$. We would then construct a vector $\varphi_2 \in V_2$ such that $\|\varphi_2\| = 1$ and $T(\varphi_2) = c_2\varphi_2$, where $|c_2| = r_2$. Then $\varphi_2 = T(\varphi_2)/c_2 \in W$, and $\{\varphi_1, \varphi_2\}$ is an orthonormal subset of W. Note also that $|c_1| \geqslant |c_2|$.

Suppose we repeat this argument $n - 1$ times obtaining an orthonormal sequence $\{\varphi_1, \ldots, \varphi_{n-1}\} \subseteq W$, a sequence $r_1 \geqslant r_2 \geqslant \cdots \geqslant r_{n-1}$ in $\mathbb{R}$, and subspaces $V_i = L(\{\varphi_1, \ldots, \varphi_{i-1}\})^\perp$ such that $r_i = \|T|_{V_i}\|$ and $T(\varphi_i) = c_i\varphi_i$, where $|c_i| = r_i$. Here $i = 1, \ldots, n-1$, and $V_1 = V$. We suppose $r_{n-1} > 0$. Set $V_n = L(\{\varphi_1, \ldots, \varphi_{n-1}\})^\perp$, and $r_n = \|T|_{V_n}\|$. Again if $r_n = 0$, then $V_n \subseteq \ker(T) = W^\perp \subseteq V_n$. Thus, applying Theorem 1.21, $V = V_n \oplus L(\{\varphi_1, \ldots, \varphi_{n-1}\})$, and $W = L(\{\varphi_1, \ldots, \varphi_{n-1}\})$. So, the argument is complete whenever our construction process produces an $r_n = 0$. In this case, $\dim(W) < \infty$. In any case, our construction produces orthonormal sequences $\{\varphi_1, \ldots, \varphi_n\}$ in W. Thus, if $\dim(W) < \infty$, we produce an orthonormal basis of W consisting of eigenvectors of T in a finite number of steps. Therefore, the proof is now complete when $\dim(W) < \infty$.

Suppose $\dim(W) = \infty$. Then no r_n can be zero, and our construction process continues ad infinitum. Inductively, we obtain an orthonormal sequence $\{\varphi_i\} \subseteq W$, and a sequence $\{c_i\} \subseteq \mathbb{R}$ such that $T(\varphi_i) = c_i\varphi_i$, $|c_i| = \|T|_{V_i}\|$, and $|c_1| \geqslant |c_2| \geqslant \cdots$ for all $i \in \mathbb{N}$. Here $V_i = L(\{\varphi_1, \ldots, \varphi_{i-1}\})^\perp$ with $V_1 = V$. We also know that $c_i \neq 0$ for any i. We claim that $\lim_{i \to \infty} c_i = 0$, and the set $\{\varphi_i | i \in \mathbb{N}\}$ is a "basis" of W.

Suppose $\{c_i\}$ does not converge to 0. Then there exists a positive number b such that $|c_i| \geqslant b$ for all $i \in \mathbb{N}$. We have $\|T(\varphi_i) - T(\varphi_j)\|^2 = \|c_i\varphi_i - c_j\varphi_j\|^2 = \|c_i\varphi_i\|^2 + \|c_j\varphi_j\|^2 = |c_i|^2 + |c_j|^2 \geqslant 2b^2$ whenever $i \neq j$. This inequality implies the sequence $\{T(\varphi_i)\}$ has no convergent subsequence. Since $\overline{T(S)}$ is sequentially compact, this is impossible. Hence, $\{c_i\} \to 0$.

It remains to show that $\{\varphi_i \mid i \in \mathbb{N}\}$ is a "basis" of W. Recall that this means $\beta = \sum_{i=1}^{\infty} \langle \beta, \varphi_i \rangle \varphi_i$ for every $\beta \in W$. So, let $\beta \in W$. Then $\beta = T(\alpha)$ for some $\alpha \in V$. Set $b_i = \langle \beta, \varphi_i \rangle$, and $a_i = \langle \alpha, \varphi_i \rangle$ for every i. Then $b_i = \langle \beta, \varphi_i \rangle = \langle T\alpha, \varphi_i \rangle = \langle \alpha, T\varphi_i \rangle = \langle \alpha, c_i\varphi_i \rangle = c_i a_i$. In particular, $T(a_i\varphi_i) = a_i T(\varphi_i) = a_i c_i \varphi_i = b_i \varphi_i$. Thus, $T(\alpha - \sum_{i=1}^{n} a_i\varphi_i) = \beta - \sum_{i=1}^{n} b_i\varphi_i$. Since the a_i are the Fourier coefficients of α (with respect to $\{\varphi_i\}$), $\alpha - \sum_{i=1}^{n} a_i\varphi_i$ is a vector in V_{n+1} by Theorem 1.33. The norm of T on V_{n+1} is $|c_{n+1}|$. Thus, we have $\|T(\alpha - \sum_{i=1}^{n} a_i\varphi_i)\| \leqslant |c_{n+1}| \|\alpha - \sum_{i=1}^{n} a_i\varphi_i\|$. Since $\alpha - \sum_{i=1}^{n} a_i\varphi_i$ is orthogonal to $\sum_{i=1}^{n} a_i\varphi_i$, we have $\|\alpha\|^2 = \|\alpha - \sum_{i=1}^{n} a_i\varphi_i\|^2 + \|\sum_{i=1}^{n} a_i\varphi_i\|^2$. In particular, $\|\alpha - \sum_{i=1}^{n} a_i\varphi_i\| \leqslant \|\alpha\|$. Putting this all together, we have $\|\beta - \sum_{i=1}^{n} b_i\varphi_i\| = \|T(\alpha - \sum_{i=1}^{n} a_i\varphi_i)\| \leqslant |c_{n+1}| \|\alpha\|$. Since $\{c_i\} \to 0$, we conclude that $\{\sum_{i=1}^{n} b_i\varphi_i\} \to \beta$. Thus, $\beta = \sum_{i=1}^{\infty} \langle \beta, \varphi_i \rangle \varphi_i$. This completes the proof of Theorem 2.26. $\square$

EXERCISES FOR SECTION 2

(1) Let $T \in \text{Hom}_{\mathbb{R}}(\mathbb{R}^n, \mathbb{R}^n)$. Show that T is orthogonal if and only if $T(\varphi)$ is an orthonormal basis of $\mathbb{R}^n$. Here the inner product is assumed to be the standard inner product on $\mathbb{R}^n$. φ is an orthonormal basis of $\mathbb{R}^n$.

(2) If T is a self-adjoint operator on a Hilbert space V and $p(X) \in \mathbb{R}[X]$, show $p(T)$ is self-adjoint.

(3) Show that the subspaces $Y_1, \dots, Y_r$ in Corollary 2.15 are unique up to permutations.

(4) Prove Corollary 2.16.

(5) Let

$$A = \begin{bmatrix} 4 & -1 & 1 \\ -1 & 4 & -1 \\ 1 & -1 & 4 \end{bmatrix}$$

Find an orthogonal matrix P such that PAP^{-1} is diagonal.

(6) Prove Corollary 2.23.

(7) Write a specific formula for the UDV factorization of a nonsingular matrix A. The formula should involve A and the eigenvectors of $A^t A$.

(8) Show that any nilpotent, self-adjoint operator is zero.

(9) Let V be a Hilbert space, and let f: $V \times V \to \mathbb{R}$ be a bounded bilinear form. Thus, there exists a positive constant b such that $|f(\alpha, \beta)| \leqslant b\|\alpha\| \|\beta\|$ for all $\alpha, \beta \in V$. Show that there exists a unique $T \in \mathscr{B}(V, V)$ such that $f(\alpha, \beta) = \langle \alpha, T\beta \rangle$. Show T is self-adjoint if and only if f is symmetric.

(10) Show that the factorization $T = RS$ given in Corollary 2.22 is unique.

(11) If $A \in M_{n \times n}(\mathbb{R})$ is orthogonal, show the rows of A are an orthonormal basis of $\mathbb{R}^n$.

(12) In 2.25, show T is compact if and only if for every bounded sequence $\{\alpha_n\} \subseteq V$, $\{T(\alpha_n)\}$ has a convergent subsequence.

(13) Consider the inner product space in Exercise 13 of Section 1. Let $C \in M_{n \times n}(\mathbb{R})$. Compute the adjoint of the linear map $T(A) = CA$.

(14) Suppose C is nonsingular in Exercise 13. Compute the adjoint of the linear transformation $S(A) = C^{-1}AC$.

(15) Let $V = \mathbb{R}[X]$. Define an inner product $\langle \ , \ \rangle$ on V by setting $\langle f, g \rangle = \int_0^1 f(t)g(t) \, dt$. Let f be a fixed polynomial in $\mathbb{R}[X]$. Define a linear transformation T on V by setting $T(g) = fg$. Show that T has an adjoint, that is, there exists a map $S \in \mathscr{E}(V)$ such that $\langle Tg, h \rangle = \langle g, Sh \rangle$ for all $g, h \in V$.

(16) Let $(V, \langle \ , \ \rangle)$ be the inner product space in Exercise 15. Let D: $V \to V$ be differentiation. Show that D has no adjoint. Note that V is not a Hilbert space.

(17) If V is a finite dimensional Hilbert space and $T \in \mathscr{E}(V)$ is an isomorphism, prove that T^* is an isomorphism, and $(T^{-1})^* = (T^*)^{-1}$. Is this true if $\dim V = \infty$?

(18) Suppose T and S are self-adjoint operators on a Hilbert space V. Show that ST is self-adjoint if and only if $ST = TS$.

(19) Let V be a Hilbert space. An operator $T \in \mathscr{E}(V)$ is said to be positive if T is self-adjoint and $\langle T\alpha, \alpha \rangle > 0$ for all $\alpha \in V - \{0\}$. If S and T are two positive operators on V, show that $S + T$ is positive, but ST need not be positive.

(20) Let $(V, \langle \ , \ \rangle)$ be a finite-dimensional inner product space. Let $T \in \mathscr{E}(V)$. Prove that T is positive if and only if $T = S^*S$ for some isomorphism $S \in \mathscr{E}(V)$.

(21) In Exercise 20, suppose $\underline{\alpha}$ is an orthonormal basis of V. Let T be a positive operator on V. Show that every entry on the main diagonal of $\Gamma(\underline{\alpha}, \underline{\alpha})(T)$ is positive.

(22) A variant of Corollary 2.23 is as follows: Let $A \in M_{m \times n}(\mathbb{R})$ be nonzero. Show there exists orthogonal matrices V and U such that

$$V^t A U = \begin{bmatrix} D & 0 \\ 0 & 0 \end{bmatrix}$$

where D is a diagonal matrix of the form $D = \text{diag}(x_1, \ldots, x_r)$ with $x_1 \geqslant x_2 \geqslant \cdots \geqslant x_r > 0$. [*Hint*: Apply Theorem 2.14 to the matrix $A^t A$]. This factorization is called the *singular value decomposition* of A.

(23) Let $A \in M_{m \times n}(\mathbb{R})$ be nonzero, and suppose A has singular value decomposition as in Exercise 22. Let $\beta \in M_{m \times 1}(\mathbb{R})$. Show that the vector

$$X = U \begin{bmatrix} D^{-1} & 0 \\ 0 & 0 \end{bmatrix} V^t \beta$$

is a least-squares solution to the equation $AX = \beta$. Thus, $\| \beta - AX \|$ is as small as possible. (Notice that the sizes of the zero matrices have been changed here.)

3. COMPLEX INNER PRODUCT SPACES

In this section, we want to extend our discussion of inner product spaces to include vector spaces over the complex numbers $\mathbb{C}$. Suppose V is a vector space over $\mathbb{C}$. We want a complex inner product on V to be a function $\langle \ , \ \rangle : V \times V \to \mathbb{C}$ that satisfies conditions similar to those in Definition 1.1. It is obvious that some changes in the definition are going to have to be made. $\mathbb{C}$, for instance, is not an ordered field. (There is no order relation $\leqslant$ on $\mathbb{C}$ that behaves nicely with respect to addition and multiplication.) Therefore, Definition 1.1(d) makes no sense unless we demand that $\langle \alpha, \alpha \rangle \in \mathbb{R}$ for every $\alpha \in V$. Whatever definition we decide on, we should like $\| \alpha \| = \langle \alpha, \alpha \rangle^{1/2}$ to behave like a norm on V.

To motivate the definition we shall use, consider the complex vector space $\mathbb{C}^n$. The analog of the standard inner product (on $\mathbb{R}^n$) for $\mathbb{C}^n$ is the bilinear map $\langle \alpha, \beta \rangle = \sum_{k=1}^n z_k w_k$. Here $\alpha = (z_1, \ldots, z_n)$, $\beta = (w_1, \ldots, w_n)$, and $z_k, w_k \in \mathbb{C}$ for all $k = 1, \ldots, n$. This bilinear form does not work well as a candidate for an inner product on $\mathbb{C}^n$. If $\alpha = (1, i, 0, \ldots, 0)$, for example, then $\alpha \neq 0$, but $\langle \alpha, \alpha \rangle = 0$. We can fix this problem by defining $\langle \alpha, \beta \rangle = \sum_{k=1}^n z_k \bar{w}_k$. Here $\bar{w}_k$ denotes the complex conjugate of w_k. Now if $\alpha = (z_1, \ldots, z_n) \in \mathbb{C}^n$, then $\langle \alpha, \alpha \rangle = \sum_{k=1}^n z_k \bar{z}_k = \sum_{k=1}^n |z_k|^2$. Here the notation $|z|$ indicates the modulus of the complex number z. The reader will recall that the modulus of a complex number $z = a + bi$ [$a, b \in \mathbb{R}$] is defined to be the positive square root of $a^2 + b^2$. Thus, $|z| = (a^2 + b^2)^{1/2}$. The modulus is a function from $\mathbb{C}$ to the nonnegative real numbers. It agrees with the ordinary absolute value on $\mathbb{R}$. Hence, we have chosen the same notation $| \ |$ for the absolute value on $\mathbb{R}$ and the modulus on $\mathbb{C}$. The reader can easily check that the modulus $| \ | : \mathbb{C} \to \mathbb{R}$ is a norm on the real vector space $\mathbb{C}$. We also have $|zz'| = |z| |z'|$, and $z\bar{z} = |z|^2$ for all $z \in \mathbb{C}$. In particular, $\langle \alpha, \alpha \rangle$ is a nonnegative real number for every $\alpha \in \mathbb{C}^n$. Also, it is clear that $\langle \alpha, \alpha \rangle = 0$ if and only if $\alpha = 0$.

We give up something here with this new definition of $\langle \ , \ \rangle$. The function

$\langle \alpha, \beta \rangle = \sum_{k=1}^{n} z_k \bar{w}_k$ is not a symmetric, bilinear form on $\mathbb{C}^n \times \mathbb{C}^n$. Instead, $\langle \, , \, \rangle$ satisfies the following conditions:

3.1: (a) $\langle z\alpha + z'\beta, \gamma \rangle = z\langle \alpha, \gamma \rangle + z'\langle \beta, \gamma \rangle$.

(b) $\langle \gamma, z\alpha + z'\beta \rangle = \bar{z}\langle \gamma, \alpha \rangle + \bar{z}'\langle \gamma, \beta \rangle$.

(c) $\langle \alpha, \beta \rangle = \overline{\langle \beta, \alpha \rangle}$.

(d) $\langle \alpha, \alpha \rangle$ is a positive real number for all $\alpha \in \mathbb{C}^n - \{0\}$.

The reader can easily verify that these equations hold for all $\alpha, \beta,$ and γ in $\mathbb{C}^n$, and all z, z' in $\mathbb{C}$. This function $\langle \, , \, \rangle$ has the desired length properties and furthermore reduces to the standard inner product on $\mathbb{R}^n$ when restricted to $\mathbb{R}^n \times \mathbb{R}^n$. Finally, we note that the conditions listed in 3.1 make sense for any complex vector space V, and any function f: $V \times V \to \mathbb{C}$. Hence, we adopt these conditions as our definition of a complex inner product.

Definition 3.2: Let V be a vector space over $\mathbb{C}$. By a complex inner product on V, we shall mean a complex valued function $\langle \, , \, \rangle \colon V \times V \to \mathbb{C}$ that satisfies the following conditions:

(a) $\langle z\alpha + z'\beta, \gamma \rangle = z\langle \alpha, \gamma \rangle + z'\langle \beta, \gamma \rangle$.

(b) $\langle \alpha, \beta \rangle = \overline{\langle \beta, \alpha \rangle}$.

(c) $\langle \alpha, \alpha \rangle$ is a positive, real number for every $\alpha \in V - \{0\}$.

These conditions are to hold for all $\alpha, \beta, \gamma \in V$, and all $z, z' \in \mathbb{C}$.

A complex vector space V together with an inner product $\langle \, , \, \rangle$ on V will be called a complex inner product space and written $(V, \langle \, , \, \rangle)$. As with real inner product spaces, a given complex vector space may admit several complex inner products. Thus, a complex inner product space is an ordered pair consisting of a vector space V over $\mathbb{C}$ and a complex valued function $\langle \, , \, \rangle \colon V \times V \to \mathbb{C}$ that satisfies the conditions in 3.2.

Suppose $(V, \langle \, , \, \rangle)$ is a complex inner product space. The reader has undoubtedly noted that 3.1(b) follows from 3.1(a) and 3.1(c). The same remark can be made in any V. We have $\langle \gamma, z\alpha + z'\beta \rangle = \bar{z}\langle \gamma, \alpha \rangle + \bar{z}'\langle \gamma, \beta \rangle$ for all $\alpha, \beta, \gamma \in V$, and all $z, z' \in \mathbb{C}$. This readily follows from 3.2(a) and 3.2(b). A function f: $V \to \mathbb{C}$ is said to be *conjugate linear* if $f(z\alpha + z'\beta) = \bar{z}f(\alpha) + \bar{z}'f(\beta)$ for all $\alpha, \beta \in V$ and $z, z' \in \mathbb{C}$. Thus, for any $\beta \in V$, the function $\langle \beta, \, \rangle$ is conjugate linear on V, and the function $\langle \, , \beta \rangle$ is linear. Note also that $\langle 0, \beta \rangle = \langle \beta, 0 \rangle = 0$ for all $\beta \in V$. Thus, for every $\alpha \in V$, we have $\langle \alpha, \alpha \rangle \geqslant 0$ with equality only if $\alpha = 0$.

The first four examples in Section 1 all have complex analogs. We begin with the example that motivates the definition:

Example 3.3: Set $V = \mathbb{C}^n$. Define $\langle \alpha, \beta \rangle = \sum_{k=1}^{n} z_k \bar{w}_k$, where $\alpha = (z_1, \ldots, z_n)$ and $\beta = (w_1, \ldots, w_n)$. Then $(\mathbb{C}^n, \langle \, , \, \rangle)$ is a complex inner product space. We shall refer to this particular inner product on $\mathbb{C}^n$ as the *standard inner product*. $\square$

Notice that when $n = 1$ in Example 3.3, $\langle z, z \rangle = z\bar{z} = |z|^2$. Hence, the modulus function $|\,|: \mathbb{C} \to \mathbb{R}$ is the norm given by the standard inner product on $\mathbb{C}$. We had mentioned that a given V might support more than one inner product. Here is a second inner product on $\mathbb{C}^2$.

Example 3.4: Let $V = \mathbb{C}^2$. Define $\langle\ ,\ \rangle'$ by the following formula: $\langle(z_1, z_2), (w_1, w_2)\rangle' = 2z_1\bar{w}_1 + z_1\bar{w}_2 + z_2\bar{w}_1 + z_2\bar{w}_2$. The reader can easily verify that $\langle\ ,\ \rangle'$ satisfies conditions (a)–(c) in Definition 3.2. Hence, $(\mathbb{C}^2, \langle\ ,\ \rangle')$ is a complex inner product space. □

Example 3.5: Let V be the set of all continuous, complexed valued functions on the closed interval $[a, b] \subseteq \mathbb{R}$. Clearly, V is a complex vector space via pointwise addition and scalar multiplication. V becomes a complex inner product space when we define $\langle f, g \rangle = \int_a^b f(t)\overline{g(t)}\, dt$. □

Example 3.6: Let $V = \bigoplus_{k=1}^{\infty} \mathbb{C}$. V becomes an infinite-dimensional, complex inner product space via $\langle \alpha, \beta \rangle = \sum_{k=1}^{\infty} z_k\bar{w}_k$. Here $\alpha = (z_1, z_2, \dots)$ and $\beta = (w_1, w_2, \dots)$. Since α and β have at most finitely many nonzero components, the formula for $\langle \alpha, \beta \rangle$ makes perfectly good sense. □

A more general example is the complex analog of l^2.

Example 3.7: Let $V = \{\{z_k\} \in \mathbb{C}^{\mathbb{N}} \mid \sum_{k=1}^{\infty} |z_k|^2 < \infty\}$. V is a complex vector space via componentwise addition and scalar multiplication. We can define an inner product on V by setting $\langle \alpha, \beta \rangle = \sum_{k=1}^{\infty} z_k\bar{w}_k$. Here $\alpha = \{z_k\}$, and $\beta = \{w_k\}$.

This space is the complex analog of l^2 in Example 1.5 of Section 1. In the literature, it is also called l^2. □

The reader will note that several of these examples are the same as the corresponding real inner product spaces. We have simply enlarged the base field from $\mathbb{R}$ to $\mathbb{C}$. There is a standard way to produce complex inner product spaces from real ones. Namely, pass to the complexification.

Let $(W, \langle\ ,\ \rangle)$ be a real inner product space. Consider the complexification $W^{\mathbb{C}} = W \otimes_{\mathbb{R}} \mathbb{C}$ of W (see Section 2 of Chapter II). Recall that if we identify a vector $\alpha \in W$ with its image $\alpha \otimes 1$ in $W^{\mathbb{C}}$, then $W^{\mathbb{C}}$ is spanned by W. Thus, every vector in $W^{\mathbb{C}}$ can be written in the form $z_1\alpha_1 + \cdots + z_n\alpha_n$, where $\alpha_1, \dots, \alpha_n \in W$ and $z_1, \dots, z_n \in \mathbb{C}$. Any basis $\underline{\alpha}$ of W over $\mathbb{R}$ is also a basis of $W^{\mathbb{C}}$ over $\mathbb{C}$. Also, any vector in $W^{\mathbb{C}}$ can be written uniquely in the form $\alpha + i\beta$ for vectors α and $\beta \in W$. In terms of this representation, addition and scalar multiplication in $W^{\mathbb{C}}$ are given by the following formulas:

3.8:
$$(\alpha + i\beta) + (\mu + i\lambda) = (\alpha + \mu) + i(\beta + \lambda)$$

$$(a + bi)(\alpha + i\beta) = (a\alpha - b\beta) + i(a\beta + b\alpha)$$

In equation 3.8, $\alpha, \beta, \mu, \lambda \in W$ and $a, b \in \mathbb{R}$.

It is also useful to note that the complexification $W^{\mathbb{C}}$ has a natural $\mathbb{R}$-isomorphism on it given by $\alpha + i\beta \to \alpha - i\beta$. If $\gamma = \alpha + i\beta$, then $\alpha - i\beta$ is called the *conjugate* of γ and written $\bar{\gamma}$. The map $\gamma \to \bar{\gamma}$ is a conjugate linear isomorphism of $W^{\mathbb{C}}$.

Now the real inner product $\langle \ , \ \rangle$ on W can be extended to a complex inner product $\langle \ , \ \rangle_1$ on $W^{\mathbb{C}}$ in a natural way such that the following diagram is commutative:

3.9:

$$
\begin{array}{ccc}
W^{\mathbb{C}} \times W^{\mathbb{C}} & \xrightarrow{\langle \ , \ \rangle_1} & \mathbb{C} \\
\uparrow{\scriptstyle i} & & \uparrow{\scriptstyle i} \\
W \times W & \xrightarrow{\langle \ , \ \rangle} & \mathbb{R}
\end{array}
$$

The formula for $\langle \ , \ \rangle_1$ is as follows:

3.10: $\langle \mu_1 + i\lambda_1, \mu_2 + i\lambda_2 \rangle_1 = \langle \mu_1, \mu_2 \rangle - i\langle \mu_1, \lambda_2 \rangle + i\langle \lambda_1, \mu_2 \rangle + \langle \lambda_1, \lambda_2 \rangle$

Equation 3.10 is the only definition of $\langle \ , \ \rangle_1$ possible if $\langle \ , \ \rangle_1$ is to be a complex inner product on $W^{\mathbb{C}}$ making diagram 3.9 commute. The fact that 3.9 is indeed commutative is obvious. The fact that $\langle \ , \ \rangle_1$ satisfies the conditions in Definition 3.2 is a tedious but straightforward exercise. Let us summarize what we have said here in the following theorem:

Theorem 3.11: Let $(W, \langle \ , \ \rangle)$ be a real inner product space. Then there is a unique complex inner product $\langle \ , \ \rangle_1$ on the complexification $W^{\mathbb{C}}$ of W such that diagram 3.9 is commutative. Furthermore, $\langle \ , \ \rangle_1$ is defined by equation 3.10. $\square$

Since $\langle \ , \ \rangle_1$ is the natural extension of $\langle \ , \ \rangle$, we shall drop any notational differences in the symbols and simply write $\langle \ , \ \rangle$ for the inner product on both W and $W^{\mathbb{C}}$. As an illustration of Theorem 3.11, the reader should check that the standard inner product on $\mathbb{C}^n$ is the complexification of the standard inner product on $\mathbb{R}^n$. Also, Example 3.6 is the complexification of Example 1.3.

Again suppose $(V, \langle \ , \ \rangle)$ is a complex inner product space. We claim that V is a normed linear vector space over $\mathbb{R}$ relative to $\|\alpha\| = \langle \alpha, \alpha \rangle^{1/2}$. Here and elsewhere in this section, we use the symbols $(x)^{1/2}$ to mean the nonnegative real number whose square is x. We need Schwarz's inequality for complex inner product spaces.

Lemma 3.12: Let $(V, \langle \ , \ \rangle)$ be a complex inner product space. Then $|\langle \alpha, \beta \rangle| \leqslant \langle \alpha, \alpha \rangle^{1/2} \langle \beta, \beta \rangle^{1/2}$ for all $\alpha, \beta \in V$.

Before proving Lemma 3.12, let us make a couple of comments about the quantities appearing in the inequality. If α and β are vectors in V, then $\langle \alpha, \beta \rangle$ is a complex number. Thus, the left side of 3.12 is the modulus of the complex number $\langle \alpha, \beta \rangle$. By 3.2(c), $\langle \alpha, \alpha \rangle$ and $\langle \beta, \beta \rangle$ are nonnegative real numbers. Thus, the right side of 3.12 is the product of the (nonnegative) square roots of these quantities. In the lemma, we compare these three real numbers.

Proof of 3.12: Fix α and β in V. If $\langle \alpha, \beta \rangle$ is a real number, then the proof of the inequality is the same as in Lemma 1.6.

So, we assume $z = \langle \alpha, \beta \rangle \in \mathbb{C} - \mathbb{R}$. Then $z \neq 0$, and $z^{-1}\alpha \in V$. Since $\langle z^{-1}\alpha, \beta \rangle = z^{-1}\langle \alpha, \beta \rangle = 1$, a real number, Schwarz's inequality is true for the vectors $z^{-1}\alpha$ and β. Thus, $1 = |\langle z^{-1}\alpha, \beta \rangle| \leqslant \langle z^{-1}\alpha, z^{-1}\alpha \rangle^{1/2} \langle \beta, \beta \rangle^{1/2}$. But $\langle z^{-1}\alpha, z^{-1}\alpha \rangle = (z\bar{z})^{-1}\langle \alpha, \alpha \rangle = |z|^{-2}\langle \alpha, \alpha \rangle$. Therefore, $\langle z^{-1}\alpha, z^{-1}\alpha \rangle^{1/2} = |z|^{-1}\langle \alpha, \alpha \rangle^{1/2}$. Thus, $|z| \leqslant \langle \alpha, \alpha \rangle^{1/2} \langle \beta, \beta \rangle^{1/2}$, and the proof is complete. $\square$

We can now define a norm on any complex inner product space.

Corollary 3.13: Let $(V, \langle \ , \ \rangle)$ be a complex inner product space. Then $\|\alpha\| = \langle \alpha, \alpha \rangle^{1/2}$ defines a real valued function on V that satisfies the following conditions:

(a) $\|\alpha\| > 0$ for all $\alpha \in V - \{0\}$.

(b) $\|z\alpha\| = |z| \|\alpha\|$ for all $\alpha \in V$, and $z \in \mathbb{C}$.

(c) $\|\alpha + \beta\| \leqslant \|\alpha\| + \|\beta\|$ for all $\alpha, \beta \in V$.

Proof: (a) Definition 3.2(c) implies that $\|\alpha\| > 0$ for every $\alpha \in V - \{0\}$. Note also that $\|\alpha\| = 0$ if and only if $\alpha = 0$.

(b) $\|z\alpha\| = \langle z\alpha, z\alpha \rangle^{1/2} = (z\bar{z}\langle \alpha, \alpha \rangle)^{1/2} = (|z|^2 \langle \alpha, \alpha \rangle)^{1/2} = |z| \|\alpha\|$.

(c) In order to prove the triangle inequality, we need to recall the definition of the real part, Re(z), of a complex number z. If $z = a + bi$, then Re(z) = a. Note that $z + \bar{z} = 2\,\text{Re}(z)$ and $\text{Re}(z) \leqslant |z|$.

Now suppose α and β are vectors in V. Then $\|\alpha + \beta\|^2 = \langle \alpha + \beta, \alpha + \beta \rangle = \langle \alpha, \alpha \rangle + \langle \alpha, \beta \rangle + \langle \beta, \alpha \rangle + \langle \beta, \beta \rangle = \|\alpha\|^2 + \langle \alpha, \beta \rangle + \overline{\langle \alpha, \beta \rangle} + \|\beta\|^2 = \|\alpha\|^2 + 2\,\text{Re}(\langle \alpha, \beta \rangle) + \|\beta\|^2 \leqslant \|\alpha\|^2 + 2|\langle \alpha, \beta \rangle| + \|\beta\|^2$. By Schwarz's inequality, $\|\alpha\|^2 + 2|\langle \alpha, \beta \rangle| + \|\beta\|^2 \leqslant \|\alpha\|^2 + 2\|\alpha\| \|\beta\| + \|\beta\|^2 = (\|\alpha\| + \|\beta\|)^2$. Thus, $\|\alpha + \beta\|^2 \leqslant (\|\alpha\| + \|\beta\|)^2$. Taking square roots gives us (c). $\square$

Any complex inner product space is of course a vector space over $\mathbb{R}$ since $\mathbb{R} \subseteq \mathbb{C}$. Corollary 3.13 says that any complex inner product space is a normed linear vector space over $\mathbb{R}$. Actually, 3.13(b) is a stronger statement than what is required in Definition 1.1(b) of Chapter IV. A complex vector space V together with a real valued function $\| \ \|: V \to \mathbb{R}$ satisfying conditions (a)–(c) in 3.13 is called a *complex, normed linear vector space*. Thus, Corollary 3.13 says that any complex inner product space $(V, \langle \ , \ \rangle)$ is a complex, normed linear vector space relative to $\|\alpha\| = \langle \alpha, \alpha \rangle^{1/2}$.

As usual, we shall call the norm $\|\alpha\| = \langle \alpha, \alpha \rangle^{1/2}$ defined by the inner product on V the norm associated with $\langle \ , \ \rangle$. Topological statements about a complex inner product space will always be relative to the associated norm on V. We can rewrite Schwarz's inequality as follows:

3.14: $$|\langle \alpha, \beta \rangle| \leqslant \|\alpha\| \, \|\beta\| \qquad \text{for all} \quad \alpha, \beta \in V$$

We now introduce the same definition discussed in Section 1 for real inner product spaces.

Definition 3.15: Let (V, $\|\ \|$) be a complex, normed linear vector space. We say (V, $\|\ \|$) is a pre-Hilbert space if there exists a complex inner product on V such that $\|\alpha\| = \langle \alpha, \alpha \rangle^{1/2}$ for all $\alpha \in V$. If the pre-Hilbert space (V, $\|\ \|$) is complete, then (V, $\|\ \|$) is called a Hilbert space.

The inner product spaces $\mathbb{C}^n$ and l^2 are (complex) Hilbert spaces. The space $\bigoplus_{k=1}^{\infty} \mathbb{C}$ is a pre-Hilbert space, but not a Hilbert space.

It is not our intention at this point to give the complex analog of every theorem in Section 1 for complex pre-Hilbert spaces. We shall say just a few words about some of these results. Let (V, $\langle \ , \ \rangle$) be a complex inner product space. Two vectors α and β in V are said to be *orthogonal* if $\langle \alpha, \beta \rangle = 0$. If α and β are orthogonal, we shall write $\alpha \perp \beta$. Since $\langle \alpha, \beta \rangle = \overline{\langle \beta, \alpha \rangle}$, we see $\alpha_\perp \beta$ if and only if $\beta_\perp \alpha$. We shall use the same terminology introduced in Definition 1.16 for complex inner product spaces.

The parallelogram law, Corollary 1.18, the Gram–Schmidt theorem, Bessel's inequality, and so on are all true in any complex inner product space (with only minor changes in their statements and proofs). For example, Bessel's inequality becomes the following statement in a complex inner product space: Let $\{\varphi_k\}$ be an orthonormal sequence in V, and set $z_k = \langle \alpha, \varphi_k \rangle$ for all $k \in \mathbb{N}$. Then $\sum_{k=1}^{\infty} |z_k|^2 \leqslant \|\alpha\|^2$. We shall cover these results in the exercises at the end of this section.

We shall be interested in finite-dimensional, complex inner product spaces in the next section. For these spaces, the complex analog of Theorem 1.21 can be proved purely algebraically.

Theorem 3.16: Let (V, $\langle \ , \ \rangle$) be a finite-dimensional, complex inner product space. Let W be a subspace of V. Then $V = W \oplus W^\perp$.

Proof: Applying the Gram–Schmidt theorem to W, we can find an orthonormal basis $\underline{\alpha} = \{\alpha_1, \ldots, \alpha_r\}$ of W. If $\alpha \in V$, then $\text{Proj}_W(\alpha) = \sum_{k=1}^{r} \langle \alpha, \alpha_k \rangle \alpha_k$ is a vector in W such that $\alpha\text{-Proj}_W(\alpha)$ is orthogonal to W. Thus, $\alpha = \text{Proj}_W(\alpha) + (\alpha - \text{Proj}_W(\alpha)) \in W + W^\perp$. Therefore, $V = W + W^\perp$. Since we always have $W \cap W^\perp = (0)$, $V = W \oplus W^\perp$. $\square$

It easily follows from Theorem 3.17 that $W^{\perp\perp} = W$ for any subspace W of V. We leave this as an exercise at the end of this section.

EXERCISES FOR SECTION 3

(1) Verify that the standard inner product on $\mathbb{C}^n$ indeed satisfies the conditions listed in 3.1.

(2) Show that the map $\langle \ , \ \rangle'$ given in Example 3.4 is a complex inner product on $\mathbb{C}^2$.

(3) Do the same for the map given in Example 3.7.

(4) Show that $\langle \ , \ \rangle_1$ given in equation 3.10 is a complex inner product on $W^{\mathbb{C}}$.

(5) Show that Examples 3.3 and 3.6 are the complexifications of Examples 1.2 and 1.3, respectively.

(6) Let $(V, \langle \ , \ \rangle)$ be a complex inner product space. Prove the Gram–Schmidt theorem in this setting. Thus, if $\{\alpha_k\}$ is a finite or infinite sequence of linearly independent vectors in V, show that there exists an orthonormal sequence $\{\varphi_k\}$ in V such that $L(\{\alpha_1, \ldots, \alpha_j\}) = L(\{\varphi_1, \ldots, \varphi_j\})$ for all j.

(7) Show that the parallelogram law is valid in any complex inner product space V. Thus, $\|\alpha + \beta\|^2 + \|\alpha - \beta\|^2 = 2(\|\alpha\|^2 + \|\beta\|^2)$.

(8) The complex analog of Lemma 1.17(b) is not true. If α and β are orthogonal, show $\|\alpha + \beta\|^2 = \|\alpha\|^2 + \|\beta\|^2$. Give an example that shows that the converse of this statement is false.

(9) Show that two vectors α and β in a complex inner product space are orthogonal if and only if $\|z\alpha + z'\beta\|^2 = \|z\alpha\|^2 + \|z'\beta\|^2$ for all $z, z' \in \mathbb{C}$.

(10) If two vectors α and β in a real inner product space have the property that $\|\alpha\| = \|\beta\|$, then $\alpha + \beta$ and $\alpha - \beta$ are orthogonal (prove!). Discuss the corresponding statement for complex inner product spaces.

(11) State and prove the complex analogs of Bessel's inequality and Parseval's equation.

(12) Suppose $(V, \langle \ , \ \rangle)$ is a complex inner product space. Let W be a subspace of V, and $\alpha \in V$. If $\langle \alpha, \beta \rangle + \langle \beta, \alpha \rangle \leqslant \langle \beta, \beta \rangle$ for all $\beta \in W$, prove that α is orthogonal to W.

(13) Suppose $(V, \| \ \|)$ is a complex pre-Hilbert space. Let W be a finite-dimensional subspace of V. Suppose $\beta \in V$. Show that there exists a vector in W closest to β in the $\| \ \|$-norm. Find a formula for such a vector.

(14) Let $(V, \| \ \|)$ be a complex pre-Hilbert space. Show that the map $\langle \ , \ \rangle \colon V \times V \to \mathbb{C}$ is continuous.

(15) Describe in detail the complexification of the real inner product space given in Exercise 13 of Section 1.

(16) Find an orthonormal basis of the complex inner product space given in Exercise 15.

(17) What do all complex inner products on $\mathbb{C}$ look like?

(18) Let $\langle \ , \ \rangle$ denote the standard inner product on $\mathbb{C}^2$. Show there is no nonzero matrix $A \in M_{2 \times 2}(\mathbb{C})$ such that $\langle \alpha, \alpha A \rangle = 0$ for all $\alpha \in \mathbb{C}^2$.

4. NORMAL OPERATORS

In this section, we shall assume that $(V, \langle \ , \ \rangle)$ is a finite-dimensional, complex inner product space. We shall let $\mathscr{E}(V) = \text{Hom}_{\mathbb{C}}(V, V)$. The reader will recall from Chapter I that $\mathscr{E}(V)$ is a finite-dimensional algebra over $\mathbb{C}$. In the literature, a map in $\mathscr{E}(V)$ may be called a linear transformation, an endomorphism of V, or a linear operator on V. All three names are used in various places. In this section, we shall usually refer to a map $T \in \mathscr{E}(V)$ as a linear operator on V. Our first order of business is to construct an adjoint of T. We begin with the following lemma:

Lemma 4.1: Let $(V, \langle \ , \ \rangle)$ be a finite-dimensional, complex inner product space. Then for every $f \in V^*$, there exists a unique vector $\gamma \in V$ such that $f(\alpha) = \langle \alpha, \gamma \rangle$ for all $\alpha \in V$.

Proof: Let $f \in V^* = \text{Hom}_{\mathbb{C}}(V, \mathbb{C})$. Let $\underline{\alpha} = \{\alpha_1, \ldots, \alpha_n\}$ be an orthonormal basis of V. Set $\gamma = \sum_{k=1}^{n} \overline{f(\alpha_k)}\alpha_k$. Define a complex valued function g on V by $g(\alpha) = \langle \alpha, \gamma \rangle$. We had seen in Section 3 that $g \in V^*$.

We want to argue that $f = g$. This is true if and only if f and g agree on $\underline{\alpha}$. So, fix $j \in \{1, \ldots, n\}$. Then $g(\alpha_j) = \langle \alpha_j, \gamma \rangle = \langle \alpha_j, \sum_{k=1}^{n} \overline{f(\alpha_k)}\alpha_k \rangle = \sum_{k=1}^{n} f(\alpha_k)\langle \alpha_j, \alpha_k \rangle = f(\alpha_j)$. Thus, $f = g$, and, in particular, $f(\alpha) = \langle \alpha, \gamma \rangle$ for all $\alpha \in V$.

To show that γ is unique, suppose $f(\alpha) = \langle \alpha, \gamma' \rangle$ for all $\alpha \in V$. Then $\langle \alpha, \gamma - \gamma' \rangle = 0$. In particular, $\|\gamma - \gamma'\|^2 = \langle \gamma - \gamma', \gamma - \gamma' \rangle = 0$. Hence, $\gamma = \gamma'$. $\square$

Corollary 4.2: Let $(V, \langle \ , \ \rangle)$ be a finite-dimensional, complex inner product space. Then for every $T \in \mathscr{E}(V)$, and every $\beta \in V$, there exists a unique vector $\gamma \in V$ such that $\langle T\alpha, \beta \rangle = \langle \alpha, \gamma \rangle$ for all $\alpha \in V$.

Proof: Let $T \in \mathscr{E}(V)$ and $\beta \in V$. Then the map $f(\alpha) = \langle T\alpha, \beta \rangle$ is clearly a linear transformation from V to $\mathbb{C}$. Thus, $f \in V^*$. By Lemma 4.1, there exists a unique vector $\gamma \in V$ such that $f(\alpha) = \langle \alpha, \gamma \rangle$ for all $\alpha \in V$. Thus, $\langle T\alpha, \beta \rangle = \langle \alpha, \gamma \rangle$ for all α. $\square$

We can use Corollary 4.2 to construct an adjoint of a linear operator T.

Theorem 4.3: Let $(V, \langle \ , \ \rangle)$ be a finite-dimensional, complex inner product space. Let $T \in \mathscr{E}(V)$. Then there exists a unique linear operator $T^* \in \mathscr{E}(V)$ such that $\langle T\alpha, \beta \rangle = \langle \alpha, T^*\beta \rangle$ for all $\alpha, \beta \in V$.

Proof: Let $T \in \mathscr{E}(V)$. By Corollary 4.2, there exists a function T^* from V to V such that $\langle T\alpha, \beta \rangle = \langle \alpha, T^*\beta \rangle$ for all $\alpha, \beta \in V$. For a fixed β, $T^*(\beta)$ is the γ in 4.2 for which $\langle T\alpha, \beta \rangle = \langle \alpha, \gamma \rangle$ for all $\alpha \in V$. We claim that T^* is a linear operator on V.

Suppose β_1 and β_2 are vectors in V. Then for any $\alpha \in V$, we have $\langle T\alpha, \beta_1 \rangle = \langle \alpha, T^*\beta_1 \rangle$, and $\langle T\alpha, \beta_2 \rangle = \langle \alpha, T^*\beta_2 \rangle$. Thus, $\langle T\alpha, \beta_1 + \beta_2 \rangle = \langle T\alpha, \beta_1 \rangle + \langle T\alpha, \beta_2 \rangle = \langle \alpha, T^*\beta_1 \rangle + \langle \alpha, T^*\beta_2 \rangle = \langle \alpha, T^*\beta_1 + T^*\beta_2 \rangle$. On the other hand, $T^*(\beta_1 + \beta_2)$ is the unique vector in V such that $\langle T\alpha, \beta_1 + \beta_2 \rangle = \langle \alpha, T^*(\beta_1 + \beta_2) \rangle$ for all α. We conclude that $T^*(\beta_1 + \beta_2) = T^*(\beta_1) + T^*(\beta_2)$. Thus, T^* is an additive map on V.

Let $\beta \in V$, and $z \in \mathbb{C}$. Then for every $\alpha \in V$, $\langle T\alpha, z\beta \rangle = \langle \alpha, T^*(z\beta) \rangle$, and $\langle T\alpha, \beta \rangle = \langle \alpha, T^*\beta \rangle$. Thus, $\langle \alpha, zT^*(\beta) \rangle = \bar{z}\langle \alpha, T^*\beta \rangle = \bar{z}\langle T\alpha, \beta \rangle = \langle T\alpha, z\beta \rangle = \langle \alpha, T^*(z\beta) \rangle$. Since α is arbitrary, we conclude that $T^*(z\beta) = zT^*(\beta)$.

We have now shown that $T^* \in V^*$. The fact that T^* is unique is obvious. $\square$

The operator T^* constructed in Theorem 4.3 satisfies the same functional equation as the adjoint of a real operator, that is, equation 2.7. In complex theory, the map T^* is called the Hermitian adjoint of T. Thus, we have the following definition:

Definition 4.4: Let $(V, \langle \ , \ \rangle)$ be a finite-dimensional, complex inner product space, and $T \in \mathscr{E}(V)$. The unique $T^* \in \mathscr{E}(V)$ for which $\langle T\alpha, \beta \rangle = \langle \alpha, T^*\beta \rangle$ for all $\alpha, \beta \in V$ is called the Hermitian adjoint of T.

Thus, T^* is the complex analog of the adjoint of a real operator on a real inner product space. In fact, we have the following relation between the Hermitian adjoint and the adjoint:

Theorem 4.5: Suppose $(W, \langle \ , \ \rangle)$ is a finite-dimensional, real inner product space. Let $T \in \mathrm{Hom}_{\mathbb{R}}(W, W)$. Let T^* denote the adjoint of T. Then the Hermitian adjoint of $T^{\mathbb{C}}$ on $W^{\mathbb{C}}$ is $(T^*)^{\mathbb{C}}$.

Proof: Recall from Chapter II that if $T \in \mathrm{Hom}_{\mathbb{R}}(W, W)$, then $T^{\mathbb{C}} = T \otimes_{\mathbb{R}} I_{\mathbb{C}}$ is the $\mathbb{C}$-linear transformation on $W^{\mathbb{C}}$ given by $T^{\mathbb{C}}(\sum_{k=1}^n z_k\alpha_k) = \sum_{k=1}^n z_k T(\alpha_k)$. In this equation, $z_1, \ldots, z_n \in \mathbb{C}$ and $\alpha_1, \ldots, \alpha_n \in W$. The adjoint T^* of T is the unique map in $\mathrm{Hom}_{\mathbb{R}}(W, W)$ for which $\langle T\alpha, \beta \rangle = \langle \alpha, T^*\beta \rangle$ for all $\alpha, \beta \in V$. The inner product $\langle \ , \ \rangle$ on $W^{\mathbb{C}}$ is given by equation 3.10. Thus, to prove the theorem, we need to show that the following equation in $\mathbb{C}$ is valid:

4.6: $\qquad \langle T^{\mathbb{C}}\alpha_1, \alpha_2 \rangle = \langle \alpha_1, (T^*)^{\mathbb{C}}\alpha_2 \rangle \qquad$ for all $\quad \alpha_1, \alpha_2 \in W^{\mathbb{C}}$

To see this, write $\alpha_j = \mu_j + i\lambda_j$ with μ_1, μ_2, λ_1 and λ_2 in W. Using equations 3.10 and 2.7, we have $\langle T^{\mathbb{C}}\alpha_1, \alpha_2 \rangle = \langle T\mu_1 + iT\lambda_1, \alpha_2 \rangle = \langle T\mu_1, \alpha_2 \rangle$

$+ i\langle T\lambda_1, \quad \alpha_2\rangle = \langle T\mu_1, \quad \mu_2\rangle - i\langle T\mu_1, \quad \lambda_2\rangle + i\langle T\lambda_1, \quad \mu_2\rangle + \langle T\lambda_1, \quad \lambda_2\rangle = \langle \mu_1, T^*\mu_2\rangle - i\langle \mu_1, T^*\lambda_2\rangle + i\langle \lambda_1, T^*\mu_2\rangle + \langle \lambda_1, T^*\lambda_2\rangle = \langle \mu_1 + i\lambda_1, T^*\mu_2 + iT^*\lambda_2\rangle = \langle \alpha_1, (T^*)^C\alpha_2\rangle.$ □

Equation 4.6 says the Hermitian adjoint of the complexification of a real operator T is the complexification of the adjoint of T. This statement can be represented in symbols as follows: $(T^C)^* = (T^*)^C$.

Let us return to the general setting of a finite-dimensional, complex inner product space $(V, \langle \ , \ \rangle)$. The map on $\mathscr{E}(V)$ which sends T to T^* is a conjugate linear isomorphism. This follows from the first three properties listed in our next lemma.

Lemma 4.7: Let $(V, \langle \ , \ \rangle)$ be a finite-dimensional, complex inner product space. The map on $\mathscr{E}(V)$ which sends T to T^* satisfies the following properties:

(a) $(T^*)^* = T$.

(b) $(S + T)^* = S^* + T^*$.

(c) $(zT)^* = \bar{z}T^*$.

(d) $(ST)^* = T^*S^*$.

(e) If $\underline{\alpha}$ is an orthonormal basis of V, then $\Gamma(\underline{\alpha}, \underline{\alpha})(T^*)$ is the conjugate transpose of $\Gamma(\underline{\alpha}, \underline{\alpha})(T)$.

Proof: All the above assertions follow immediately from the functional equation:

4.8: $\qquad\qquad \langle T\alpha, \beta\rangle = \langle \alpha, T^*\beta\rangle \qquad$ for all $\quad \alpha, \beta \in V$

We prove (e) and leave the rest as exercises at the end of this section. Suppose $\underline{\alpha} = \{\alpha_1, \ldots, \alpha_n\}$ is an orthonormal basis of V. Set $\Gamma(\underline{\alpha}, \underline{\alpha})(T) = (z_{kj}) \in M_{n \times n}(\mathbb{C})$. Then $T(\alpha_j) = \sum_{k=1}^n z_{kj}\alpha_k$ for all $j = 1, \ldots, n$. If $T^*(\alpha_j) = \sum_{k=1}^n w_{kj}\alpha_k$, then we have $z_{pj} = \langle T\alpha_j, \alpha_p\rangle = \langle \alpha_j, T^*\alpha_p\rangle = \langle \alpha_j, \sum_{k=1}^n w_{kp}\alpha_k\rangle = \bar{w}_{jp}$. Thus, the conjugate transpose of $\Gamma(\underline{\alpha}, \underline{\alpha})(T)$ is $\Gamma(\underline{\alpha}, \underline{\alpha})(T^*)$. □

Although we did not make it explicit in Section 2, the map sending $T \rightarrow T^*$ for real inner product spaces also satisfies the same properties (a)–(e) (without the conjugate) in Lemma 4.7.

We can now introduce the definitions of normal, Hermitian, and unitary operators.

Definition 4.9: Let $(V, \langle \ , \ \rangle)$ be a finite-dimensional, complex inner product space. Let $T \in \mathscr{E}(V)$.

(a) We say T is *normal* if $TT^* = T^*T$.

(b) T is *Hermitian* if $T^* = T$.

(c) T is *unitary* if $T^*T = I_V$.

Obviously a Hermitian operator is normal. Since $\dim_{\mathbb{C}}(V) < \infty$, $T^*T = I_V$ if and only if $TT^* = I_V$. Thus, a unitary operator is normal also. Hermitian and unitary operators are the complex analogs of self-adjoint and orthogonal operators on real inner product spaces. In fact, we have the following theorem:

Theorem 4.10: Let $(W, \langle \ , \ \rangle)$ be a finite-dimensional, real inner product space. Let $T \in \mathrm{Hom}_{\mathbb{R}}(W, W)$.

(a) If T is self-adjoint, then $T^{\mathbb{C}}$ is Hermitian.

(b) If T is orthogonal, then T^* is unitary.

Proof: These results follow immediately from Theorem 4.5 and the fact that the complexification of a product of two endomorphisms is the product of their complexifications. Thus, if T is self-adjoint, then $T = T^*$. Hence, $(T^{\mathbb{C}})^* = (T^*)^{\mathbb{C}} = T^{\mathbb{C}}$. Therefore, $T^{\mathbb{C}}$ is Hermitian. If T is orthogonal, then $T^*T = I_W$. Let I denote the identity map on the complexification $W^{\mathbb{C}}$. Then $I = (I_W)^{\mathbb{C}} = (T^*T)^{\mathbb{C}} = (T^*)^{\mathbb{C}}T^{\mathbb{C}} = (T^{\mathbb{C}})^*T^{\mathbb{C}}$. Therefore, $T^{\mathbb{C}}$ is unitary. $\square$

In terms of the complex inner product on V, the definitions of Hermitian and unitary can be rewritten as follows:

4.11: (a) T is Hermitian if and only if $\langle T\alpha, \beta \rangle = \langle \alpha, T\beta \rangle$ for all $\alpha, \beta \in V$.

(b) T is unitary if and only if $\langle T\alpha, T\beta \rangle = \langle \alpha, \beta \rangle$ for all $\alpha, \beta \in V$.

The proof of 4.11 is completely analogous to 2.8′ in Section 2.

Since Hermitian and unitary operators satisfy the same functional relations as their real analogs, self-adjoint and orthogonal operators, they should have the same names. Unfortunately, they do not. Here is a handy chart to help you remember the names and definitions of the real objects and their corresponding complex analogs:

4.12:

Real Inner Product Spaces	*Complex Inner Product Spaces*
(a) $\langle T\alpha, \beta \rangle = \langle \alpha, T^*\beta \rangle$ T^* is the adjoint of T	$\langle T\alpha, \beta \rangle = \langle \alpha, T^*\beta \rangle$ T^* is the Hermitian adjoint of T
(b) $\langle T\alpha, \beta \rangle = \langle \alpha, T\beta \rangle$ T is self-adjoint	$\langle T\alpha, \beta \rangle = \langle \alpha, T\beta \rangle$ T is Hermitian
(c) $\langle T\alpha, T\beta \rangle = \langle \alpha, \beta \rangle$ T is orthogonal	$\langle T\alpha, T\beta \rangle = \langle \alpha, \beta \rangle$ T is unitary

In the last part of this section, we discuss normal operators. Since Hermitian and unitary operators are both normal, whatever we say applies to both types of linear transformations. Throughout the rest of this section, $(V, \langle \, , \, \rangle)$ will denote a finite-dimensional, complex inner product space. T is a linear operator on V.

Lemma 4.13: T is unitary if and only if there exists an orthonormal basis $\underline{\alpha}$ of V, such that $T(\underline{\alpha})$ is an orthonormal basis of V.

Proof: Let $\underline{\alpha} = \{\alpha_1, \ldots, \alpha_n\}$ be an orthonormal basis of V. Suppose T is unitary. Then $\langle T\alpha_k, T\alpha_j \rangle = \langle \alpha_k, \alpha_j \rangle = 1$ if $j = k$, and 0 otherwise. It ready follows that $T(\underline{\alpha}) = \{T(\alpha_1), \ldots, T(\alpha_n)\}$ is an orthonormal basis of V.

Conversely, suppose $\underline{\alpha} = \{\alpha_1, \ldots, \alpha_n\}$ is an orthonormal basis of V such that $\{T(\alpha_1), \ldots, T(\alpha_n)\}$ is also an orthonormal basis of V. Let $\alpha, \beta \in V$, and write $\alpha = \sum_{k=1}^n z_k \alpha_k$ and $\beta = \sum_{k=1}^n w_k \alpha_k$. Then $\langle \alpha, \beta \rangle = \langle \sum_{k=1}^n z_k \alpha_k, \sum_{k=1}^n w_k \alpha_k \rangle = \sum_{k=1}^n z_k \bar{w}_k$. This last equation follows from the fact that $\underline{\alpha}$ is a set of orthonormal vectors. Since $T(\underline{\alpha})$ is also a set of orthonormal vectors in V, we have $\langle T\alpha, T\beta \rangle = \langle \sum_{k=1}^n z_k T(\alpha_k), \sum_{k=1}^n w_k T(\alpha_k) \rangle = \sum_{k=1}^n z_k \bar{w}_k$. Thus, $\langle T\alpha, T\beta \rangle = \langle \alpha, \beta \rangle$ for all $\alpha, \beta \in V$, and T is unitary. $\square$

An operator $T \in \mathscr{E}(V)$ is said to be skew-Hermitian, if $T^* = -T$. Clearly, a skew-Hermitian operator is normal. A typical example of a skew-Hermitian operator is $(T - T^*)$. Here T is any operator. A nonzero, skew-Hermitian operator cannot be Hermitian. In general, skew-Hermitian operators are not unitary either. In particular, a normal operator need not be Hermitian or unitary. If T is an arbitrary endomorphism of V, then $T = (T + T^*)/2 + (T - T^*)/2$. Since $(T + T^*)/2$ is Hermitian, and $(T - T^*)/2$ is skew-Hermitian, we have the following lemma:

Lemma 4.14: Every operator $T \in \text{Hom}_{\mathbb{C}}(V, V)$ is the sum of two normal operators. $\square$

Since $\mathbb{C}$ is an algebraically closed field, any operator T on V has a characteristic polynomial of the following form:

4.15:
$$c_T(X) = \prod_{k=1}^r (X - z_k)^{n_k}$$

Here $\{z_1, \ldots, z_r\} = \mathscr{S}_{\mathbb{C}}(T)$, the spectrum of T. We also have $\sum_{k=1}^r n_k = n = \dim_{\mathbb{C}}(V)$.

Lemma 4.16: If T is Hermitian, then $\mathscr{S}_{\mathbb{C}}(T) \subseteq \mathbb{R}$.

Proof: Let z be an eigenvalue of T. Then there exists a nonzero vector $\alpha \in V$ such that $T(\alpha) = z\alpha$. We have $z\langle \alpha, \alpha \rangle = \langle z\alpha, \alpha \rangle = \langle T\alpha, \alpha \rangle = \langle \alpha, T\alpha \rangle = \langle \alpha, z\alpha \rangle = \bar{z}\langle \alpha, \alpha \rangle$. Thus, $(z - \bar{z})\|\alpha\|^2 = 0$. Since $\alpha \neq 0$, we conclude that $z = \bar{z}$. Hence, z is a real number. $\square$

Thus, the eigenvalues of a Hermitian operator are always real. Our next lemma says that nilpotent Hermitian operators are zero.

Lemma 4.17: If T is Hermitian, and $T^k(\alpha) = 0$ for some vector $\alpha \in V$, then $T(\alpha) = 0$.

Proof: We can assume that $T^{2^m}(\alpha) = 0$ for some $m \geqslant 1$. Set $S = T^{2^{m-1}}$. Then S is Hermitian by Lemma 4.7(d). Since $SS^* = S^2 = T^{2^m}$, we see that $SS^*(\alpha) = 0$. But then, $0 = \langle SS^*\alpha, \alpha \rangle = \langle S^*\alpha, S^*\alpha \rangle = \langle S\alpha, S\alpha \rangle = \|S\alpha\|^2$. We conclude that $S(\alpha) = 0$. We can now repeat this argument. We finally get $T^2(\alpha) = 0$. Then $0 = \langle T^2\alpha, \alpha \rangle = \langle T\alpha, T\alpha \rangle = \|T\alpha\|^2$. Therefore, $T(\alpha) = 0$. $\square$

We can extend this result to normal operators in general.

Corollary 4.18: If T is normal, and $T^k(\alpha) = 0$ for some vector α in V, then $T(\alpha) = 0$.

Proof: Set $S = T^*T$. Then $S^* = (T^*T)^* = T^*T^{**} = T^*T = S$. Thus, S is Hermitian. Since T is normal, T commutes with T^*. In particular, we have $S^k(\alpha) = (T^*T)^k(\alpha) = (T^*)^k T^k(\alpha) = 0$. By Lemma 4.17, $S(\alpha) = 0$. But then, we have $0 = \langle S\alpha, \alpha \rangle = \langle T^*T\alpha, \alpha \rangle = \langle T\alpha, T\alpha \rangle = \|T(\alpha)\|^2$. We conclude that $T(\alpha) = 0$. $\square$

Corollary 4.18 is important for what is to follow. If T is a nilpotent, normal operator, then $T = 0$. More generally, suppose T is a normal operator on V. If $p(X) \in \mathbb{C}[X]$, then clearly $p(T)$ is a normal operator on V. Thus, if $p(T)$ is nilpotent, we can conclude that $p(T) = 0$.

Lemma 4.19: If T is a normal operator on V, then $\|T(\alpha)\| = \|T^*(\alpha)\|$ for all α in V.

Proof: $\|T(\alpha)\|^2 = \langle T\alpha, T\alpha \rangle = \langle \alpha, T^*T\alpha \rangle = \langle \alpha, TT^*\alpha \rangle = \langle T^*\alpha, T^*\alpha \rangle = \|T^*(\alpha)\|^2$. Taking square roots, gives us the desired result. $\square$

One immediate application of Lemma 4.19 is the fact that $\ker(T) = \ker(T^*)$ for any normal operator T. An important special case of this equality is the following corollary:

Corollary 4.20: Suppose T is a normal operator on V. Then $T(\alpha) = z\alpha$ for some $\alpha \in V$, and $z \in \mathbb{C}$ if and only if $T^*(\alpha) = \bar{z}\alpha$.

Proof: Since T is normal, $T - zI_V$ is also normal. The Hermitian adjoint of $T - zI_V$ is given by $(T - zI_V)^* = T^* - \bar{z}I_V$. By Lemma 4.19, $\ker(T - zI_V) = \ker(T^* - \bar{z}I_V)$. Thus, $T(\alpha) = z\alpha$ if and only if $T^*(\alpha) = \bar{z}\alpha$. $\square$

Thus, if T is a normal operator on V with spectrum given by $\mathscr{S}_{\mathbb{C}}(T) = \{z_1, \ldots, z_r\}$, then T* is a normal operator on V with spectrum given by $\mathscr{S}_{\mathbb{C}}(T^*) = \{\bar{z}_1, \ldots, \bar{z}_r\}$. Furthermore, α is an eigenvector for T associated with z_k if and only if α is an eigenvector for T* associated with $\bar{z}_k$.

We can use these results to say something about the spectrum of a unitary operator.

Corollary 4.21: If T is a unitary operator on V and $z \in \mathscr{S}_{\mathbb{C}}(T)$, then $|z| = 1$.

Proof: Let α be an eigenvector of T associated with z. Then $T(\alpha) = z\alpha$. Since unitary operators are normal, Corollary 4.20 implies $T^*(\alpha) = \bar{z}\alpha$. Therefore, $\alpha = T^*T(\alpha) = T^*(z\alpha) = zT^*(\alpha) = z\bar{z}\alpha$. Since $\alpha \neq 0$, we conclude that $z\bar{z} = 1$. Thus, $|z| = 1$. $\square$

Corollary 4.21 of course says that a unitary operator has all of its eigenvalues lying on the unit circle in the complex plane $\mathbb{C}$.

We need one more set of ideas before presenting the spectral theorem for normal operators. The reader will recall that an endomorphism $T \in \mathscr{E}(V)$ is called *idempotent* if $T^2 = T$. A typical example of an idempotent map is the projection $\text{Proj}_W(\ )$ of V onto a subspace W. If T is idempotent, then $V = \ker(T) \oplus \text{Im}(T)$. To see this, first note that any vector $\alpha \in V$ can be written in the form $\alpha = T(\alpha) + (\alpha - T(\alpha))$. Clearly, $T(\alpha) \in \text{Im}(T)$. Since $T(\alpha - T(\alpha)) = T(\alpha) - T^2(\alpha) = T(\alpha) - T(\alpha) = 0$, we see $\alpha - T(\alpha) \in \ker(T)$. Therefore, $V = \ker(T) + \text{Im}(T)$. If $\alpha \in \ker(T) \cap \text{Im}(T)$, then $\alpha = T(\beta)$ for some $\beta \in V$. Then $\alpha = T(\beta) = T^2(\beta) = T(\alpha) = 0$. This last equality comes from the fact that $\alpha \in \ker(T)$. Thus, $\ker(T) \cap \text{Im}(T) = (0)$, and V is the direct sum of $\ker(T)$ and $\text{Im}(T)$.

Now, in general, when T is idempotent, the subspaces $\ker(T)$ and $\text{Im}(T)$ are not orthogonal. Thus, V is not in general an orthogonal direct sum of $\ker(T)$ and $\text{Im}(T)$. Those idempotent T for which $\ker(T)_\perp\text{Im}(T)$ are called *orthogonal projections*. We can construct examples of orthogonal projections by following the same procedure as in Section 2. Suppose W is a subspace of V. Let $\underline{\alpha} = \{\alpha_1, \ldots, \alpha_m\}$ be an orthonormal basis of W. We can define an operator $\text{Proj}_W(\ )$ on V by setting $\text{Proj}_W(\alpha) = \sum_{k=1}^m \langle \alpha, \alpha_k \rangle \alpha_k$. The reader can easily check that $\text{Proj}_W(\text{Proj}_W(\alpha)) = \text{Proj}_W(\alpha)$. Thus, the map $\text{Proj}_W(\ )$ is an idempotent operator on V. The image of $\text{Proj}_W(\ )$ is clearly W. One can also check that $\alpha - \text{Proj}_W(\alpha)$ is orthogonal to W. It easily follows from this fact that $\ker(\text{Proj}_W(\ ))$ is orthogonal to $\text{Im}(\text{Proj}_W(\ ))$. Thus, $\text{Proj}_W(\ )$ is an orthogonal projection of V onto W.

We shall need the following result about orthogonal projections:

Lemma 4.22: Suppose T is idempotent and normal. Then T is a Hermitian, orthogonal projection.

Proof: We first note that T* is idempotent. We have $\langle \alpha, T^*\beta \rangle = \langle T\alpha, \beta \rangle = \langle T^2\alpha, \beta \rangle = \langle \alpha, (T^*)^2\beta \rangle$. We conclude from these equations that $T^* =$

$(T^*)^2$. Our comments before this lemma now imply that $V = \ker(T) \oplus \text{Im}(T) = \ker(T^*) \oplus \text{Im}(T^*)$.

Since T is normal, Lemma 4.19 implies that $\ker(T) = \ker(T^*)$. We claim that $\text{Im}(T) = \text{Im}(T^*)$. Since T commutes with T^*, we have $T^*(\text{Im}(T)) \subseteq \text{Im}(T)$. Thus, $\text{Im}(T^*) = T^*(V) = T^*(\ker(T) + \text{Im}(T)) = T^*(\ker(T^*) + \text{Im}(T)) = T^*(\text{Im}(T)) \subseteq \text{Im}(T)$. Reversing the roles of T and T^*, gives us the other inclusion $\text{Im}(T) \subseteq \text{Im}(T^*)$. Thus, $\text{Im}(T) = \text{Im}(T^*)$.

We now claim that $T = T^*$. Since both maps are idempotent, they are both the identity map on $\text{Im}(T) = \text{Im}(T^*)$. Let $\alpha \in V$. We can write α as $\alpha = \alpha_1 + \alpha_2$ where $\alpha_1 \in \ker(T) = \ker(T^*)$, and $\alpha_2 \in \text{Im}(T) = \text{Im}(T^*)$. Then $T(\alpha) = T(\alpha_1) + T(\alpha_2) = T(\alpha_2) = \alpha_2 = T^*(\alpha_2) = T^*(\alpha_1) + T^*(\alpha_2) = T^*(\alpha)$. We have now established that T is Hermitian.

For any Hermitian operator on V, we have $\ker(T) = \text{Im}(T)^{\perp}$. This argument is exactly the same as the self-adjoint argument. If $\alpha \in \ker(T)$, and $\beta \in \text{Im}(T)$, then $\beta = T(\gamma)$ for some $\gamma \in V$. In particular, we have $\langle \alpha, \beta \rangle = \langle \alpha, T\gamma \rangle = \langle T\alpha, \gamma \rangle = 0$. Therefore, $\ker(T) \subseteq \text{Im}(T)^{\perp}$. If $\alpha \in \text{Im}(T)^{\perp}$, and β is arbitrary, then $0 = \langle \alpha, T\beta \rangle = \langle T\alpha, \beta \rangle$. We conclude that $T(\alpha) = 0$. Thus, $\text{Im}(T)^{\perp} \subseteq \ker(T)$.

Since $\ker(T) = \text{Im}(T)^{\perp}$, in particular, $\ker(T) \perp \text{Im}(T)$. Hence, T is an orthogonal projection. $\square$

We can now state the spectral theorem for normal operators. With the lemmas we have proved in this section, the proof of the Spectral Theorem is a simple consequence of Theorem 4.23 of Chapter III. The reader is advised to review Theorem 4.23 before proceeding further.

Theorem 4.23 (Spectral Theorem): Let $(V, \langle \ , \ \rangle)$ be a finite-dimensional, complex inner product space. Let $T \in \text{Hom}_C(V, V)$ be a normal operator. Suppose the characteristic polynomial of T is $c_T(X) = \prod_{k=1}^{r}(X - z_k)^{n_k}$. Then there exists a set of pairwise orthogonal idempotents $\{P_1, \ldots, P_r\} \subseteq \text{Hom}_C(V, V)$ having the following properties:

(a) $\sum_{k=1}^{r} P_k = I_V$.

(b) $\sum_{k=1}^{r} z_k P_k = T$.

(c) For each $k = 1, \ldots, r$, $\text{Im}(P_k) = \{\alpha \in V \mid T(\alpha) = z_k \alpha\}$.

If we set $V_k = \text{Im}(P_k)$ for each $k = 1, \ldots, r$, then we also have

(d) $\dim_C(V_k) = n_k$.

(e) $V = V_1 \oplus \cdots \oplus V_r$.

(f) $V_k \perp V_j$ for all $1 \leqslant j < k \leqslant r$.

(g) $m_T(X) = \prod_{k=1}^{r}(X - z_k)$.

Proof: Let $P_1, \ldots, P_r$ and $N_1, \ldots, N_r$ be the endomorphisms in $\mathscr{E}(V)$ given by Theorem 4.23 of Chapter III. Each N_k is a nilpotent operator on V and is also a polynomial in T. Since T is normal, our comments after Corollary 4.18 imply $N_1 = \cdots = N_r = 0$. It now follows from Theorem 4.23(d) of Chapter III that

$\sum_{k=1}^{r} z_k P_k = T$. We had also proved in Theorem 4.23 that $P_1, \ldots, P_r$ are pairwise orthogonal idempotents whose sum is I_V. Hence, we have established (a) and (b).

In Theorem 4.23 of Chapter III, we had also established that $V_k = \text{Im}(P_k) = \ker(T - z_k)^{n_k}$ for each $k = 1, \ldots, r$. Since $T - z_k$ is normal, Corollary 4.18 implies $\ker(T - z_k)^{n_k} = \ker(T - z_k)$. This proves (c). The assertions in (d) and (e) were also established in 4.23 of Chapter III. The assertion in (g) follows from Corollary 4.22 of Chapter III. The only thing that remains to be proved is the statement in (f).

Each P_k is a polynomial in T. Thus, each P_k is a normal operator on V. Lemma 4.22 implies each P_k is a Hermitian, orthogonal projection of V onto V_k. Suppose $k \neq j$, and let $\alpha \in V_k$ and $\beta \in V_j$. Then $\alpha = P_k(\alpha')$, and $\beta = P_j(\beta')$ for some $\alpha', \beta' \in V$. Since $P_k P_j = 0$, we have $\langle \alpha, \beta \rangle = \langle P_k(\alpha'), P_j(\beta') \rangle = \langle \alpha', P_k^* P_j \beta' \rangle = \langle \alpha', P_k P_j \beta' \rangle = \langle \alpha', 0 \rangle = 0$. Thus, $V_k \perp V_j$ and (f) is proved. $\square$

The spectral theorem says that V decomposes into an orthogonal direct sum of the eigenspaces V_k. If we choose an orthonormal basis (by Gram–Schmidt) of each V_k and take their union, we get an orthonormal basis of V consisting entirely of eigenvectors of T. Hence, we can restate Theorem 4.23 as follows:

Corollary 4.24: Let T be a normal operator on a finite-dimensional, complex inner product space V. Then V has an orthonormal basis consisting of eigenvectors of T. $\square$

The matrix version of Theorem 4.23 is easy to state. The Hermitian adjoint A^* of a complex matrix A is its conjugate transpose. Thus, $A^* = (\bar{A})^t$. A matrix U is unitary if $U^*U = I$. A matrix A is Hermitian if $A^* = A$. A change in orthonormal bases in $\mathbb{C}^n$ is given by a unitary matrix. Thus, the matrix version of 4.23 is as follows:

Corollary 4.25: Let $A \in M_{n \times n}(\mathbb{C})$ be a normal matrix, that is, $AA^* = A^*A$. Then there exists a unitary matrix U such that UAU^{-1} is diagonal. $\square$

The two most important examples of normal operators on V are Hermitian and unitary operators. Thus, the most important applications of the spectral theorem are the following two special cases:

Corollary 4.26: If T is a Hermitian operator on a finite-dimensional, complex inner product space V, then V has an orthonormal basis consisting of eigenvectors of T. Equivalently, if $A \in M_{n \times n}(\mathbb{C})$ is Hermitian, then there exists a unitary matrix U such that UAU^{-1} is diagonal. $\square$

Corollary 4.27: If T is a unitary operator on a finite-dimensional, complex inner product space V, then V has an orthonormal basis consisting of eigenvectors of T. Equivalently, if $A \in M_{n \times n}(\mathbb{C})$ is a unitary matrix, then there exists a second unitary matrix U such that UAU^{-1} is diagonal. $\square$

EXERCISES FOR SECTION 4

(1) Prove assertions (a)–(d) in Lemma 4.7.

(2) Show that the assertions in 4.11 are correct.

(3) Exhibit a normal operator on $\mathbb{C}^n$ which is neither Hermitian nor unitary.

(4) If T is a normal operator on V and $p(X) \in \mathbb{C}[X]$, compute the Hermitian adjoint of p(T). Use this answer to show that p(T) is normal.

(5) Prove Corollary 4.25.

(6) Generalize Corollary 4.25 as follows: If $A_1, \ldots, A_s$ are a finite number of normal, commuting matrices [in $M_{n \times n}(\mathbb{C})$], then there exists a unitary matrix U such that $UA_k U^{-1}$ is diagonal for all $k = 1, \ldots, s$.

(7) Show that Exercise 6 is false if A_1 and A_2 do not commute.

(8) Use the spectral theorem to show that for every symmetric matrix $A \in M_{n \times n}(\mathbb{R})$, there exists an orthogonal matrix $P \in M_{n \times n}(\mathbb{R})$ such that PAP^{-1} is diagonal. (*Hint*: Pass to the complexification, and argue that the eigenvalues and eigenvectors you need are all real.)

(9) Let V be a complex inner product space, and let $T \in \mathscr{E}(V)$. Suppose T is normal. Show that
(a) T is Hermitian if and only if $\mathscr{S}_\mathbb{C}(T) \subseteq \mathbb{R}$.
(b) T is unitary if and only if $\mathscr{S}_\mathbb{C}(T) \subseteq \{z \in \mathbb{C} \,|\, |z| = 1\}$.

(10) Return to the example you gave in Exercise 3, and find an orthonormal basis of $\mathbb{C}^n$ consisting of eigenvectors for your normal operator.

(11) Suppose T is a nonzero, skew-Hermitian operator on V. Show the eigenvalues of T are purely imaginary.

(12) Suppose T is a normal operator on V. Show that the Hermitian adjoint T* of T is a polynomial in T.

(13) Let

$$A = \begin{pmatrix} 1 & 0 & 0 & 0 \\ 0 & 0 & 1 & 0 \\ 0 & 1 & 0 & 0 \\ 0 & 0 & 0 & 1 \end{pmatrix}$$

Find a unitary matrix U such that UAU^{-1} is diagonal.

(14) Let $A \in M_{n \times n}(\mathbb{C})$. Show there exists a unitary matrix U such that UAU^{-1} is upper triangular.

(15) Show that an n × n matrix A is normal if and only if A commutes with AA*.

(16) Let $A \in M_{n \times n}(\mathbb{C})$. Suppose the characteristic polynomial of A is given by $c_A(X) = \prod_{k=1}^{r}(X - z_k)^{n_k}$. Let $w_1, \ldots, w_n$ be a list of the z_k, each repeated n_k times. Thus, z_1 appears n_1 times, z_2 appears n_2 times etc. Show that $\sum_{k=1}^{n}|w_k|^2 \leqslant \text{Tr}(A^*A)$. Show that A is normal if and only if we have equality here. In this problem, $\text{Tr}(A^*A)$ denotes the trace of the matrix A*A.

(17) If $A, B \in M_{n \times n}(\mathbb{C})$ and AB = 0, then BA need not be zero [Example!]. What happens when A and B are both normal? Does AB = 0 imply BA = 0?

(18) Write down the complex analog of Exercise 15 of Section 2. This is a little more interesting than the real case.

(19) Return to the complex inner product space given in Exercise 15 of Section 3. Compute the Hermitian adjoint of the map T(A) = CA for any $C \in M_{n \times n}(\mathbb{C})$.

(20) In Exercise 19, suppose C is nonsingular. Compute the Hermitian adjoint of the map $S(A) = C^{-1}AC$.

(21) Let $(V, \langle \, , \, \rangle)$ be a finite-dimensional, complex inner product space. Let $T \in \mathscr{E}(V)$. Prove that the following assertions are equivalent:
(a) T is normal.
(b) $T^* = g(T)$ for some $g(X) \in \mathbb{C}[X]$.
(c) $\|T(\alpha)\| = \|T^*(\alpha)\|$ for all $\alpha \in V$.
(d) Every T-invariant subspace of V is T*-invariant.

Glossary of Notation

F	an arbitrary field, 1
$\mathbb{Q}$	the field of rationals numbers, 2
$\mathbb{R}$	the field of real numbers, 2
$\mathbb{C}$	the field of complex numbers, 2
$\mathbb{F}_p$	the field of p elements, 2
$\mathbb{Z}$	the integers, 2
V	an arbitrary vector space, 2
$\mathbb{N}$	the natural numbers, 3
F^n	n-tuples of elements from F, 3
B^A	all functions from A to B, 3
V^n	all functions from $\{1, \ldots, n\}$ to V, 3
(a_{ij})	a matrix, 3
$M_{m \times n}(F)$	the set of $m \times n$ matrices over F, 3
$F[X]$	polynomials in X over F, 3
C(I)	the set of continuous functions on $I \subseteq \mathbb{R}$, 4
$C^k(I)$	k-times differentiable functions on I, 4
$\mathscr{R}(A)$	Riemann integrable functions on A, 4
A^t	the transpose of a matrix A, 4
$\sum_{i \in \Delta} W_i$	the sum of the W_i, 5
L(S)	the linear span of S, 6
$\mathscr{P}(V)$	the set of all subsets of V, 6

254

$\mathscr{S}(V)$	the set of all subspaces of V, 6
$\|A\|$	the cardinality of the set A, 8
$(A, \tilde{\ })$	a set A and relation $\tilde{\ }$ on A, 8
dim V	the dimension of V, 12
$\underline{\alpha} = \{\alpha_1, \ldots, \alpha_n\}$	a basis of V, 14
$[\cdot]_{\underline{\alpha}}$	the $\underline{\alpha}$-skeleton, 14
$M(\underline{\delta}, \underline{\alpha})$	change of basis matrix, 14
$Col_i(M)$	ith column of M, 14
T	a linear transformation 17
Hom(V, W)	the linear transformations from V to W, 17
ker T	the kernel of T, 18
Im T	the image of T, 18
$\cong$	an isomorphism, 18
$(\cdot)_{\underline{\alpha}}$	transpose of $[\cdot]_{\underline{\alpha}}$, 19
I_n	the n × n identity matrix, 20
$V \times \cdots \times V$	a finite product of V, 20
$\Gamma(\underline{\alpha}, \underline{\beta})(T)$	the matrix representation of T relative to $\underline{\alpha}$, $\underline{\beta}$, 22
$rk(A)$	the rank of a matrix A, 24
$C = \{(V_i, d_i)\}$	a chain complex, 26
I_V	the identity map on V, 28
$\prod_{i \in \Delta} V_i$	the product of the V_i, 30
$(f(i))_{i \in \Delta}$	a Δ-tuple, 30
$\bigoplus_{i \in \Delta} V_i$	the direct sum of the V_i, 33
$V_1 \oplus \cdots \oplus V_n$	a finite direct sum, 34
$\mathscr{E}(V)$	the algebra of endomorphisms of V, 36
$\equiv$	an equivalence relation, 39
$\bar{x}$	the equivalence class containing x, 39
V/W	the set of equivalence classes of V modulo W, 39
$\alpha + W$	a coset of W, 40
$\mathscr{A}(V)$	the set of all affine subspaces of V, 40
S_α	translation through α, 41
$Aff_F(V, V')$	the set of affine transformations from V to V′, 41
V^*	the dual of V, 46
$\underline{\alpha}^* = \{\alpha_1^*, \ldots, \alpha_n^*\}$	the dual basis of $\underline{\alpha} = \{\alpha_1, \ldots, \alpha_n\}$, 46
$\omega(\alpha, \beta)$	a bilinear map, 47
$A^\perp$	the annihilator of A, 48

T^*	the adjoint of T, 49
$rk\{T\}$	the rank of T, 50
$Tr(A)$	the trace of A, 53
$q(\xi)$	a quadratic form, 54
ϕ	a multilinear mapping, 59
$C^\infty(I)$	infinitely differentiable functions on I, 60
$Mul_F(V_1 \times \cdots \times V_n, V)$	the set of multilinear maps from $V_1 \times \cdots \times V_n$ to V, 61
$V_1 \otimes_F \cdots \otimes_F V_n$	the tensor product of $V_1, \ldots, V_n$, 64
$\alpha_1 \otimes \cdots \otimes \alpha_n$	a coset in $V_1 \otimes_F \cdots \otimes_F V_n$, 65
$A \otimes B$	the Kronecker product of A and B, 67
$V^{\otimes n}$	$V \otimes \cdots \otimes V$ (n times), 68
$\mathcal{T}(V)$	the tensor algebra of V, 68
$T_1 \otimes \cdots \otimes T_n$	the tensor product of the maps T_i, 71
$V^{\mathbb{C}}$	the complexification of V, 79
$T^{\mathbb{C}}$	the complexification of the map T, 79
S_n	the group of permutations of $\{1, \ldots, n\}$, 83
$(i_1, \ldots, i_r)$	an r-cycle, 85
$sgn(\sigma)$	the sign of the permutation σ, 86
$Alt(\phi)$	the alternating map formed from ϕ, 86
$Alt_F(V^n, W)$	the space of alternating maps from V^n to W, 87
$\Lambda_F^n(V)$	the n-th exterior power of V, 89
$\alpha_1 \wedge \cdots \wedge \alpha_n$	a coset in $\Lambda_F^n(V)$, 89
$\binom{N}{n}$	the binomial coefficient N over n, 91
$\Lambda^n(T)$	the n-th exterior power of T, 92
$\Lambda(V)$	the exterior algebra of V, 93
$Sym_F(V^n, W)$	symmetric multilinear maps from V^n to W, 94
$S(\phi)$	the symmetric map formed from ϕ, 94
$S_F^n(V)$	the n-th symmetric power of V, 95
$[\alpha_1] \cdots [\alpha_n]$	a coset in $S_F^n(V)$, 95
$S_F^n(T)$	the n-th symmetric power of T, 96
$F[X_1, \ldots, X_n]$	the polynomial algebra in n variables, 97
$S(V)$	the symmetric algebra of V, 97
$\partial(f)$	the degree of f(X), 99
$f \mid g$	f divides g, 99
$g.c.d.(f_1, \ldots, f_n)$	a greatest common divisor of $f_1, \ldots, f_n$, 100
$R(f)$	the roots of f, 102
$\overline{F}$	the algebraic closure of F, 103

V^F	$V^F = V \otimes_F \bar{F}$, 103
T^F	$T^F = T \otimes_F I_F$, 103
l.c.m.$(f_1, \ldots, f_n)$	a least common multiple of $f_1, \ldots, f_n$, 104
$m_T(X)$	the minimal polynomial of T, 105
$\hookrightarrow$	the inclusion map, 107
V^K	$V^K = V \otimes_F K$, 107
T^K	$T^K = T \otimes_F I_K$, 107
$M_{m \times n}(F[X])$	m × n matrices with coefficients in F[X], 110
adj(A)	the adjoint of A, 110
$c_A(X)$	the characteristic polynomial of A, 111
$\mathscr{S}_F(T)$	the eigenvalues of T in F, 118
diag$(a_1, \ldots, a_n)$	n × n diagonal matrix, 121
N_k	a k × k subdiagonal matrix, 123
$\bar{\beta}$	the conjugate of β in V^C, 143
e^A	the exponential of A, 153
C(g(X))	the companion matrix of g(X), 161
|x|	the absolute value of x, 171
‖ ‖	a norm, 171
(V, ‖ ‖)	a normed linear vector space, 172
ℓ^2	the Hilbert space of square summable sequences, 173
d(α, β)	the distance between α and β, 173
$B_r(\alpha)$	a ball of radius r about α, 173
d(β, A)	the distance between β and A, 173
A^0	the interior of A, 173
A^c	the complement of A, 173
$\lim_{\xi \to \alpha} f(\xi)$	the limit of f as ξ approaches α, 174
$\mathscr{B}(V, W)$	the set of bounded linear operators from V to W, 175
‖T‖	the uniform norm on $\mathscr{B}(V, W)$, 177
$\bar{A}$	the closure of the set A, 179
A^∂	the boundary of the set A, 179
‖ ‖$_s$	the sum norm, 181
$\{\alpha_n\}$	a sequence in V, 186
$\{\alpha_n\} \to \alpha$	the sequence $\{\alpha_n\}$ converges to α, 186
$\{\beta_n\} \hookrightarrow \{\alpha_n\}$	$\{\beta_n\}$ is a subsequence of $\{\alpha_n\}$, 189
$\langle \, , \, \rangle$	an inner product, 206
(V, $\langle \, , \, \rangle$)	an inner product space, 207

$\alpha \perp \beta$	a and β are orthogonal, 211
$A \perp B$	A and B are orthogonal, 211
$A^{\perp}$	the vectors orthogonal to A, 211
$\text{Proj}_W(\beta)$	the orthogonal projection of β onto W, 213
$|z|$	the modulus of a complex number z, 236

References

1. E. Kamke, *Theory of Sets*, Dover, New York, 1950.
2. I. Kaplansky, *Linear Algebra and Geometry*, Allyn & Bacon, Boston, Massachusetts, 1969.
3. J. L. Kelley, *General Topology*, Van Nostrand, Princeton, New Jersey, 1955.
4. S. Lang, *Algebra*, Addison-Wesley, Reading, Massachusetts, 1965.
5. D. G. Northcott, *Multilinear Algebra*, Cambridge University Press, London, 1984.
6. O. Zariski and P. Samuel, *Commutative Algebra*, Vol. I, Van Nostrand, Princeton, New Jersey, 1958.

Subject Index